Aerosol Science and Technology

Aerosol Science and Technology

Parker C. Reist
Professor of Air and Industrial Hygiene Engineering
Department of Environmental Sciences and Engineering
University of North Carolina
Chapel Hill, North Carolina

Second Edition

McGraw-Hill, Inc.
New York St. Louis San Francisco Auckland Bogotá
Caracas Lisbon London Madrid Mexico Milan
Montreal New Delhi Paris San Juan São Paulo
Singapore Sydney Tokyo Toronto

Library of Congress Cataloging-in-Publication Data

Reist, Parker C.
  Aerosol science and technology / Parker C. Reist—2nd ed.
    p.   cm.
  Includes index.
  ISBN 0-07-051882-3
  1. Aerosols. I. Title.
QC882.42.R45 1993
541'.3'4515—dc20   93-14765
                                                                    CIP

Copyright © 1993 by McGraw-Hill, Inc. All rights reserved. Printed in the United States of America. Except as permitted under the United States Copyright Act of 1976, no part of this publication may be reproduced or distributed in any form or by any means, or stored in a data base or retrieval system, without the prior written permission of the publisher.

1 2 3 4 5 6 7 8 9 0   DOC/DOC   9 8 7 6 5 4 3 2

ISBN 0-07-051882-3

The sponsoring editor for this book was Gail Nalven, the editing supervisor was Caroline Levine, and the production supervisor was Suzanne W. Babeuf. This book was set in Century Schoolbook by McGraw-Hill's Professional Book Group composition unit.

Printed and bound by R. R. Donnelley & Sons Company.

The first edition of this book was published by Macmillan, Inc. in 1984.

Information contained in this work has been obtained by McGraw-Hill, Inc., from sources believed to be reliable. However, neither McGraw-Hill nor its authors guarantee the accuracy or completeness of any information published herein and neither McGraw-Hill nor its authors shall be responsible for any errors, omissions, or damages arising out of use of this information. This work is published with the understanding that McGraw-Hill and its authors are supplying information but are not attempting to render engineering or other professional services. If such services are required, the assistance of an appropriate professional should be sought.

Contents

Preface   xi
Preface to the First Edition   xiii

Chapter 1. Introduction and Definitions   1
  Units   1
  Definitions   2
  Morphological Properties of Aerosols   3
    Shape   3
    Size   4
    Structure   8
    Fractal Properties   8
  Surface Properties   11

Chapter 2. Particle Size Distributions   13
  Introduction   13
  Mean and Median Diameter   13
  Histograms   15
  Mathematical Representation of Distribution   19
    Normal Distribution   20
    Log-normal Distribution   22
  Log-Probability Paper   24
  Other Definitions of Means   25

Chapter 3. Fluid Properties   31
  Kinetic Theory   32
  Gas Behavior   33
    Molecular Speeds (Bernoulli)   33
    Mean Free Path   38
  Gas Viscosity, Heat Conductivity, and Diffusion   40

Chapter 4. Macroscopic Fluid Properties   45
  Reynolds Number   45
  Drag   49

Chapter 5. Viscous Motion and Stokes' Law — 59
- Continuous Medium — 60
- Incompressible Medium — 63
- Viscous Medium — 63
- Infinite Medium — 64
- Rigid Particles — 65
- Spherical Particle — 68

Chapter 6. Particle Kinetics: Settling, Acceleration, and Deceleration — 75
- Equation of Motion of an Aerosol Particle — 76
- Particle Motion in the Absence of External Forces Except Gravity — 77
- Terminal Settling Velocity — 81
- Stop Distance — 83
- Particle Acceleration or Deceleration — 83
- Limitations — 84
- One-Dimensional Motion at High Reynolds Numbers — 84
- Ideal Stirred Settling — 86

Chapter 7. Particle Kinetics: Impaction — 91
- Curvilinear Motion — 91
- Impaction of Particles — 92
  - Impactor Operation — 96
  - Particle Bounce — 101
  - Impactors for Very Small Particle Sizes — 102
  - Pressure Drop in Impactors — 104
  - Analysis of Impactor Data — 105
  - Errors Associated with Impactor Data — 107
  - Impactor Analysis Using Phase Trajectories — 108

Chapter 8. Particle Kinetics: Centrifugation, Isokinetic Sampling, and Respirable Sampling — 113
- Centrifugation of Particles — 115
- Cyclones — 117
- Isokinetic Sampling — 120
- Respirable Sampling — 124

Chapter 9. Brownian Motion and Simple Diffusion — 131
- Brownian Motion — 131
- Fick's Laws of Diffusion — 132
- Einstein's Theory of Brownian Motion — 133
- Brownian Displacement — 136
- Brownian Motion of Rotation — 138
- "Barometric" Distribution of Particles — 139
- Effect of Aerosol Mass on the Diffusion Coefficient — 140
- Aerosol Apparent Mean Free Path — 142

Chapter 10. Particle Diffusion — 145
- Steady-State Diffusion — 145
- Non-Steady-State Diffusion — 146
  - Infinite Volume, Plane Vertical Wall — 146
  - Two Vertical Walls a Distance H Apart — 148
- Diffusion in Flowing Air Streams—Convective Diffusion — 150
- General Equations of Convective Diffusion — 150
  - Convective Diffusion Defined by the Peclet Number — 151
- Tube Deposition — 152
- Laminar Boundary Layer — 155
  - Turbulent Boundary Layer — 157
  - Concentration Boundary Layer — 157
  - The Diffusion Velocity — 158
  - Application of Diffusion Velocity — 159

Chapter 11. Thermophoresis — 163
- Early Observations of Thermophoresis — 165
- Theory — 166
- Thermophoresis in the Free Molecule Region (Kn ≫ 1) — 166
- Thermal Forces in the Slip-Flow Regime (Kn ≤ 0.2) — 169
- Epstein's Equation — 170
- Brock's Equation — 171
- Derjaguin and Yalamov's Equation — 172
- Thermophoretic Velocity — 173
- Thermophoretic Velocity for All Particle Sizes — 175
- The Dust-free Space — 175

Chapter 12. Aerosols Charging Mechanisms — 179
- Definition of Force — 179
- Particle Mobility — 180
- Particle Charge, q — 181
  - Direct Ionization of the Particle — 181
  - Static Electrification — 182
  - Collisions with Ions or Ion Clusters — 185
  - Diffusion Charging—Unipolar Ions — 186
  - Field Charging — 189
  - Combined Diffusion and Field Charging — 195
  - Ion Production by Corona Discharge — 195
  - Maximum Attainable Particle Charge — 198
  - Charge Equilibrium — 200
  - Steady-State Theory of Charge Equilibrium — 201
  - Transient Approach to Charge Equilibrium — 207

Chapter 13. Electrostatic Controlled Aerosol Kinetics — 209
- Electric Fields — 209
- Field Strength of a Point Charge — 210

Coulomb's Law	211
Electrical Units	211
General Equations for Field Strength	212
Constant Field Strength	213
Computation of the Electric Field for Simple Geometries	213
Negligible Ionic Space Charge	213
Ionic Space Charge Present	214
Electric Field—Particles Present	216
Perturbations in the Electric Field Caused by a Particle or Other Object	218
Particle Drift in an Electric Field	219
Efficiency of an Electrostatic Precipitator	221

Chapter 14. Condensation and Evaporation Phenomena in Aerosols   225

Early Observations	225
Types of Nucleation	226
Saturation Ratio	227
Homogeneous Nucleation—Kelvin's Equation	228
Rate of Formation of Critical Nuclei	232
Ions as Nuclei	233
Heterogenous Nucleation	238
Condensation Nuclei	238
Sources of Condensation Nuclei	240
Composition of Condensation Nuclei	241
Utilization of Nuclei	241
Insoluble Nuclei	242
Soluble Nuclei	242
Hysteresis in Evaporation and Condensation	246

Chapter 15. Evaporation and Growth   251

Maxwell's Equation	251
Growth or Lifetime of Drops—Langmuir's Equation	256
Modifications to Langmuir's Equation	258
Evaporation Time in a Saturated Medium	259
Growth and Evaporation of Moving Droplets	260

Chapter 16. Optical Properties: Extinction   263

Definition of Terms	264
Extinction of Light—Bouguer's Law	266
Assumptions Implicit in Bouguer's Law	270
Computation of Extinction Coefficient	270
Receptor—Contrast	276
Alteration of Contrast	276

Chapter 17. Optical Properties: Angular Scattering   281

Definitions	281

Mie Scattering—The Mie Theory	283
Approximations to Mie Theory	284
Polydisperse Aerosol	289
Rayleigh Scattering	289
Scattering Patterns with Increasing α	291
Radiative Transfer	293
Applications	294
Diffraction Rings	294
Higher-order Tyndall Spectra	295
Use of the Forward Scattering Lobe	296
Single Particle Scattering Measurements	297
Chapter 18. Coagulation of Particles	**301**
Coagulation of Monodisperse Spherical Particles	301
Coagulation of Particles of Two Different Sizes	306
Coagulation of Many Sizes of Particles	306
Differential Equation Form	309
Limitations of the Differential Equation Form	310
Use of a Nonlinear Integro-Differential Equation	310
Terms for Gravity and Deposition Effects	311
The "Self-Preserving" Size Distribution	312
Coagulation of Nonspherical Particles	312
External Factors in Coagulation	313
Electrical Effects in Coagulation	313
Coagulation in Moving Atmospheres	314
Chapter 19. Viable Aerosols	**319**
Types of Viable Aerosols	320
Units of Measure	320
Factors Influencing Viable Aerosol Concentrations	321
Estimates of Viable Aerosol Concentrations	323
Chapter 20. Explosive Aerosols	**327**
Severity of Explosions	328
Types of Explosive Dusts	329
Ignition Sources	331
Particle Size	332
Control of Dust Explosions	336
Appendix A. Corrected Sedimentation Velocities	**339**
Appendix B. Stokes' Law	**341**
Appendix C. Error Function	**344**
Appendix D. Units, Definitions, and Conversions	**346**

Appendix E. Adiabatic Expansion 350

Appendix F. Psychrometric Chart 352

Appendix G. Bessel Functions of Order 1a 355

References 357

Index 365

Preface

This book has had a strange history. It was originally started in 1969 based on lecture notes from an aerosol course I was teaching at the Harvard School of Public Health. In 1972 I moved to the University of North Carolina and brought the course and my work on the book along with me. The new setting and new responsibilities did their part to delay things, and it was not until 1984 that the book was finally published by Macmillan. It was well received. However, in the intervening years there has been a great spurt in the growth of the field of aerosol science: There are a number of universities offering courses in aerosol science, and at one English university it is possible to receive a master's degree in aerosol science; a whole new field has grown up around the concept of using controlled aerosol-producing reactions to create exotic new materials; aerosols are being seen as an effective method for administration of some drugs and may someday replace many intravenous procedures; the ultimate answers to the greenhouse effect appear to be intimately associated with the property of aerosols to absorb some radiation wavelengths better than others; and finally the relative importance of aerosols to the microelectronics industry has been widely recognized, in both a positive and a negative sense.

Accordingly I felt that an update of the 1984 book was in order. New developments in sampling equipment design, refinements in fundamental background information for aerosols, the emergence of fractal geometry as an aerosol tool, and my recognition that several important areas were completely ignored in the first edition all made compelling reasons for this revision. New chapters have been added covering thermophoresis, viable aerosols, and dust explosions; several other chapters have been substantially rewritten.

Finally, at the request of many of my former students, more information on units has been added, many of the worked examples have been clarified, and a number of the figures have been replaced with better illustrations.

In the meantime, much of what was good about the earlier book has been retained, including the original introduction, which still, I think, says it all.

Parker C. Reist
Chapel Hill
March 5, 1992

Preface to the First Edition

From dust we came and to dust we shall return.

This book is about dust, dust and all the myriad tiny things that hang suspended in the air. These clouds of fine particles, or aerosols, can cheer us up when we look at a spectacular sunset, or they can be depressing, such as on a gray day in a smoky town. Particles suspended in air act as sites on which water can condense and thus play a principal role in the water cycle and the formation of rain. Dust clouds on a back road allow us to follow a vehicle at great distances, and smoke screens promise protection in an electronic war. We use fine particles suspended in air to kill mosquitoes, treat allergies, control underarm odor, and even oil machinery. High concentrations of some particles are extremely explosive, and low concentrations of other particles are extremely toxic. Whether we realize it or not, we are at all times surrounded by literally thousands of small particles, and their importance to the natural functioning of the earth is incalculable.

Considering the importance of airborne particles, one might think that they would have attracted the attention of modern scientists and that fundamental knowledge of particle behavior would be widespread and well known by now. This is not the case. Rather, aerosol science is a much neglected stepdaughter of physics or perhaps physical chemistry and is only now beginning to blossom and provoke the interest it deserves.

Systematic study of the fundamental properties of airborne particles has been intermittent in the past. For some reason we, as a society, tend to look on everyday phenomena with blind acceptance, regarding what we see as so common that it never occurs to us to ask why. Why does a cloud remain airborne—and where does it come from and where does it go? What is "smoke"—a solid or a gas? (When asked this question on the first day of class, many of my students erroneously think that smoke is a gas.) Why are some dusts harmful and others not? Or similarly, why is the same dust sometimes harmful while at other times it is not?

For centuries people have suspected that dust could be harmful. At least, early writers indicated in their works a general connection between lung diseases and dust inhalation, even though they didn't distinguish between the various types of respiratory diseases. For example, Pliny refers to inhalation of "fatal dust," and Agricola speaks of the "pestilential air" and "the corrosive dust." In his book published in 1700, Ramazzini describes the effect of dust on the respiratory organs and describes numerous cases of fatal dust disease.

With the industrial revolution in the 19th century and the advent of high-speed machinery, dust exposure increased dramatically, as did dust-caused diseases. In the latter part of the 19th century, interest focused on dust exposure of miners, especially in the gold mines of South Africa and the tin mines of Cornwall. As a result of these studies and others, it was found that high exposure concentrations gave rise to more cases of lung disease.

Even with evidence showing the relationship of dust levels in the air to disease, only the simplest effort was made by the medical profession to study the properties of dust in the air—how to sample it, how to control it, what its important physical properties were, how it was produced, or where it ultimately went. The focus of the medical profession was primarily on gross effects.

In the natural sciences, however, aerosols were in the forefront in the 19th century because these small particles represented the smallest divisions of matter known at the time. Many individuals whom we now consider the intellectual giants of that time contributed to our understanding of aerosols, and the names Tyndall, Lister, Kelvin, Maxwell, Aitken, and Einstein, to name a few, are familiar in the aerosol literature as well as in the fields for which they are most famous.

However, with the discovery of radioactivity and the development of quantum mechanics, the passion for finding the smallest division of all matter drove scientists away from studies of aerosols, and the field as a scientific discipline lay dormant, despite continuing discoveries in medicine regarding the relationship between dust and disease. Only in the area of occupational health were aerosol studies continued, and these were of an applied nature. The use of aerosols in warfare and screening smokes led to some effort to study their properties between World War I and World War II, but it was not until World War II that aerosol problems again began to attract the attention of the main scientific community. The reasons for this increased interest were several. First, production of fissionable materials involved working with radioactive aerosols, potentially dangerous materials. Second, the advent of radar created the need for understanding the effect of clouds on the transmitted and reflected signals and how this effect could be either minimized or maximized, depending on whether one wanted to hide or seek. Finally, the threat of chemical and biological warfare needed to be dealt with on the basis of knowledge, not guesswork, and since aerosols represent the chief means for dispensing these agents, study of aerosol behavior was essential.

In the past 20 years, work relating directly to the study of aerosols has increased greatly with at least two journals specifically devoted to aerosol studies and numerous others regularly publishing articles on various aspects of particulates in air. Aerosols appear to play a major role in the removal of pollutant gases from the atmosphere either by absorbing them on existing particles or through the creation of new particles. A knowledge of aerosol properties is useful in studying the atmosphere of planets other than earth. Many air pollutants originate in particulate form or become particulates soon after discharge and must be dealt with as such. Acid rain is an example of an aerosol problem where gas is transformed to a liquid—in this case sulfur dioxide is transformed in the air to sulfuric acid.

As many frustrated investigators have noted again and again, the study of aerosols is by no means easy. Particles in air behave differently from the air in which they are suspended and behave differently among themselves depending on their size, shape, and composition. Collecting a representative sample of an aerosol for any purpose can be a frustrating and time-consuming task, and a knowledge of aerosol properties and behavior is essential to maximize chances for adequate sample collection. This is especially true when many of the automated sampling devices available today are used. The device generates the numbers, whether they are accurate or not, and it is up to the investigator to interpret and understand what is being generated.

This book is an attempt to present, in a rigorous but illustrative manner, introductory information on the study of aerosol properties and behavior so that an individual desiring to learn the mysteries of the field will not be completely discouraged. The text has evolved out of more than 15 years' experience in teaching an introductory course on aerosol science to numerous first-year graduate students, some of whom picked at the edges of the course and were sufficed, others who digested all the material and developed an insatiable appetite for more. I hope in this book to reach both groups. Many examples are given of aerosol studies which can be applied almost directly to other situations without much attention being paid to the underlying theories. However, for the more inquiring mind, equations have been developed to attempt to illustrate the thought process used to arrive at a particular solution. Some solutions may not be the most accurate or up to date. I have no apologies. In learning the simpler approximate solution, one develops the terminology, conventions, and methods of thinking which lead to greater understanding of the more rigorous complex solutions.

This book is a textbook. If it helps individuals to better read and understand the current aerosol literature and to extend the field on their own, then it will have served its purpose.

Acknowledgments

I would like to thank my friends and students at the University of North Carolina for the assistance they have provided during preparation of the manuscript for the current edition, especially Doris Mitchell, Delores Plummer, and Don Fox; my patient editors at McGraw-Hill, Gail Nalven and Carol Levine; and of course my family, particularly my wife Jan, for her constant encouragement even when nothing seemed to be going right.

In addition, no book can be written without much assistance from others who may never realize the help they give. Consider the vast number of researchers whose efforts often go unacknowledged, aside from a citation in a scientific journal. Without the labor of these individuals, a book such as this could not be written—we build, after all, on the work of others. Realizing this, I would like to thank all those scientists and engineers, both past and present, whose interest in aerosols and related subjects founded, developed, and enlightened a new field of study. May those who follow continue to add to our store of information and understanding of the world around us.

Parker C. Reist
Chapel Hill
April 4, 1983

Aerosol Science and Technology

Chapter 1

Introduction and Definitions

Aerosols are ubiquitous in our environment. Haze particles are formed over vegetation; dust clouds are blown up by the wind; volcanoes erupt, spewing dense smoke into the atmosphere; and, of course, in their many activities people mark their way by the particles they discharge into the air. This book is about aerosol particles, their physical properties, and the scientific basis that has been developed for predicting their behavior.

Units

Aerosol sizes are usually referred to in terms of the micrometer (μm) (previously called the micron μ). One micrometer is equal to 10^{-4} centimeters (cm), 10^{-6} meters (m), or 10^4 angstrom units, abbreviated Å. In working problems it is necessary to use a consistent set of units. Since most physical constants are available either in cgs or mks units (English units are too cumbersome to use), aerosol sizes given in micrometers very often must be converted to either centimeters or meters for computations (depending on the system of units chosen). When you are working problems involving ratios of particle size, this conversion is not necessary.

Example 1.1 A basketball is 12 in in diameter. Express its diameter in micrometers.

$$1 \text{ in} = 2.54 \text{ cm}$$

$$1 \text{ cm} = 10^4 \text{ }\mu\text{m}$$

$$\text{Diameter} = 12 \text{ in} \times 2.54 \text{ cm/in} \times 10^4 \text{ }\mu\text{m/cm}$$

$$= 3.05 \times 10^5 \text{ }\mu\text{m}$$

Definitions

To begin the systematic study of particles, it is first necessary to consider several commonly used definitions of various types of aerosols.

Aerosol A suspension of solid or liquid particles in a gas, usually air; a colloid. Included in this definition would be:

Dust Solids formed by disintegration processes such as crushing, grinding, blasting, and drilling. The particles are small replicas of the parent material, and the particle sizes can range from submicroscopic to microscopic. Very often sizes are specified by screen mesh size. For example, the percentage passing or retained on a given mesh is indicative of size.

> **Example 1.2** How many spherical particles just passing through a 200-mesh screen are required to equal the mass of a single spherical particle that just passes through a 50-mesh screen? Assume that the diameter of the particle passing through the mesh equals the mesh opening and a particle density of 2.65 g/cm^3.
>
> $$\text{Mass of particle passing 50-mesh screen} = \frac{\pi}{6}d^3\rho = \frac{3.14}{6}(0.0297)^3(2.65)$$
>
> $$= 3.64 \times 10^{-5}\,\text{g}$$
>
> $$\text{Mass of particle passing 200-mesh screen} = \frac{\pi}{6}d^3\rho = \frac{3.14}{6}(0.0074)^3(2.65)$$
>
> $$= 5.62 \times 10^{-7}\,\text{g}$$
>
> $$\text{No. particles required} = \frac{3.64 \times 10^{-5}}{5.62 \times 10^{-7}} = 64.7, \text{ say 65 particles}$$

Fumes Solids produced by physicochemical reactions such as combustion, sublimation, or distillation. Typical fumes are the metallurgical fumes of PbO, Fe_2O_3, or ZnO. Particles making up fumes are quite small, below 1 μm in size, and thus cannot be sized on screens. The particles appear to flocculate readily.

Smoke A cloud of particles produced by some sort of oxidation process such as burning. The optical density is presupposed. Generally, smokes are considered to have an organic origin and typically come from coal, oil, wood, or other carbonaceous fuels. Smoke particles are in the same size range as fume particles.

TABLE 1.1 Openings of Some Typically Small Mesh Sizes*

Mesh	Opening, mm
50	0.297
100	0.150
200	0.074
400	0.038

*From *Handbook Chem. Phys.*, 54th ed., CRC Press, Cleveland, 1973, p. F147.

Mists and fog Aerosols produced by the disintegration of liquid or the condensation of vapor. Because liquid droplets are implied, the particles are spherical. They are small enough to appear to float in moderate air currents. When these droplets coalesce to form larger drops of about 100 μm or so, they can then appear as rain.

Haze Particles with some water vapor incorporated into them or around them, as observed in the atmosphere.

Smog A combination of smoke and fog, usually containing photochemical reaction products combined with water vapor to produce an irritating aerosol. Smog particle sizes are usually quite small, being somewhat less than 1 μm in diameter.

These definitions have arisen from popular usage, so there is little wonder that they overlap. What one person might call smog someone else could call haze, and both would be correct. Therefore we should generally use the more precise, if less colorful, definition of aerosol and then fill in the details on a more qualitative basis.

Since an aerosol is a collection of particles, it is often desirable to indicate whether the particles are all alike or are dissimilar. Thus there are several other descriptions of aerosols that must also be taken into account.

Monodisperse All particles exactly the same size. A *monodisperse aerosol* contains particles of only a single size. As might be expected, this condition is extremely rare in nature.

Polydisperse Containing particles of more than one size.

Homogeneous Chemical similarity. A *homogeneous aerosol* is one in which all particles are chemically identical. In an *inhomogeneous aerosol* different particles have different chemical compositions.

Morphological Properties of Aerosols

Shape

It is convenient to think of all aerosol particles as spheres for calculation, and this also helps visualize the processes taking place. But, with the exception of liquid droplets, which are always spherical, many shapes are possible. These shapes can be divided into three general classes.

1. *Isometric particles* are those for which all three dimensions are roughly the same. Spherical, regular polyhedral, or particles approximating these shapes belong in this class. Most knowledge regarding aerosol behavior pertains mainly to isometric particles.

2. *Platelets* are particles that have two long dimensions and a small third dimension. Leaves or leaf fragments, scales, and disks fall into this class. Very little is known about platelet behavior in air, and care must be exercised in applying knowledge derived from studying isometric particles to platelets.

3. *Fibers* are particles with great length in one dimension compared to much smaller lengths in the other two dimensions. Examples are prisms, needles, and threads or mineral fibers such as asbestos. Recent concern over the health hazard posed by inhalation of asbestos fibers has prompted study of fiber properties in air. There is still not as much known about fibers as isometric particles.

Example 1.3 An asbestos fiber is 10 μm in length with a circular cross-section of 0.5-μm diameter. Find the diameter of a sphere that has the same volume as the fiber.

$$\text{Volume of fiber} = \frac{\pi}{4}(0.5)^2(10)$$

$$= 1.96 \text{ μm}^3$$

$$\text{Volume of sphere} = \frac{\pi}{6}d^3$$

$$d^3 = 1.96\frac{6}{\pi} = 3.75$$

$$d = 1.55 \text{ μm}$$

Particle shape can vary with the formation method and the nature of the parent material. Particles formed by the condensation of vapor molecules are generally spherical, especially if they go through a liquid phase during condensation. Particles formed by breaking or grinding larger particles, termed *attrition*, are seldom spherical, except in the case where liquid droplets are broken up to form smaller liquid droplets.

Size

A particle is generally imagined to be spherical or nearly spherical. Either particle radius or particle diameter can be used to describe particle size. In theoretical discussions of particle properties, the radius is most commonly used, whereas in more practical applications the diameter is the descriptor of choice. Thus one should carefully ascertain which definition is being used when the term *particle size* is used. In this text particle diameter is used throughout.

Once a choice of diameter or radius is made, there are a number of ways that this diameter or radius can be defined which reflect particle properties other than physical size. For a monodisperse aerosol, a single measure describes the diameters of all the particles. But with polydisperse aerosols a single diameter is not sufficient to describe all particle diameters, and certain presumptions must be made as to the distribution of sizes. Other parameters besides diameter alone must be used. This is discussed in more detail in Chap. 2.

Two commonly encountered definitions of particle size are Feret's diameter and Martin's diameter. These refer to estimates of approximate particle size when determined from viewing the projected images of a number of irregularly shaped particles. *Feret's diameter* is the maximum distance from edge to edge of each particle, and *Martin's diameter* is the length of the line that separates each particle into two equal portions. Since these measures could vary depending on the orientation of the particle, they are valid only if averaged over a number of particles and if all measurements are made parallel to one another. Then, by assuming random orientation of the particles, an average diameter is measured.

This measurement problem can be simplified somewhat by using the *projected area diameter* instead of Feret's or Martin's diameter. This is defined as the diameter of a circle having the same projected area as the particle in question. Figure 1.1 illustrates these three definitions. In general, Feret's diameter will be larger than the projected area diameter which will be larger than Martin's diameter.

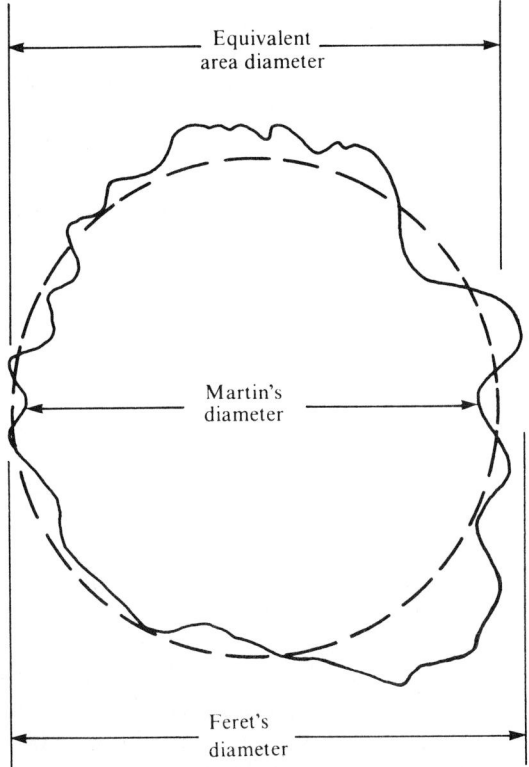

Figure 1.1 Illustration of three common definitions of particle diameter. In general, Martin's diameter is less than the equivalent area diameter, which in turn is less than Feret's diameter.

Example 1.4 Figure 1.2 shows a collection of five irregularly shaped particles. By measuring along lines parallel to the scale line, determine Martin's, Feret's, and the projected area diameter for this collection of particles.

The measured values are

$$\text{Feret's diameter} = 15 \text{ scale units}$$

$$\text{Martin's diameter} = 10 \text{ scale units}$$

$$\text{Projected area diameter} = 13 \text{ scale units}$$

Sometimes a diameter is defined in terms of particle settling velocity. All particles having similar settling velocities are considered to be the same size, regardless of their actual size, composition, or shape. Two such definitions which are most common are

Aerodynamic diameter Diameter of a unit density sphere (density = 1 g/cm^3) having the same aerodynamic properties as the particle in question. This means that particles of any shape or density will have the same aerodynamic diameter if their settling velocity is the same.

Stokes' diameter Diameter of a sphere of the same density as the particle in question having the same settling velocity as that particle. Stokes' diameter and aerodynamic diameter differ only in that Stokes' diameter includes the particle density whereas the aerodynamic diameter does not.

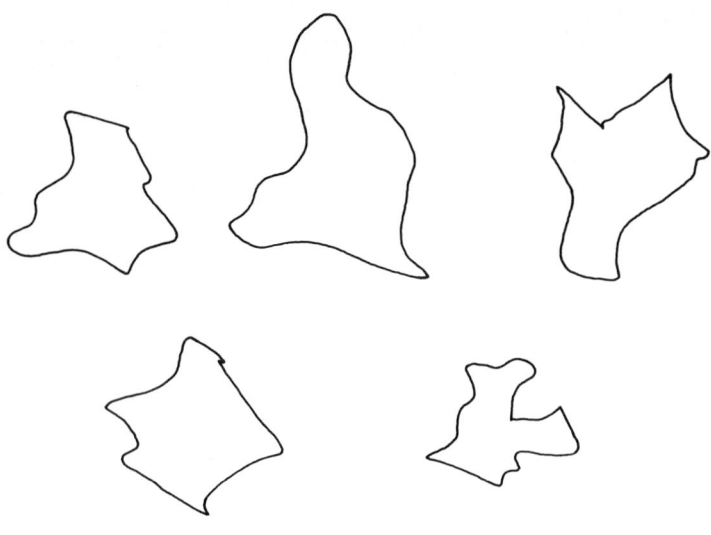

Figure 1.2 Illustration for Example 1.4.

Example 1.5 A sodium chloride cube (density = 2.165 g/cm^3) settles at a rate of 0.3 cm/s. Find the aerodynamic diameter of this cube.

Appendix A gives a corrected sedimentation velocity of 0.306 cm/s for a 10-μm-diameter unit-density sphere. Hence 10 μm is the aerodynamic diameter of this particular salt cube.

Particle diameters of interest in aerosol science cover a range of about four orders of magnitude, from 0.01 μm as a lower limit to approximately 100 μm as the upper limit. The lower limit approximates roughly the point where the transition from molecule to particle takes place. Particles much greater than about 100 μm or so do not normally remain suspended in the air for a sufficient length of time to be of much interest in aerosol science. There are occasions where particles that are either smaller or larger than these limits are important, but usually most particle diameters will fall within the limits of 0.01 to 100 μm.

Particles much greater than 5 to 10 μm in diameter are usually removed by the upper respiratory system, and those smaller than 5 μm can penetrate deep into the alveolar spaces of the lung. Thus 5 to 10 μm is often considered to be the upper diameter for aerosols of physiological interest.

Within the size range of 0.01 to 100 μm lie a number of physical dimensions which have a significant effect on particle properties. For example, the mean free path of an "air" molecule is about 0.07 μm. This means that the air in which a particle is suspended exhibits different properties, depending on particle size. Also the wavelengths of visible light lie in the narrow band of 0.4 to 0.7 μm. Particles smaller than the wavelength of light scatter light in a distinctly different manner than do larger particles.

Particle size is the most important descriptor for predicting aerosol behavior. This is apparent from the above discussion and will become even more apparent in later chapters. Typical particle sizes of selected materials are given in Table 1.2.

TABLE 1.2 Typical Particle Diameters, μm

Tobacco smoke	0.25	Lycopodium	20
Ammonium chloride	0.1	Atmospheric fog	2–50
Sulfuric acid mist	0.3–5	Pollens	15–70
Zinc oxide fume	0.05	"Aerosol" spray products	1–100
Flour dust	15–20	Talc	10
Pigments	1–5	Photochemical aerosols	0.01–1

Structure

Aerosol particles may occur by themselves or may be formed into chains of spheres or cubes. These are called *agglomerates* or *flocs*. Agglomerates are usually formed from highly charged small particles such as are found in dense smokes or metal fumes.

Particles may also occur as gas-filled hollow drops or as particle-filled hollow particles. Fly ash is an example of this latter type of material. Thus particle density can be significantly different from the density of the parent material.

Fractal properties

Since Mandelbrot's original description (1977, 1983), fractal geometry has found relevance in a number of scientific disciplines including aerosol technology and science [e.g., see Lovejoy (1982), Meakin (1983), Kaye (1984), Sheaffer (1987), Reist et al. (1989)]. Many applications are covered in some detail by Kaye (1989), so only a brief description of fractals is given here.

There are some geometric shapes which have an infinite boundary even though the area enclosed by that boundary is finite. These anomalies were dismissed by most 19th-century mathematicians as insignificant oddities, but Mandelbrot, drawing on the work of Richardson (1961) and others, showed that these shapes were extremely common in nature and were part of a generalized geometric system which he termed *fractal geometry*.

Clouds, trees, plant root systems, and even the human lung are all examples of structures which can be described as fractals, as are random agglomerates of many small spherical particles which form into one large particle of a highly irregular shape. Many metal fumes are observed to be large numbers of these large, irregular particles. Because these fume particles lack a specific definable shape, it is hard to describe them quantitatively by diameter or area and even more difficult to predict their aerodynamic properties. Fractal geometry offers a method whereby descriptors can be assigned to these particles, thus permitting their quantitative study.

There are several ways to consider the definition of a fractal. For example, consider the line, square, and cube shown in Fig. 1.3. The total length of the line T is

$$T = na^1$$

where n is the number of segments and a the length of one segment. For the square, the total area T is

$$T = na^2$$

Line $T = na^1$

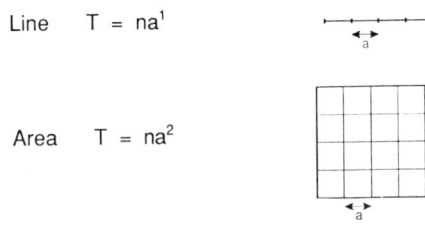

Area $T = na^2$

Volume $T = na^3$

Figure 1.3 Simple shapes and fractals.

and for the cube, the total volume T is

$$T = na^3$$

In each case the shape can be considered to be completely "filled"—the line filled with line segments, the area with squares, and the volume with cubes. These three equations could be written as

$$T = na^\delta$$

where δ could be 1, 2, or 3.

But now consider what a noninteger value for δ means. This implies that the shape is only partially filled, the degree of filling being greater as the value of δ becomes greater. Thus an irregular particle with many internal interstices could have its volume described by the factor δ, which would imply something about how loosely or tightly packed the particle was. The factor δ in this case is called by Mandelbrot the *fractal dimension* of the particle.

A perfectly filled geometric shape such as a sphere or cube has a fractal dimension of 3, whereas an irregular shape such as the agglomerate shown in Fig. 1.4 might have a fractal dimension of 2.43, indicating that there is some openness in the particle.

A second method of defining fractal dimensions is to consider an irregular boundary around a finite area (the coastline of Britain is often used as an example). If the coastline is measured with a ruler 1 m long, it will measure longer than if it were measured with a ruler 1 km long. This is because in using the longer ruler many little twists and kinks in the coastline will be missed. Thus the length of the coast-

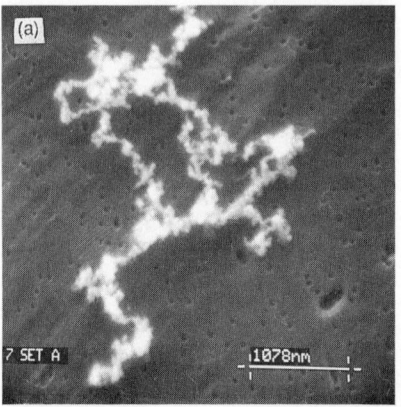
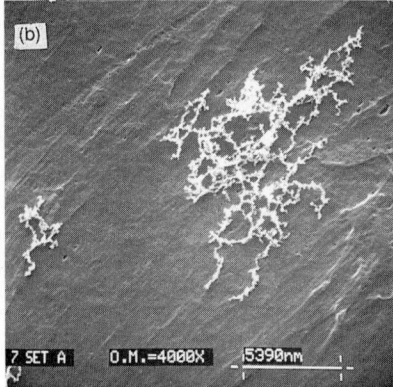
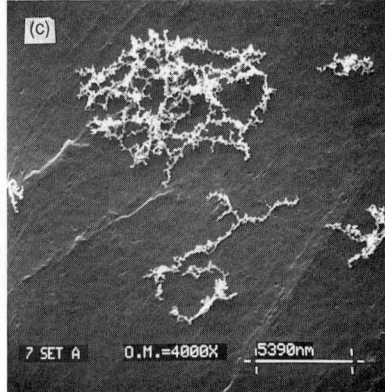

Figure 1.4 Electron micrograph of agglomerates, 0.1 atm. Sample taken at 30 min after explosion. (*a*) 20,000X, (*b*) (*c*) 4000X.

line is dependent on the length of the ruler. Plotting the logarithm of the coastline length as a function of the logarithm of the ruler length gives a straight line, as shown in Fig. 1.5. And 1 minus the slope of this line gives the fractal dimension, as defined above. Again the fractal dimension δ is indicative of the space-filling ability of the curve.

Mandelbrot defines the surface area fractal dimension δ as

$$(A_s)^{1/\delta} \propto V^{1/3}$$

where A_s is the surface area of a fractal object of volume V. For natural objects this relationship holds only over some range or "structural range," with the upper limit related to the finite size of the structure of the agglomerate and the lower limit related to the size of

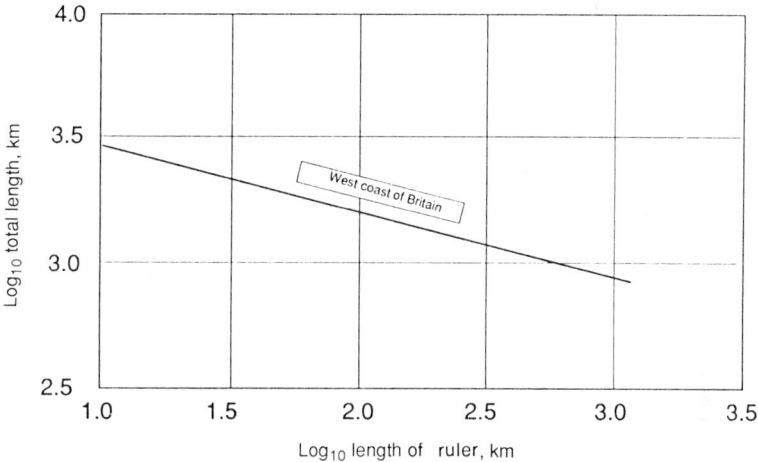

Figure 1.5 Richardson's empirical data. (*From Mandelbrot, 1983.*)

the fundamental units that make up the object. For agglomerated aerosols that are natural fractals, the lower size limit is thought to be the size of the primary particles (Kay, 1984).

One important quality of fractal objects with the same fractal dimension is that they are all self-similar; i.e., they possess no descriptive length scale at all (Herrmann, 1986). Thus for a set of fractal objects having the same source and same fractal dimension, it is impossible to say which are the big ones or which are the small ones without some other frame of reference. Thus fractal dimensions must be used in conjunction with some other, more familiar particle measure.

Surface Properties

Aerosol particles, because of their small size, present a large amount of surface for chemical reactions such as burning, adsorption, absorption, or other chemical reactions or for such physical properties as wettability or electrostatic effects. The amount of area per gram of material increases as the particle size decreases, and for a given average size, increasing polydispersity decreases the surface area per gram. As particle size becomes very small, the boundary conditions between the particle and the air around it become confused, but also become more important.

Example 1.6 What is the surface area of 1 g of a monodisperse water aerosol if the particle diameter is 10 and 1 μm?
Let n = particles per gram

$$1\text{ g} = \left(\frac{\pi}{6}\right)(d^3)(1)(n)$$

12 Chapter One

If A = surface area per gram

Then
$$A = \pi d^2 n$$

$$A = \frac{(1)(6/\pi)}{d^3}(\pi d^2) = \frac{6}{d}$$

For 10-μm particles

$$A = \frac{6}{10^{-3}} = 6000 \text{ cm}^2$$

For 1-μm particles

$$A = 60{,}000 \text{ cm}^2$$

Problems

1 What is the ratio of the volume of a spherical particle that will just pass through a 200-mesh screen compared to a sphere that will just pass through a 400-mesh screen?

2 It is given that 0.2 g of particles is passed through a 325-mesh sieve but retained on a 400-mesh sieve. Assuming the particles are spheres and are all the same size, estimate the maximum and minimum number of particles present. Assume a particle density of 2.65 g/cm^3.

3 Express the earth's equatorial diameter in micrometers (d = 7912 mi). Express the diameter of an electron in micrometers (d = 10^{-12} cm). Express the diameter of a hydrogen molecule in micrometers (d = 2.9 Å).

4 Compare relative dimensions of a sphere, platelet, and fiber, assuming that the fiber element diameter and platelet thickness are one-tenth the sphere diameter and that the volumes of the sphere, platelet, and fiber are equal. Assume a circular cross-section.

5 If the sphere in Prob. 4 is a 1-μm-diameter silica particle (ρ = 2.65 g/cm^3), what are the equivalent platelet dimensions and fiber length?

6 The settling velocity of a 5-μm-diameter sand particle can be estimated from the expression

$$v_g = (3 \times 10^{-3}) d^2 \rho$$

where v_g is the settling velocity in centimeters per second, d the particle diameter in micrometers, and ρ the density in grams per cubic centimeter. Find the aerodynamic diameter of this particle.

7 Show that for a constant mass of particles, decreasing the particle size by a factor of 10 increases the surface area by a factor of 10.

8 How many 0.1-μm-diameter H_2SO_4 droplets can be produced by splitting up one 10-μm-diameter H_2SO_4 droplet?

Chapter 2

Particle Size Distributions

Introduction

As mentioned in Chap. 1, most frequently aerosol particles are present in a variety of sizes; i.e., the aerosol is *polydisperse*.

Most aerosols are polydisperse when formed, some more than others. For example, an examination of sawdust would reveal particles of various sizes, as would that of any material formed by attrition. Since raindrops could grow by condensation or by a series of collisions with other drops, they would also be expected to be polydisperse. In fact, monodisperse aerosols are very rare in nature, and when they do appear, generally they do not last very long. Some high-altitude clouds are monodisperse, as are some materials formed by condensation. Sometimes it is satisfactory to represent all the particle sizes by only a single size. Other times more information is needed about the distribution of all particle sizes. Of course, a simple plot of particle frequency versus size gives a picture of the sizes present in the aerosol, but this may not be enough for a complete quantitative analysis.

Polydisperse aerosols can be described in a number of ways using mathematical or visual methods. Some of the more common methods are discussed in this chapter.

Mean and Median Diameter

The simplest way of treating a group of different particle diameters is to add all the diameters and divide by the total number of particles. This gives the average diameter. Mathematically this can be expressed as

$$\bar{d} = \frac{\sum n_i d_i}{\sum n_i} \qquad (2.1)$$

This is known as the *mean particle diameter*.

The *median particle diameter* can be determined by listing all diameters in order from the smallest to the largest and then finding the particle diameter that splits the list into two equal halves.

Example 2.1 Given the following particle diameter data, determine the mean and median diameters of the aerosol.

Interval, μm	No. n_i
1–2	30
2–3	90
3–5	50
5–10	20
10–20	10

By using Eq. 2.1 the following table can be formed. The midpoint of the size interval is chosen as the best estimate of the size of all particles in that interval.

Midpoint d_i	n_i	$n_i d_i$
1.5	30	45
2.5	90	225
4.0	50	200
7.5	20	150
15.0	10	150
	200	770

The mean value is 770/200 = 3.85 μm.

By inspection of the table, the median value can be seen to lie somewhere between 2 and 3 μm in diameter. With the given data a more precise evaluation of this number is not possible.

Although they are simple in concept, neither the mean nor the median diameter alone conveys much information about the general range of particle diameters present. Usually more information is required describing the spread of the particle size distribution. This gives some indication of how well the mean or median value represents all particles in the aerosol.

It is common practice to describe an aerosol solely by some average value, completely ignoring considerations of particle size distribution.

When this is done, estimates of aerosol properties are much less accurate than they would have been if all particle sizes had been taken into account.

Histograms

Besides their use in determining a mean or median value, the numbers of particles in various size intervals can be plotted as bar charts or line charts. These plots are pictures of the size distribution of the aerosol. This is useful in envisioning the range and frequency of the sizes present.

Example 2.2 Plot the data given in Example 2.1 first by plotting the midpoints of the size intervals as a function of particle diameter and then by plotting a bar chart of number of particles per unit size interval against each size interval.

Interval, μm	Midpoint d	Interval size, μm	n_i	n_i per micrometer
1–2	1.5	1	30	30
2–3	2.5	1	90	90
3–5	4.0	2	50	25
5–10	7.5	5	20	4
10–20	15.0	10	10	1

Figure 2.1a shows a line chart of the midpoints of the data. Although the particle diameter distribution is plainly shown, it is possible to alter the shape of the distribution by changing the interval size.

When a bar chart is plotted instead of a line chart, as in Fig. 2.1b, this problem is not as severe. The ordinate or height of each bar is normalized by dividing the number of particles in an interval by the width of that interval. The width of each bar represents the actual width of each size interval. Then the area of each block represents the relative frequency of particles in that particular size interval.

Charts or graphs of this sort have the advantage of showing at a glance what the particle size distribution of an aerosol looks like and is perhaps the best way of visually representing complex size-distribution data.

For atmospheric aerosols, a wide range of particle sizes may be present in numbers which can vary by several orders of magnitude. In these cases the typical bar graph will not be satisfactory since the large numbers of small particles can completely overwhelm the display of other sizes, even though the larger sizes may be most signifi-

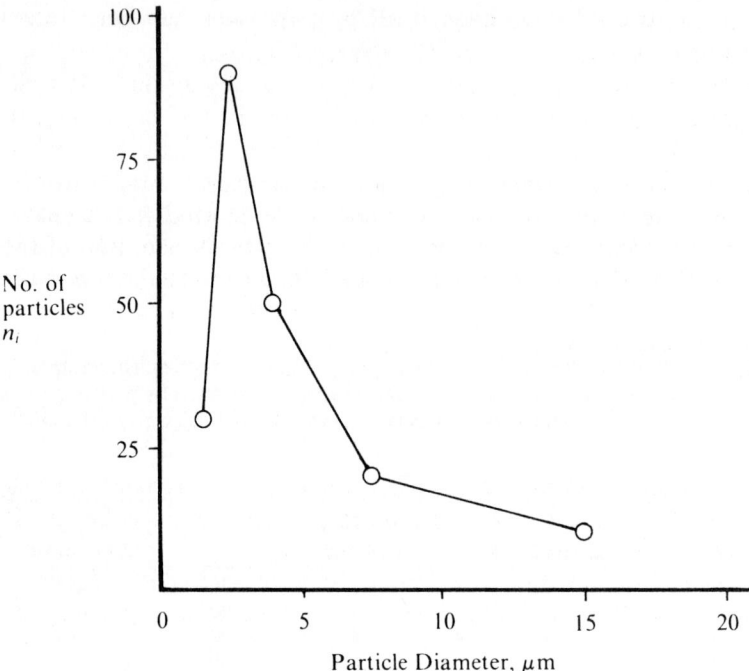

Figure 2.1a Simple plot of distribution data.

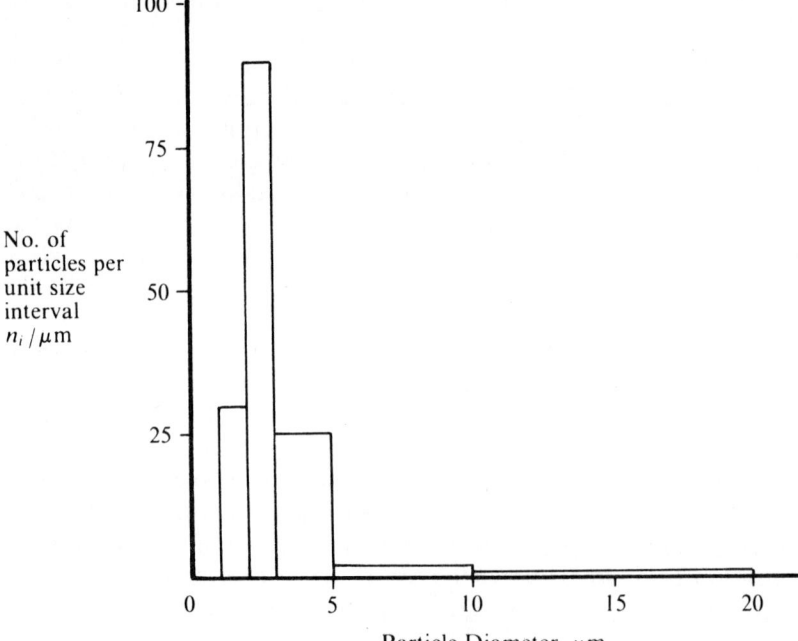

Figure 2.1b Bar graph of particle distribution data.

cant in terms of mass or surface area. Or the larger particle diameters will be displayed more prominently than the smaller ones, even though the smaller sizes may be of primary interest.

One solution is to plot the logarithms of particle diameter on the abscissa instead of the diameters themselves. This spreads out the presentation of distribution data so that a much broader range of particle sizes can be visualized. However, to maintain the relationship that the area between two particle size intervals is proportional to the total number of particles present, the ordinate scale must be altered. This is done by dividing the number of particles in each interval by the difference in the logarithms of the largest and smallest particle sizes of that interval, or, in mathematical terms,

$$\text{Ordinate value} = \frac{\Delta n}{\Delta \log d} \qquad (2.2)$$

This relationship is found for each size interval. Similar expressions can be written for particle surface area or particle mass or volume. (It should be stressed again that particle volume converts directly to particle mass by multiplication of volume and particle density. Hence in plotting size distribution data, either one can be used to represent the other.)

Example 2.3 Plot the data given in Example 2.1 in the form of $\Delta n/\Delta \log d$ versus $\log d$.

log 2 - log 1

Interval	Midpoint	No. Δn	$\Delta \log d$	$\Delta n/\Delta \log n$
1–2	1.5	30	0.30	100
2–3	2.5	90	0.18	511
3–5	4.0	50	0.22	225
5–10	7.5	20	0.30	66
10–20	15.0	10	0.30	33

The data are plotted in Fig. 2.2. A continuous distribution is assumed in order to develop a smooth plot.

Continuous curves of the type illustrated in Fig. 2.2 are often used to show the difference in size distributions of aerosol number, surface area, or mass, with the same aerosol. These differences arise when there are large numbers of small particles present in an aerosol. These particles contribute greatly to total particle count but little to total particle mass or surface area.

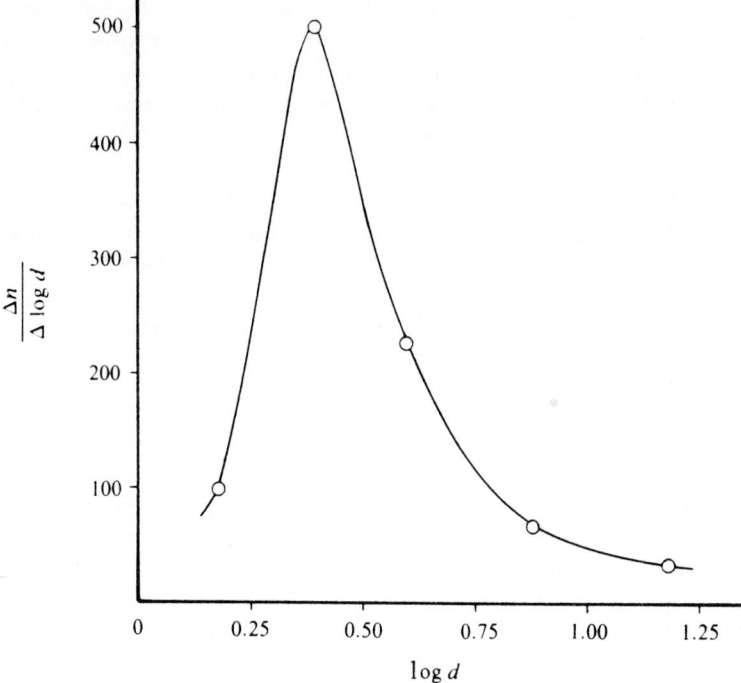

Figure 2.2 Plot of data from Example 2.3. A continuous distribution is assumed.

Example 2.4 Plot surface and volume distribution for the aerosol given in Example 2.1 in the same fashion as plotted in Example 2.3. However, in this case normalize the linear ordinate so that the relative areas and volumes under similar interval limits will be comparable.

For particle surface area S, values in the ΔS column are determined by multiplying the number of particles in each interval by the square of the midpoint diameter of that interval. For particle volume V, values in the ΔV column are found by multiplying the number of particles in each interval by the cube of the midpoint particle diameter. It is not necessary to multiply the ΔS values by π or the ΔV values by $\pi/6$ since these constants will cancel when the ΔS and ΔV quantities are normalized by dividing by the sum of all values.

Interval	ΔS	$\Delta S/(S_T \Delta \log d)$	ΔV	$\Delta V/(V_T \Delta \log d)$
1–2	67.5	0.047	101.3	0.007
2–3	562.5	0.650	1,406.3	0.170
3–5	800.0	0.757	3,200.0	0.308
5–10	1,125.0	0.778	8,437.5	0.598
10–20	2,250.0	1.556	33,750.0	2.391
	$S_T = 4,805.0$		$V_T = 46,895.1$	

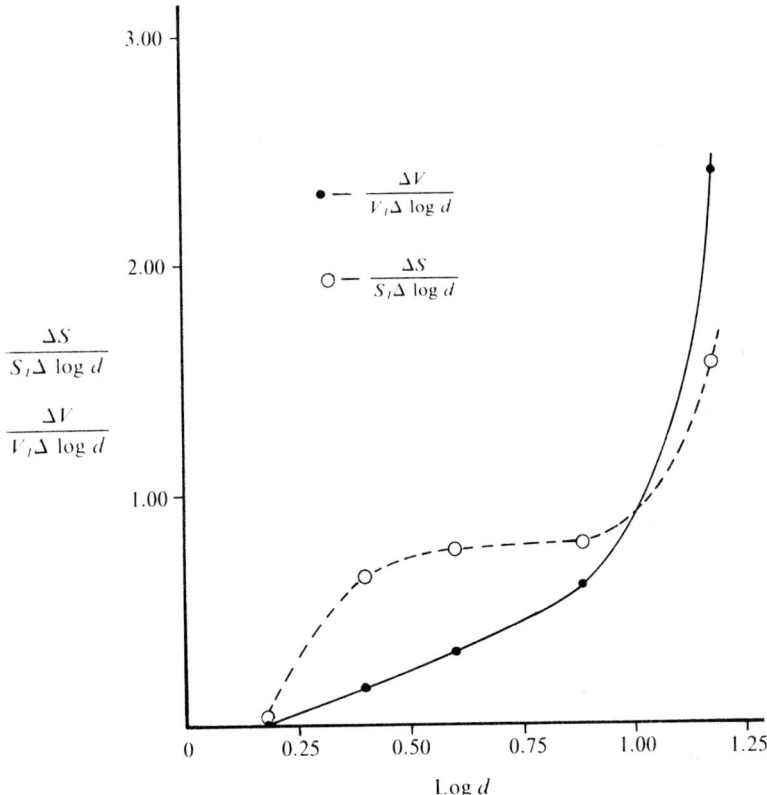

Figure 2.3 Similar plot as in Fig. 2.2 except particle surface area S and particle volume V are plotted. These curves are normalized by dividing by S_T and V_T, respectively.

Comparison of the data plotted in Fig. 2.3 shows how surface area and volume (mass) tend to be associated mainly with the larger-size particles whereas in general the smaller particles contribute mainly to the total numbers present. Therefore in presenting size distribution data it is important to consider the purpose of the presentation and which feature [number, surface area, or volume (mass)] is to be stressed.

Mathematical Representation of Distribution

If the size interval of the aerosol is permitted to become very small, the resulting histogram begins to approximate a smooth curve. Then it is possible to represent the distribution by a smooth curve or, better still, by some mathematical function, i.e.,

$$dn_i = f(d)dd \qquad (2.3)$$

where dn_i is the number of particles lying in the interval between sizes dd_{i-1} and dd_i.

Obviously, to plot this sort of curve requires analysis of the sizes of a great number of particles. Or, if it were possible to specify some identifying parameters of the distribution, a functional form could be used to represent a whole family of curves. There have been many attempts to find such a functional form. Usually these equations have been satisfactory for aerosols from the same specific sources but are not generally applicable to all aerosols.

One widely used form which is applicable to many different aerosols from a variety of sources is the *lognormal distribution*. To understand the utility of the lognormal distribution, it is first necessary to review the concept of a "normal" distribution.

Normal distribution

Many phenomena which appear to occur on a more or less random basis exhibit certain characteristics which can be used to predict future trends. For example, although it is impossible to tell on any single toss whether a coin will come up heads or tails, if the coin is unbiased, heads will come up approximately 50 percent of the time. The more tosses made, the closer one usually comes to this approximation.

Suppose 100 tosses were made and the number of heads was recorded, and the experiment is repeated many times. It would be observed that although usually there would be about 50 heads for every 100 tosses, occasionally there would be substantially greater or fewer. If the frequency of heads were plotted as a function of the number of heads observed in 100 tosses, a curve shape would be found that is entirely predictable. This shape, known as a *normal* distribution or *normal* curve, is shown in Fig. 2.4a. The primary virtue of a normal distribution is that because it is predictable, it can be described with two characteristic numbers, a mean value and a standard deviation. These are shown in Fig. 2.4a and are defined mathematically as

$$\bar{d} = \frac{\sum_{i=0}^{\infty} n_i d_i}{\sum_{i=0}^{\infty} n_i} \qquad (2.4)$$

and

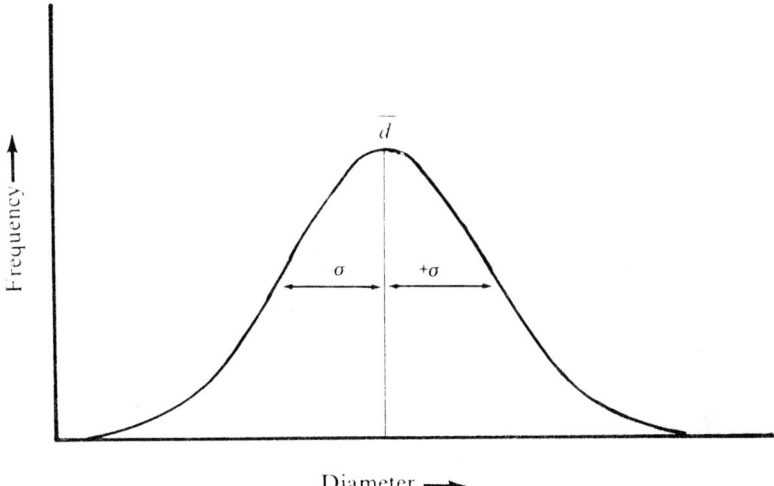

Figure 2.4a The normal distribution.

$$\sigma = \left[\frac{\sum_{i=0}^{\infty} n_i(\overline{d} - d_i)^2}{\left(\sum_{i=0}^{\infty} n\right) - 1} \right]^{1/2} \quad (2.5)$$

Example 2.5 Compute the value of the standard deviation σ for the data given in Example 2.1.
From Example 2.1 the mean value was determined to be 3.85 μm.

Interval	Midpoint	No. n_i	$\overline{d} - d_i$	$(\overline{d} - d_i)^2 n_i$
1–2	1.5	30	2.35	165.68
2–3	2.5	90	1.35	164.03
3–5	4.0	50	– 0.15	1.13
5–10	7.5	20	– 3.65	266.45
10–20	15.0	10	– 11.15	1243.23
		200		1840.52

The standard deviation $\sigma = [1840.52/(200 - 1)]^{1/2} = (9.25)^{1/2} = 3.04$ μm.

Means and standard deviations can be calculated for any set of data. For data which are normally distributed, however, the mean value lies at the midpoint of the data (hence it is also the median), and 67 percent of the distribution falls between the range of plus or minus one standard deviation.

Normal distributions occur with a variety of statistical data including average height and weight of children, grade distributions of large groups of students, and even frequency of underweight or overweight candy bars on a production line. One might guess that aerosol particle sizes would also be normally distributed.

Unfortunately, this is generally not the case. For many aerosols a plot of frequency versus size results in a graph similar to that shown in Fig. 2.4b, in which there are proportionally many more smaller particles than larger ones. The curve is said to be skewed toward the larger particle sizes.

Lognormal distribution

It was observed many years ago that particle size data which were skewed and did not fit a normal distribution would very often fit a normal distribution if frequency were plotted against the *logarithm* of particle size instead of particle size alone. This tended to spread out the smaller size ranges and compress the larger ones. If the new plot then looked like a normal distribution, the particles were said to be *lognormally* distributed and the distribution was called a *lognormal* distribution. By analogy with a normal distribution, the mean and standard deviation became

$$\log d_g = \frac{\sum n_i \log d_i}{\sum n_i} \qquad (2.6)$$

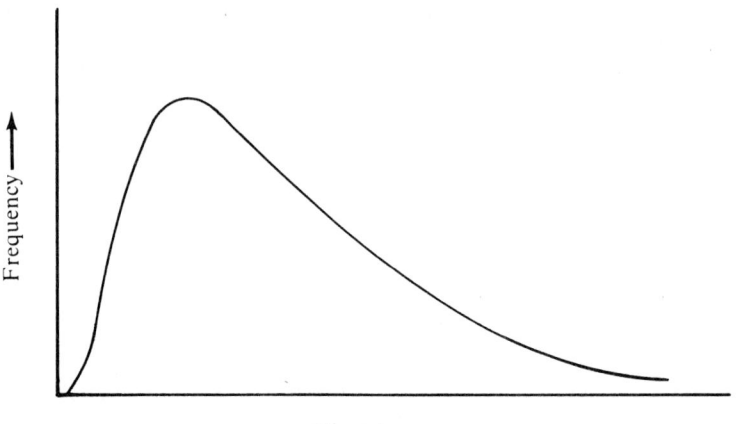

Figure 2.4b The lognormal distribution.

known as the *geometric mean diameter*, and

$$\log \sigma_g = \left[\frac{\sum n_i (\log d_g - \log d_i)^2}{\sum n_i - 1} \right]^{1/2} \quad (2.7)$$

where σ_g is known as the *geometric standard deviation*.

Example 2.6 Compute the geometric mean diameter d_g and geometric standard deviation σ_g for the data given in Example 2.1.
To compute the geometric mean

Interval	d_i	No. n_i	$\log d_i$	$n_i \log d_i$
1–2	1.5	30	0.176	5.283
2–3	2.5	90	0.398	35.815
3–5	4.0	50	0.602	30.103
5–10	7.5	20	0.875	17.501
10–20	15.0	10	1.176	11.761
		200		100.463

The geometric mean $d_g = \log^{-1}(100.463/200) = \log^{-1} 0.502 = 3.18$ μm.
To compute the geometric standard deviation

d_i	No. n_i	$\log d_g - \log d_i$	$n_i (\log d_g - \log d_i)^2$
1.5	30	0.326	3.195
2.5	90	0.104	0.983
4.0	50	−0.100	0.496
7.5	20	−0.373	2.777
15.0	10	−0.674	4.538
	200		11.989

The geometric standard deviation

$$\sigma_g = \log^{-1} \left(\frac{11.989}{199} \right)^{0.5}$$

$$= \log^{-1} 0.245 = 1.760$$

Notice that σ_g is a pure number. Unlike the regular standard deviation, it has no units. This is because it represents a ratio of diameters.

With a lognormal distribution, one geometric standard deviation represents a range of particle sizes within which lie 67 percent of all sizes. In this case the range is from d_g/σ_g to $d_g\sigma_g$, unlike the simple additive case for a normal distribution. Ninety-five percent of all particles would lie in a range d_g/σ_g^2 to $\sigma_g^2 d_g$. Thus for a monodisperse aerosol, σ_g is equal to 1 whereas σ is equal to 0 for a normal distribution.

The functional form of the lognormal distribution can be written as (Herdan, 1960)

$$f(d) = \frac{1}{d \ln \sigma_g (2\pi)^{0.5}} \exp\left[-\frac{(\ln d - \ln d_g)^2}{2 \ln^2 \sigma_g}\right] \tag{2.8}$$

where

$$\int_0^\infty f(d)\,dd = 1 \tag{2.9}$$

Example 2.7 Given $d_g = 1$ µm and $\sigma_g = 2$. Find $f(d)$ when $d = d_g$.

solution

$$f(d) = \frac{1}{d \ln \sigma_g (2\pi)^{0.5}} \exp\left[-\frac{(\ln d - \ln d_g)^2}{2 \ln^2 \sigma_g}\right]$$

$$= \frac{1}{1 \ln 2 (2\pi)^{0.5}} \exp\left[-\frac{(\ln 1 - \ln 1)^2}{2 (\ln 2)^2}\right]$$

$$= 0.576 \text{ µm}^{-1}$$

Letting $dd = 0.1$, the approximate fraction of particles lying within the range of 0.95 to 1.05 µm would be $0.576 \times 0.1 = 0.058 = 5.8$ percent.

Log Probability Paper

Because a lognormal distribution can be expressed as a distinct mathematical function, it is possible to construct graph paper on which a cumulative lognormal distribution plots as a straight line. An example of such a plot is shown in Fig. 2.5. Data are plotted as cumulative percentage of particles equal to or less than the largest size of each size interval versus the upper size of that size interval. A straight line on such a plot implies a lognormal distribution.

If a straight line can be fitted to the plot, then the median particle diameter can be determined as being the 50 percent value on the plot (remember that when you are plotting number distribution, geometric mean and median for the number distribution are the same if there is a lognormal distribution). The geometric standard deviation is determined by the ratio

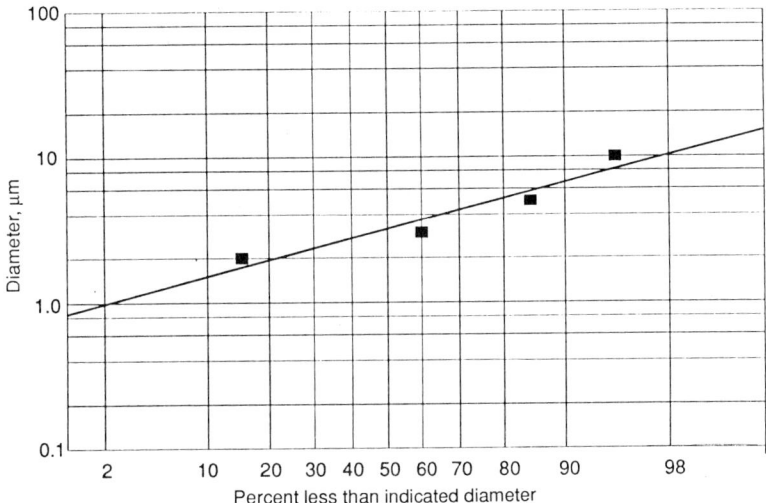

Figure 2.5 Log probability plot using the data from Example 2.1.

$$\sigma_g = \frac{84.13\% \text{ diameter}}{50\% \text{ diameter}} = \frac{50\% \text{ diameter}}{15.87\% \text{ diameter}} \quad (2.10)$$

Use of log probability paper is the simplest way to determine the mean and geometric standard deviation provided *the distribution does indeed follow a lognormal shape or at least approximates it.*

Other Definitions of Means

There are a number of different mean or median values which can be defined for a particle size distribution. These means or medians are useful depending on where the data came from or how the data are to be used. For example, the diameter of average mass (volume) can be defined as representing the diameter of a particle whose mass (volume) times the number of particles gives the total mass (volume) of all the particles. Similarly, the diameter of average surface represents the diameter of a particle whose surface times the number of particles gives the total surface.

Choice of which average diameter to use in a given situation depends on how the diameter was measured or how it is to be used. For the case where aerosol mass is measured and the fractions collected are associated with specific particle diameters, the resulting average value is the mass median diameter. In studying chemical reaction rates, the volume-surface mean diameter may be more important than just the arithmetic mean or geometric number mean.

Table 2.1 gives definitions for various "average" diameters. For a lognormally distributed aerosol the different diameters defined in Table 2.1 can be related by the equation (Raabe, 1971)

$$d_p = d_g \exp(p \ln^2 \sigma_g) \qquad (2.11)$$

TABLE 2.1 Definitions for Various "Average" Diameters

Indicated diameter	Symbol	Definition	Description
Mode	d_0 $p = -1$	d at maximum n_i	Diameter associated with the maximum number of particles in a distribution
Geometric mean	d_g $p = 0$	$\log^{-1}(\Sigma n_i \log d_i / \Sigma n_i)$	The Σnth root of the product of all particle diameters, also for a lognormal distribution the median diameter
Arithmetic mean	d $p = 0.5$	$\Sigma n_i d_i / \Sigma n_i$	The sum of all diameters divided by the total number of particles
d of average surface	d_s $p = 1$	$\sqrt{\Sigma n_i d_i^2 / \Sigma n_i}$	The diameter of a hypothetical particle having average surface area
d of average volume (mass)	d_v $p = 1.5$	$\sqrt[3]{\Sigma n_i d_i^3 / \Sigma n_i}$	The diameter of a hypothetical particle having average volume or mass
Surface median diameter	d_{smd} $p = 2$	$\log^{-1}(\Sigma n_i d_i^2 \log d_i / \Sigma n_i d_i^2)$	The geometric mean of the particle surface areas or for a lognormal distribution the area median diameter
Surface mean diameter (Sauter diameter)	d_{sm} $p = 2.5$	$\Sigma n_i d_i^3 / \Sigma n_i d_i^2$	The average diameter based on unit surface area of a particle
Volume median diameter (mass)	d_{mmd} $p = 3$	$\log^{-1}(\Sigma n_i d_i^3 \log d_i / \Sigma n_i d_i^3)$	The geometric mean of particle volumes (mass) or for a lognormal distribution the volume (mass) median diameter
Volume mean diameter (mass)	d_{vm} $p = 3.5$	$\Sigma n_i d_i^4 / \Sigma n_i d_i^3$	The average diameter based on the unit volume (mass) of a particle

p values assume a lognormal distribution.

where p is a parameter which serves to define the various possible diameters. Values of p which are associated with the different diameters are listed in Table 2.1.

Example 2.8 Given the particle size distribution shown in the table below, compare values for different diameters, using the definitions in Table 2.1 with values computed with Eq. 2.11.

Size Distribution Data and Computations

d	f(d)	(1)	(2)	(3)	(4)	(5)	(6)	(7)
0.12	2	0.240	0.029	0.003	0.000	− 1.842	− 0.027	− 0.003
0.17	5	0.850	0.144	0.025	0.004	− 3.848	− 0.111	− 0.019
0.24	14	3.360	0.806	0.194	0.046	− 8.677	− 0.500	− 0.120
0.32	60	19.200	6.144	1.966	0.629	− 29.691	− 3.040	− 0.973
0.48	100	48.000	23.040	11.059	5.308	− 31.876	− 7.344	− 3.525
0.68	190	129.200	87.856	59.742	40.625	− 31.823	− 14.715	− 10.006
1	250	250.000	250.000	250.000	250.000	0.000	0.000	0.000
1.4	160	224.000	313.600	439.040	614.656	23.380	45.826	64.156
1.9	110	209.000	397.100	754.490	1,433.531	30.663	110.693	210.317
2.6	70	182.000	473.200	1,230.320	3,198.832	29.048	196.365	510.550
3.6	28	100.800	362.880	1,306.368	4,702.925	15.576	201.871	726.736
5.1	10	51.000	260.100	1,326.510	6,765.201	7.076	184.039	938.599
7.2	1	7.200	51.840	373.248	2,687.386	0.857	44.444	319.998
	1,000	1,224.850	2,226.740	5,752.965	19,699.144	− 1.156	757.501	2,755.709

(1) $-d\, f(d)$
(2) $-d^2\, f(d)$
(3) $-d^3\, f(d)$
(4) $-d^4\, f(d)$
(5) $-f(d)\, \log d$
(6) $-f(d)\, d^2 \log d$
(7) $-f(d)\, d^3 \log d$

Results

Definition	Computed from data	From Eq. 2.11
Count mean	1.22	1.21
Geometric mean	1.00	1.00
Diameter of average mass	1.79	1.76
Diameter of average area	1.49	1.46
Area median	2.19	2.13
Mass (volume) mean	3.42	3.76

Figure 2.6 shows the relative location of each of the diameters computed in Example 2.8 on a typical lognormal distribution plot.

Equation 2.11 is a more general form of a well-known relationship used for converting particle number measurements to mass measurements and vice versa known as the *Hatch-Choate equation* (Drinker and Hatch, 1954).

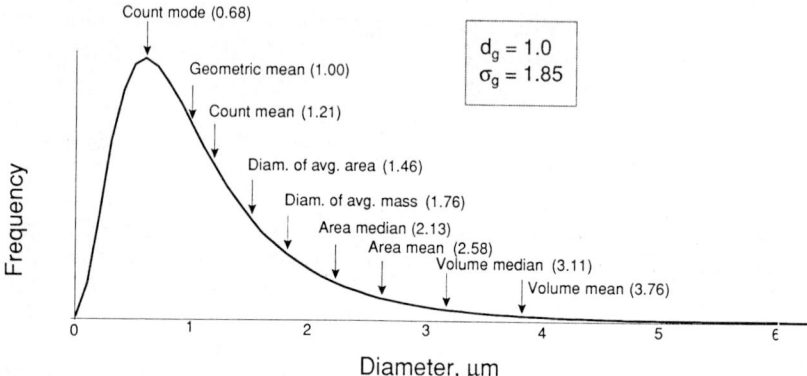

Figure 2.6 Example of lognormal distribution.

In its original form the Hatch-Choate equation for conversion of number to mass is given by

$$\log d_{mmd} = \log d_g + 6.9 \log^2 \sigma_g \qquad (2.12)$$

where d_{mmd} is the mass median diameter and for surface median diameter d_{smd}

$$\log d_{smd} = \log d_g + 4.6 \log^2 \sigma_g \qquad (2.13)$$

These equations can be derived from Eq. 2.11. It is important to note that σ_g will be the same regardless of the definition of diameter used. That is, with a lognormal distribution σ_g will be the same whether number, surface, or mass median diameters are being measured.

Example 2.9 Given a lognormally distributed aerosol with a geometric mean diameter of 1.5 μm and a σ_g of 2.3, what are the surface-area median diameter and the mass median diameter of this aerosol?

Using the Hatch-Choate equation for surface median diameter gives

$$\log d_{smd} = \log d_g + 4.6 \log^2 \sigma_g$$
$$= 0.176 + (4.6)(0.362)^2$$
$$= 0.176 + 0.602 = 0.778$$
$$d_{smd} = 6.0 \text{ μm}$$

Mass median diameter will be the same as volume median diameter (since particle density cancels in computing the means). Thus we can use the Hatch-Choate relationship directly:

$$\log d_{mmd} = \log d_g + 6.9 \log^2 \sigma_g$$
$$= 0.176 + (6.9)(0.362)^2$$
$$= 1.080$$
$$d_{mmd} = 12.0 \, \mu m$$

Note that this result can also be found by using Eq. 2.11. With a lognormal distribution, the volume or mass median diameter will always be greater than the surface median diameter which will in turn be greater than the number median diameter.

Problems

1 Given the following data:

Size interval, μm	Number
0.1–0.5	120
0.5–0.8	380
0.8–1.4	146
1.4–2.7	96
2.7–5.6	53
5.6–8.9	22
8.9–12.6	8

Construct a histogram showing number per unit size interval for each size interval. Show that the area of each block is proportional to the number of particles represented by that block.

2 Using the data in Prob. 1, compute the mean particle diameter and standard deviation of this distribution.

3 Using the same data, compute a geometric median size and geometric standard deviation. What would be the numerical value of the geometric standard deviation if the particles were all the same size?

4 With the data given in Prob. 1, plot the number distribution function $\Delta n/(n_T \Delta \log d)$ and the mass distribution function $\Delta m/(m_T \Delta \log d)$ as a function of the logarithm of the particle diameter. Assume all particles within a size interval are spheres having a diameter equal to the midpoint of the size interval. The density of the particles equals 1 g/cm^3.

5 Plot the data from Prob. 1 on log probability paper. Find the line of best fit for these data, and then determine the geometric mean and geometric standard deviation from this line.

6 Using the Hatch-Choate equation, compute the mass median diameter from the information developed in Prob. 5. If the aerosol contains 1 million particles per cubic foot and the particle density is 1 g/cm^3, find the aerosol concentration in micrograms per cubic meter.

7 Show that the Hatch-Choate equations are just special cases of the general equation for lognormal distributions, Eq. 2.11.

8 Show that the integral of $f(x)$ for a lognormal distribution (Eq. 2.8) does equal 1 when $f(x)$ is integrated over the limits of zero to infinity.

Chapter 3

Fluid Properties

An aerosol is a suspension of particles in a gaseous medium. Without the medium there would be no aerosol. The medium acts to restrain random particle motion, supports the particles against the strong pull of gravity, and in some cases acts as a buffer between particles. It is impossible to properly study aerosol behavior without first considering the medium in which the particles are suspended.

Medium behavior can be visualized in two ways. First, it can be considered to be a large collection of small spheres (molecules) that are in random motion with each other but may be in ordered motion overall. A general treatment of matter from a molecular point of view is called *statistical mechanics,* and the nonequilibrium gaseous portion is referred to as *kinetic theory.*

A second way to visualize gas behavior is by considering the gas to be a *continuous medium,* i.e., similar to some sort of interlocking syrup such as molasses or water. Study of medium properties in this case is known as *fluid dynamics* or for air *aerodynamics.* In the first case, the microscopic (small) properties of the gas are important. In the second, it is the macroscopic (large) properties which are of interest. Since aerosol particles can span the range from near-molecular sizes up to hundreds of micrometers, the gas in which the particles are suspended must be considered both from a molecular point of view and as a continuous medium.

In studying aerosols it is important to develop in one's mind's eye a picture of the process taking place. By visualizing the problem (even if it is in a simplified form) it is easier to find a method of solution, since most problems are more difficult to set up than they are to solve, once stated. To carry out this visualization, one must have an understanding of the physical phenomena that come into play and a means for estimating their effect. Thus when one is considering a pitched baseball, it is only necessary to visualize how hard the ball is thrown and whether any spin is imparted to it. If this baseball were 1 μm in diameter, it would also be extremely important to consider the properties of the air through which the baseball travels. Why? Because the medium looks different to the baseball-sized baseball than to the micrometer-sized baseball. In this chapter various properties of the medium (usually air, but it could be any other gas) are discussed from both a microscopic and a macroscopic viewpoint, so that visualization skills can be enhanced and important medium properties introduced.

Kinetic Theory

The following represents only the briefest discussion of kinetic gas theory. For more information there are many good texts on the subject [e.g., see Ladd (1986), Barrow (1973), or Daniels and Alberty (1979)].

In considering a gas from the molecular point of view, three main assumptions can be made initially (Daniels and Alberty, 1979):

1. The gas volume of interest contains a very large number of molecules.

2. The molecules are small compared to the distances between them and are in a state of continuous motion, traveling in straight lines between collisions.

3. The molecules are spherical and do not interact with each other except by elastic collisions. Elastic collisions represent no energy loss due to rearrangement of the interior of the molecule.

With these assumptions it is possible to simplify molecular behavior to a point where the gas can be treated statistically.

Example 3.1 Determine the number of molecules in 1 cm³ of air at 760-mmHg pressure and 20°C.
Let V = volume of gas occupied by 1 mol = 22.4 L at standard conditions.
For 20°C this volume must be increased in proportion to the increase in absolute temperature.
Zero degrees Celsius on the absolute scale is 273 K. Thus

$$V_{20°C} = 22.4 \left(\frac{273° + 20°}{273°} \right) = 24.04 \text{ L}$$

then, since the number of molecules in 1 mol is 6.02×10^{23}, i.e., *Avogadro's number* N_A,

$$\frac{N_A}{V} = \frac{6.02 \times 10^{23}}{24.04 \times 10^3} = 2.50 \times 10^{19} \text{ molecules/cm}^3$$

Example 3.1 illustrates the large number of molecules that are present in even a fairly small volume of gas. Thus the first assumption holds. In dealing with very low pressures (small numbers of molecules per unit volume) or small volumes, statistical assumptions may not hold; in these cases it is important to consider the medium properties quite carefully before applying any generalities about aerosol behavior.

Example 3.2 Assuming all molecules are regularly spaced within a 1-cm³ volume, determine the average distance between them.

If there are 2.50×10^{19} molecules per cubic centimeter, then there would be one molecule for each

$$\frac{1}{2.50 \times 10^{19}} \text{ cm}^3$$

$$= 4.0 \times 10^{-20} \text{ cm}^3$$

This represents a cube surrounding a single molecule. The length of one side of the cube or the distance between two molecules is

$$\text{Distance} = (4.0 \times 10^{-20})^{1/3}$$

$$= 3.42 \times 10^{-7} \text{ cm}$$

$$= 34.2 \text{ Å}$$

Typical molecular diameters for gas molecules range from about 2 to 5 Å. Hence it can be concluded that the second assumption holds, since even with this simplistic analysis the average distance between molecules is at least 10 times the molecular diameters.

For aerosols, the smallest particle diameters are about 0.005 μm, increasing to 100 μm or so (50 to 1,000,000 Å). At the smallest sizes, aerosol particles begin to approach some very large molecules in size.

Gas Behavior

Some of the basic properties of gases can be deduced by using fairly simple logic. Since this same sort of reasoning is used later to deduce aerosol properties, it is instructive here to give two examples of the types of thought processes that yield great insight into physical phenomena.

Molecular speeds (Bernoulli)

This is a very simple approach to the question of how fast the molecules are moving in a gas.

Let N molecules be enclosed in a cubical box, as illustrated in Fig. 3.1, the length of each edge being L. We assume that one-third of the molecules move back and forth so that they strike face A, one-third move in a similar manner so that they strike face B, and one-third move similarly so that they strike face C.

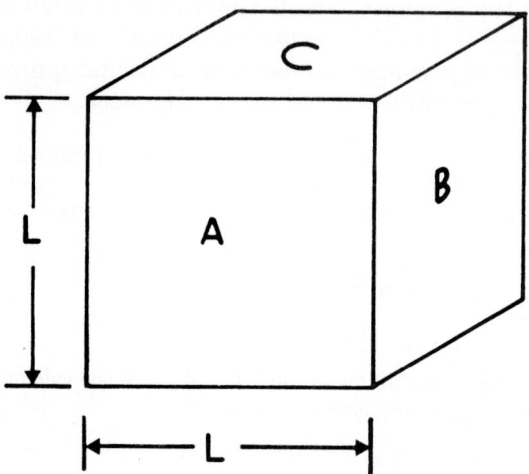

Figure 3.1 Box envisioned for molecular speed estimates.

After a molecule strikes one of the faces, say, face B, it must travel a distance $2L$ before it strikes face B again. Therefore it makes $C/(2L)$ hits per unit time, where C is the velocity of the molecule. If the mass of the molecule is m, at each hit the molecule imparts a momentum of $2mC$. (The factor 2 comes from the molecular velocity changing from $+C$ to $-C$.)

The change in momentum per unit time, dp/dt is the change per hit times the number of hits:

$$\frac{dp}{dt} = 2mC\left(\frac{C}{2L}\right) = \frac{mC^2}{L} \quad (3.1)$$

The total momentum transferred to face B in unit time is

$$\frac{N}{3}\frac{mC^2}{L} \quad (3.2)$$

Newton's second law of motion states that force is proportional to the rate of change of momentum. Therefore the total momentum transferred to the wall per unit time is equal to the force acting on that wall.

$$F = \frac{1}{3}\frac{NmC^2}{L} \quad (3.3)$$

$$\text{Pressure } p = \frac{\text{force}}{\text{area}} = \frac{NmC^2}{3L^3} = \frac{NmC^2}{3V} \quad (3.4a)$$

where V is the volume of the box.

Since Nm/V is the density ρ of the gas,

$$p = \frac{1}{3}\rho C^2 \qquad (3.4b)$$

or

$$C = \left(\frac{3p}{\rho}\right)^{1/2} \qquad (3.5)$$

Thus it is possible to compute the average speed of gas molecules merely from a knowledge of the pressure p and gas density ρ. For hydrogen under standard conditions, $C = 1696$ m/s, approximately the speed of a bullet. This simple derivation is reasonably accurate, even though the assumption is made that all molecules are traveling at the same velocity. Often simplifying assumptions permit the parameters in an equation to be identified, even if the values of the constants may be somewhat inaccurate.

Example 3.3 Compute the estimated speed of an "air" molecule at 20°C and normal pressure.

$$\text{Density air} = \frac{MW}{MV} = \frac{29}{22.4 \times 10^3} \frac{273}{293}$$

$$= 1.21 \times 10^{-3} \text{ g/cm}^3$$

Atmospheric pressure = 760 mmHg

$$= 1013.25 \text{ mbar}$$

$$C = \left[3 \frac{(1013.25 \times 10^3)}{1.21 \times 10^{-3}}\right]^{1/2}$$

$$= (2.51 \times 10^9)^{1/2}$$

$$= 502 \times 10^2 \text{ cm/s}$$

The derivation presented above gives a reasonably close estimate of actual average molecular velocities, despite its obvious simplifications.

In actuality, molecular velocities are not all the same. At any time some molecules are moving much faster than the average while others are moving more slowly than the average. For a perfect gas the velocity distribution (in one dimension) is given by the Maxwell-Boltzmann distribution function,

$$f(v_x)\, dv_x = \left(\frac{m}{2\pi kT}\right)^{1/2} \exp\left(\frac{-mv_x^2}{2kT}\right) dv_x \qquad (3.6)$$

A plot of this equation is shown in Fig. 3.2a. The k is Boltzmann's constant: $k = 1.38 \times 10^{-16}$ erg/K. $\quad erg = \frac{g\,cm^2}{s^2}$

The most probable velocity in the x direction is zero with positive and negative velocities having equal probabilities.

Example 3.4 Find the most probable value of $f(v_x)$ for an "air" molecule. Assume normal temperature and pressure.

The most probable value of $f(v_x)$ occurs when $v_x = 0$. Hence

$$f(v_x) = \left(\frac{m}{2\pi kT}\right)^{1/2} = \left[\frac{29/6.02 \times 10^{23}}{(2\pi)(1.38 \times 10^{-16})(293)}\right]^{1/2}$$

$$= 1.38 \times 10^{-5}$$

Although Eq. 3.6 represents molecular behavior in a single direction, when all three directions are taken into account simultaneously, the probability that an arbitrarily selected molecule will have a velocity between v and $v + dv$ is

$$f(v)\,dv = \left(\frac{m}{2\pi kT}\right)^{3/2} \exp\left(\frac{-mv^2}{2kT}\right) 4\pi v^2\,dv \qquad (3.7)$$

Equation 3.7 says that the probability of having zero velocity is zero. That is, there is no chance that at any time a molecule will completely stop in its motion. Figure 3.2b shows a plot of Eq. 3.7 for air.

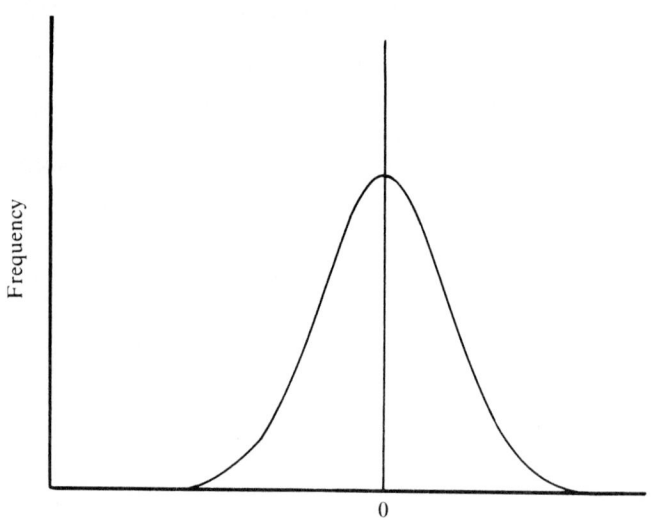

Velocity in One Direction, V_x

Figure 3.2a One-dimensional Maxwell-Boltzmann velocity distribution.

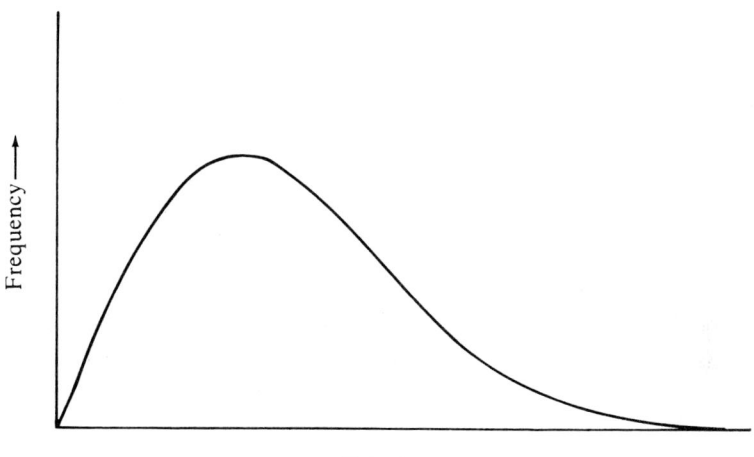

Figure 3.2b Three-dimensional Maxwell-Boltzmann velocity distribution.

There are several ways in which the representative velocity of the ensemble of molecules can be defined.

The arithmetic mean velocity is

$$\bar{v} = \int_0^\infty v f(v)\, dv = \left(\frac{8kT}{\pi m}\right)^{0.5} \tag{3.8}$$

The most probable velocity is obtained by taking the derivative of $f(v)\, dv$ with respect to v and setting it equal to zero.

$$v_p = \left(\frac{2kT}{m}\right)^{0.5} \tag{3.9}$$

The root-mean-square velocity is

$$v_{\text{rms}} = (\overline{v^2})^{0.5} = \left[\int_0^\infty f(v) v^2\, dv\right]^{0.5} = \left(\frac{3kT}{m}\right)^{0.5} \tag{3.10}$$

Example 3.5 Compute the most probable velocity of an air molecule at standard pressure and 20°C. Remember that $m = MW/N_A$.

$$v_p = \left(\frac{2kT}{m}\right)^{0.5}$$

$$= \left[\frac{2(1.38 \times 10^{-16})(293)}{4.82 \times 10^{-23}}\right]^{0.5}$$

$$= 40{,}972 \text{ cm/s}$$

Each average value of velocity can be used to best describe some particular property of the ensemble of molecular velocities. For example, in a gas all molecules have the same average kinetic energy. Hence, the root-mean-square velocity is the best estimate of velocity to use for computing parameters that are a function of kinetic energy

$$\overline{E} = \frac{mv^2}{2} = \frac{m3kT}{2m} = \frac{3kT}{2} \tag{3.11}$$

The term $\overline{E}$ is the average energy of a molecule in a gas. Interestingly, an aerosol particle suspended in the gas will acquire this same average kinetic energy from the molecules in the gas.

Example 3.6 What is the average kinetic energy of a 1.0-μm unit-density sphere which is in equilibrium with its surroundings? The air temperature is 20°C.

$$\overline{E} = \frac{3kT}{2} = \frac{(3)(1.38 \times 10^{-16})(293)}{2}$$

$$= 6.07 \times 10^{-14} \text{ erg}$$

Mean free path

The *mean free path* is defined as the average distance a molecule will travel in a gas before it collides with another molecule. This is related to molecular spacing but takes into account the fact that all molecules are in a constant state of motion and thus are more widely separated than they would be if they were firmly bound to each other. Mean free path can be estimated by using the following simple argument.

Consider a molecule traversing the centerline of a tunnel whose diameter 2σ is equal to twice the molecule diameter σ (Fig. 3.3). The molecule will collide with all molecules whose centers lie within a distance σ of the centerline of the tunnel and will miss all others.

If a molecule travels 1 cm, it sweeps out an imaginary volume of $\pi\sigma^2(1)$. With n molecules per unit volume, the number of molecules struck per centimeter is $\pi\sigma^2 n$, and the mean free path is then the reciprocal, or $1/(n\pi\sigma^2)$. If it is assumed that the molecular velocities are distributed according to maxwellian theory rather than having a sin-

Figure 3.3 Schematic of tunnel to estimate mean free path.

gle value, the mean free path equation is decreased by a factor of $\sqrt{2}$, or

$$\lambda = \frac{1}{\sqrt{2}n\pi\sigma^2} \tag{3.12}$$

Typical molecular diameters for various gases are given in Table 3.1. The value given for air represents an average value considering the relative proportions of oxygen, nitrogen, and trace gases in standard dry air.

Example 3.7

$$\lambda = \frac{1}{\sqrt{2}(2.688 \times 10^{19})(\pi)(3.617 \times 10^{-8})^2}$$

$$= 6.40 \times 10^{-6} \text{ cm} = 0.064 \text{ μm}$$

This is the mean distance between collisions at 0°C. At 20°C this distance would be

$$\lambda = \lambda_0 \frac{T}{T_0} = \frac{(0.064)(273 + 20)}{273} = 0.0687 \text{ μm}$$

The values for mean free path calculated by using Eq. 3.12 represent approximations because measurements of typical molecular diameters are not very accurate.

Accurate measurements of the mean free path of air molecules have been compiled by Jennings (1988), and these values are presented in Table 3.2. Because the mean free path is dependent on gas density,

TABLE 3.1 Typical Molecular Diameters

Gas	σ, Å
H_2	2.915
N_2	3.681
O_2	3.433
Air (dry)	3.617

SOURCE: R. B. Bird, W. E. Stewart, and E. N. Lightfoot, *Transport Phenomena*, Wiley, New York, 1960, p. 744.

TABLE 3.2 Mean Free Path in Air, μm

| Temperature, K | Relative Humidity | | |
	0	50	100
288.15	0.06391	0.06389	0.06386
293.15	0.06543	0.06544	0.06548
296.15	0.06635	0.06538	0.06647
298.15	0.06691	0.06701	0.06714

Pressure = 1.01325×10^5 Pa

SOURCE: After Jennings (1988).

Jennings considered conditions of both dry and moist air since at the same temperature and pressure there will be density differences between these two cases, depending on the amount of moisture present.

The mean free path for dry air at other temperatures and pressures can be determined from the relationship

$$\frac{\lambda}{\lambda_0} = \left(\frac{\mu}{\mu_0}\right)\left(\frac{P_0}{P}\right)\left(\frac{T}{T_0}\right)^{1/2} \quad (3.13)$$

where the subscript 0 represents a standard temperature, pressure, and viscosity. If the effect of temperature on gas viscosity is considered to be proportional to the square root of the temperature ratio, then Eq. 3.13 becomes

$$\frac{\lambda}{\lambda_0} = \left(\frac{P_0}{P}\right)\left(\frac{T}{T_0}\right) \quad (3.14)$$

i.e., the mean free path of a gas increases directly with the absolute temperature and inversely with the pressure.

Gas Viscosity, Heat Conductivity, and Diffusion

The three properties of viscosity, heat conductivity, and diffusion represent, respectively, the transfer of momentum, energy, and mass within a gas. The gas diffusion coefficient indicates the relative ability of one gas molecule to move with respect to its surroundings—the greater the value of the diffusion coefficient, the more rapid this movement. The diffusion coefficient $D_{1,2}$ for a gas of species 1 diffusing into a gas of species 2 can be estimated from the expression

$$D_{1,2} = \frac{1}{3\pi} \frac{v_{\text{rms1}}}{n_1 \sigma_{11}^2 \sqrt{2} + n_2 \sigma_{12}^2 (1 + m_1/m_2)^{1/2}} \quad (3.15)$$

where v_{rms} is the root-mean-square velocity, n_1 and n_2 are the number of molecules per cubic centimeter of species 1 and 2, and σ_{11} and σ_{12} are the collision diameters of molecule 1 with molecules 1 and 2.

Example 3.8 Show that for self-diffusion (a gas diffusing into itself) Eq. 3.15 can be simplified to

$$D = \frac{\lambda v_{\text{rms}}}{3}$$

Letting $n_1 = n/2$, $n_2 = n/2$, and $\sigma_{11} = \sigma_{12} = \sigma$, Eq. 3.15 reduces to

$$D_{1,2} = \frac{1}{3\pi} \frac{v_{\text{rms}1}}{(n/2)\sigma^2\sqrt{2} + (n/2)\sigma^2(2)^{1/2}} = \frac{v_{\text{rms}}}{3\sqrt{2}n\pi\sigma^2}$$

$$D_{1,2} = \frac{\lambda v_{\text{rms}}}{3} \tag{3.15a}$$

If three dimensions are considered, the factor $\frac{1}{3}$ above is replaced in Eqs. 3.15 and 3.15a by $3\sqrt{2}\pi/64 = 0.208$. From these equations it is seen that the diffusion coefficient of a gas varies inversely with pressure if the temperature is held constant; i.e., the diffusion coefficient varies in proportion to the mean free path.

The viscosity of a gas can be estimated from the expression (Alberty and Daniels, 1979):

$$\mu = \frac{5(\pi m k T)^{0.5}}{16\pi\sigma^2} \tag{3.16}$$

The term m represents the mass of a single molecule. The equation does not include a pressure term or depend on molecular concentration. This is confirmed with real gases at moderate pressures and normal temperatures where the viscosity is essentially independent of pressure.

Jennings (1988) reviewed the literature on viscosity with regard to compiling exact measurements for the mean free path of air molecules. He considered both dry air and moist air. At 20°C Jennings gives a value of 1.8193×10^{-4} cP for the viscosity of dry air, 1.815×10^{-4} cP for air at 50 percent relative humidity, and 1.8127×10^{-4} cP for air at 100 percent relative humidity. These figures indicate that for most aerosol work, a value for viscosity at 20°C of 1.82×10^{-4} cP is reasonably accurate regardless of the humidity.

According to Perry and Chilton (1973), the relationship of the viscosity of a gas at two different temperatures is given by

$$\frac{\mu}{\mu_0} = \left(\frac{T}{T_0}\right)^{3/2} \frac{T_0 + 1.47T_b}{T + 1.47T_b} \tag{3.17}$$

where T_b represents the normal boiling point of the gas. For air over the temperature range of 0 to 100°C, Eq. 3.17 can be approximated by the expression

$$\frac{\mu}{\mu_0} = \left(\frac{T}{T_0}\right)^{0.5} \tag{3.18}$$

with a maximum error no greater than 0.11 percent.

From Eq. 3.18 it can be seen that viscosity will increase as the temperature increases! This is just the opposite of what is observed for the behavior of typical liquids (e.g., with motor oil the viscosity increases as the temperature decreases).

Example 3.9 Determine the viscosity of helium gas at 20°C. Use 1.90 Å as the molecular diameter of helium.

Using Eq. 3.14, we get

$$\mu = \frac{5(\pi m k T)^{0.5}}{16\pi\sigma^2}$$

$$= \frac{5[(3.14)(4/6.02 \times 10^{23})(1.38 \times 10^{-16})(273 + 20)]^{0.5}}{(16)(3.14)(1.90 \times 10^{-8})^2}$$

$$= 253 \ \mu P$$

For the aerosol scientist the main point to remember about the medium from a kinetic theory point of view is that mass, energy, and momentum can be transferred within the gas—mass by diffusion, energy by heat conduction, and momentum by viscosity.

Mean free path indicates the transfer of momentum, energy, or mass a distance λ. In the steady state, the net transport equals zero. These forces, or transfer functions, always act to bring a system back to the steady state. This implies (1) diffusion from high concentration to low concentration, (2) heat conduction from hot to cold, and (3) momentum flow—mass motion energy to molecular motion (hence accompanied by a rise in temperature of the gas).

Problems

1 How many molecules of a gas are there per cubic centimeter at 20°C? At 100°C?

2 At 20°C the vapor pressure of water is 17.5 mmHg. How many molecules of H_2O are there per cubic centimeter of air when the relative humidity is 50 percent and $T = 20°C$?

3 Derive the most probable gas molecule velocity.

4 Derive the arithmetic-mean gas molecule velocity.

5 Derive the root-mean-square (rms) gas molecule velocity.

6 What is the magnitude of the rms velocity associated with a 0.1-μm-diameter unit-density spherical aerosol particle if it is in thermal equilibrium with its surroundings?

7 Compute the mean free path of a hydrogen molecule in hydrogen at 0°C, using simple theory and then using a maxwellian velocity distribution.

8 Using the equation given for viscosity, compute the viscosity of air at 20°C.

Chapter 4

Macroscopic Fluid Properties

Reynolds Number

So far the properties of the medium have been discussed from a molecular point of view. Generally, however, the medium can be thought of as a continuum, i.e., as a fluid where all molecules act in harmony with each other. This is the way one normally pictures a gas or liquid, and with this view of the medium the rules of aerodynamics can be applied.

Suppose it is desired to visualize the flow around a 1-μm sphere by studying the flow around a 1-cm sphere. One could ask, Under what conditions is it reasonable to assume that a 1-μm-diameter sphere moving in a continuous medium will behave in a manner similar to a 1-cm sphere moving in the same medium? Or more generally, under what conditions will geometrically similar flow occur around geometrically similar bodies? The answer, fundamental to fluid mechanics, is that in similar fields of flow, the forces acting on an element of either body must bear the same ratio to each other at any instant.

If the medium is considered incompressible and neglecting gravity, the main forces present are the inertial force due to the acceleration or deceleration of small fluid masses near the body and the viscous friction forces which arise due to the viscosity of the medium. For similarity these forces must be in the same ratio at any instant. Then

$$\frac{\text{Inertial force}}{\text{Viscous force}} = \frac{\rho_m v^2/d}{\mu v/d^2} = \frac{\rho_m v d}{\mu} = \text{Re} \quad (4.1)$$

(relative v)

where v is the relative velocity between the fluid and the body, ρ_m is the density of the medium, μ is the medium viscosity, and d is the body (or particle) diameter (Prandtl and Tietjens, 1957). The result is *Reynolds number*, abbreviated Re, a dimensionless number which describes the type of flow occurring around the body.

Kinematic viscosity ν can be defined as

Chapter Four

$$\nu = \frac{\mu}{\rho_m} \quad (4.2)$$

Then

$$\text{Re} = \frac{vd}{\nu} \quad (4.3)$$

is a convenient form for computing the Reynolds number in air at normal conditions. For air at normal pressure and 20°C, the kinematic viscosity ν is equal to 0.151 cm²/s.

Example 4.1 A 1-in diameter sphere moves through air with a velocity of 10 in/min. Find its Reynolds number.

$$\text{Re} = \frac{vd}{\nu} = \frac{(10 \times 2.54/60)(1 \times 2.54)}{0.151} = 7.13$$

It is also possible to derive the Reynolds number by dimensional analysis. This represents a more analytical, but less intuitive, approach to defining the condition of similar fluid flow and is essentially independent of particular shape. In this approach, variables in the Navier-Stokes equation (relative particle-fluid velocity, a characteristic dimension of the particle, fluid density, and fluid viscosity) are combined to yield a dimensionless expression. Thus

$$v^\alpha d^\beta \rho_m{}^\gamma \mu^\delta = F^0 L^0 T^0 = 1$$

Let $\alpha = 1$

$$\left(\frac{L}{T}\right)^1 (L^\beta)\left(\frac{FT^2}{L^4}\right)^\gamma \left(\frac{FT}{L^2}\right)^\delta = 1$$

Then

$$\gamma + \delta = 0$$
$$1 + \beta - 4\gamma - 2\delta = 0$$
$$2\gamma + \delta - 1 = 0$$
$$\beta = 1$$
$$\gamma = 1$$
$$\delta = -1$$

so

$$\text{Re} = \frac{vd\rho_m}{\mu}$$

Table 4.1 gives typical values for viscosity, density, and kinematic viscosity for air at 0 and 20°C.

TABLE 4.1 Useful Constants for Air*

Property	0°C	20°C	Units
Viscosity	1.76×10^{-4}	1.82×10^{-4}	$P = g/(cm \cdot s)$
Density	0.001295	0.001206	$= g/cm^3$
Kinematic viscosity	0.136	0.151	$St = cm^2/s$

*P = 760 mmHg.

The Reynolds number is useful in describing the type of flow that is taking place. At high Reynolds numbers, inertial forces will be much greater than viscous forces, while at low Reynolds numbers the opposite is true. Laminar or streamline flow is the result of the predominance of viscous forces. Thus at low Reynolds numbers the flow is laminar. Streamlines persist for great distances both upstream and downstream of the body, and little mixing takes place. When inertial forces predominate, streamlines disappear and the flow is turbulent. With turbulent flow there is rapid and random mixing downstream of the body, and streamlines are relatively undisturbed in front of the body until they almost reach the body surface. In the range where the Reynolds number increases from laminar flow to turbulence, the flow is said to be *intermediate* since at any time it can either be laminar or turbulent. Laminar flow can also be known as Stokes' flow or viscous flow.

Reynolds number can be applied to either a fluid flowing around a body or a fluid flowing inside a pipe. The transition from laminar to turbulent flow occurs at different Reynolds numbers for these two cases. The Reynolds numbers at which different flow conditions prevail are tabulated in Table 4.2. Since v is the relative velocity between the medium and the body, the Reynolds number is the same whether the body is moving through a stationary fluid or the fluid is flowing around a stationary body.

Schematic representations of these different flow conditions are illustrated in Fig. 4.1. As a gas enters a long pipe, turbulence will develop within the pipe if the Reynolds number exceeds the values given in Table 4.2.

TABLE 4.2 Values of Re for Various Conditions of Flow*

	A sphere of diameter d in a still fluid	Fluid flowing in a pipe of diameter d
Upper limit, laminar flow	1	2100
Intermediate region	1–1000	2100–4000
Turbulent flow	> 1000	> 4000

*It should be kept in mind that these values are approximate.

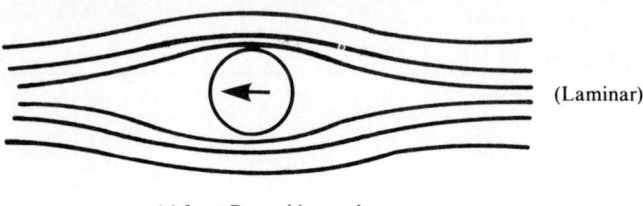
(Laminar)

(a) Low Reynolds number

(Intermediate)

(b) Intermediate Reynolds number

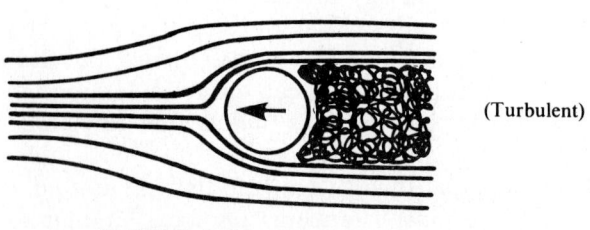
(Turbulent)

(c) High Reynolds number

Figure 4.1 Sketch of flow types.

Reynolds number is a fundamental parameter used to describe the fluid properties associated with an aerosol. Equations describing the resistance offered by a particle depend on whether the flow is laminar or turbulent, and the Reynolds number provides knowledge of the type of flow present.

Example 4.2 An aerosol comprised of 1.0-μm-diameter spheres flows through a 16-in-diameter duct with a velocity of 3500 ft/min. Determine the Reynolds number of the air flowing in the duct and of the particles in the air.

$$\text{Re in duct} = \frac{(\text{duct diameter})(\text{relative velocity of air in duct to duct})}{0.151}$$

$$= \frac{(16 \times 2.54)(3500)(30.5/60)}{0.151}$$

$$= 4.79 \times 10^5$$

This is clearly turbulent flow.

$$\text{Re of particles} = \frac{\text{(particle diameter)(relative velocity of particles to air)}}{0.151}$$

$$= \frac{(1 \times 10^{-4})(0)}{0.151} = 0$$

Since the particles are moving at the same velocity as the air in the duct, their Reynolds number is zero.

Drag

We can now consider the resistance offered by the medium to the motion of an aerosol particle. Some of the earliest interest in the motion of a body moving through a fluid arose from the desire to know where a cannonball, once fired, would land.

This problem can be treated by using the approach of Newton. Suppose the medium is composed of a large number of particles which have mass but no volume. These particles are everywhere at rest and are not connected. A body moving through this medium would experience impacts from the particles making up the medium and would impart momentum to them. The mass of particles impacting per second on the body is $\rho_m A v$, where ρ_m is the density of the particles per unit volume (and thus also the density of the medium), A is the cross-sectional area of the body normal to the direction of motion, and v the body velocity. Each impacting particle is given some velocity v' on impact which is proportional to v. Thus the momentum "created" per second is $\rho_m A v v'$.

Since the time rate of change of momentum is a force, this is also equal to the resisting force of the medium to the motion of the particle, often called the *drag*, or

$$F_D = \rho_m A v v' = k \rho_m A v^2$$

where k is a constant. The momentum transferred to the medium actually depends on whether the impacts of the gas molecules on the body are elastic or inelastic. This is reflected in the value used for the constant k. Also early estimates of k were incorrect because only the cross-sectional area of the body was considered, not the entire surface area. Depending on the type of flow, molecules can receive or impart momentum to the rear of the body, and the sides can have an influence so that the entire shape of the body is important, and not just its projected area.

There are three kinds of resistance which can be associated with the motion of a body as it passes through a medium. Deformation or viscosity drag represents the force necessary to deform the medium so that the body can pass through it. This deformation can occur at great distances both up- and downstream of the body. A second source of drag is frictional resistance which occurs at the surface of the body. The third type of resistance, pressure drag, represents compression of the medium. These latter two types make up the "skin friction" of the body. At small Reynolds numbers, deformation drag predominates, and forces that act over the entire body surface must be taken into account. At large Reynolds numbers, frictional resistance and pressure drag predominate. The drag in this case is primarily associated with the cross-sectional area normal to the fluid flow.

In cases involving high Reynolds numbers, Newton's approach (given above) agrees with experimental evidence, even though the underlying assumptions implying a constant value for k are wrong.

It is customary to write the drag equation as (Sutton, 1957)

$$F_D = \text{(some constant)} A\rho_m v^2$$

If a $v^2/2$ term (similar to the velocity head term in Bernoulli's equation) is used, then

$$F_D = C_D A \rho_m v^2/2$$

where the constant C_D is now formally known as the *coefficient of drag*, or *drag coefficient*. For a sphere of diameter d,

$$A = \frac{\pi}{4} d^2$$

and then

$$F_D = \frac{C_D \pi \rho_m d^2 v^2}{8} \tag{4.4}$$

Example 4.3 In turbulent flow the coefficient of drag is a constant with a value of about 0.4. What is the resisting force offered by air to a 6-in cannonball moving through the air with a velocity of 500 ft/s?

$$F_D = \frac{C_D \pi \rho_m d^2 v^2}{8}$$

$$= \frac{(0.4)(3.14)(0.0012)(6 \times 2.54)^2 (500 \times 30.5)^2}{8}$$

$$= 1.02 \times 10^7 \text{ dyn}$$

It was originally thought that for a given shape, body position, and

Macroscopic Fluid Properties   51

relative velocity, C_D would be a constant. This is not the case, and it is not surprising in view of the many ways in which resistance to flow can arise, depending on the Reynolds number. The coefficient of drag is a constant for a given shape and body position in those cases where the total drag is predominantly pressure drag (high Reynolds number). It is not a constant when deformation drag predominates (low Reynolds number). (laminar flow)

Similarity of flow will occur around similarly shaped bodies in those cases where the ratio of forces over the bodies' surfaces is the same; this is equivalent to saying that there will be similar resisting forces when the Reynolds numbers of the two bodies are the same. But then the drag coefficients for the two cases are also the same, i.e.,

$$C_D = f(\text{Re})$$

a statement which is true for each shape and body position. Figure 4.2 shows the relationship of C_D versus Re for spheres. In some ranges of Reynolds numbers, C_D can be determined analytically. In others, it must be estimated empirically. For laminar flow (Re < 1),

$$C_D = \frac{24}{\text{Re}} \tag{4.5}$$

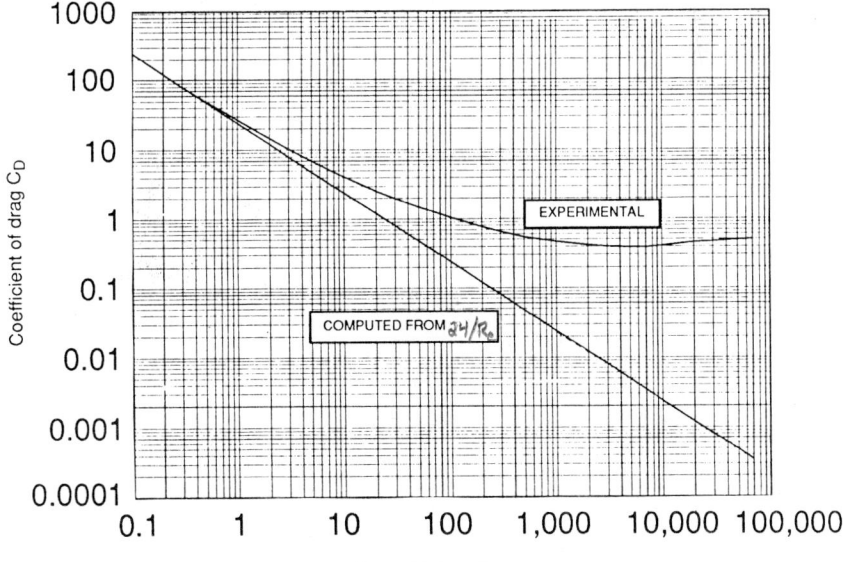

Figure 4.2   Coefficient of drag for spheres.

In the intermediate region (1 < Re < 1000) there are many empirical formulas for C_D such as (Crawford, 1976; Orr, 1966)

$$C_D = \frac{24}{Re}(1 + 0.15Re^{0.687}) \quad (4.6)$$

$$C_D = \frac{14}{Re^{0.5}} \quad \text{for } 2 < Re < 800 \quad (4.7)$$

$$C_D = \frac{24}{Re} + \frac{4}{Re^{0.33}} \quad (4.8)$$

$$C_D = \frac{18.5}{Re^{0.6}} \quad (4.9)$$

Example 4.4 Compare values of C_D as computed from Eqs. 4.6 through 4.9 for Re = 2. Which one is most nearly correct?

	Estimated Values of C_D from Eq.:			
	4.6	4.7	4.8	4.9
Re = 2	14.90	9.90	15.18	12.21

The reported measured value for C_D for spheres with Re = 2 is 14.6. Equations 4.7 and 4.9 are not very accurate, but they can be useful because of their simplicity. Where greater accuracy is needed, Eq. 4.6 or 4.8 should be used.

In the lower turbulence region ($1000 < Re < 2 \times 10^5$)

$$C_D = 0.44 \quad (4.10)$$

and in the upper turbulence region ($Re > 2 \times 10^5$)

$$C_D = 0.10 \quad (4.11)$$

Table 4.3 gives computed values for C_D for various values of Re based on these different equations, compared to actual measurements of drag coefficients for spheres.

TABLE 4.3 Experimental and Computed Values of C_D as a Function of Re

Re	Experimental	Eq. 4.6	Approximations		
			Eq. 4.7	Eq. 4.8	Eq. 4.9
0.1	240	247.4	44.3	248.6	73.6
0.2	120	126.0	31.3	126.8	48.6
0.3	80	85.3	25.6	86.0	38.1
0.5	49.5	52.5	19.8	53.0	28.0
0.7	36.5	38.3	16.7	38.8	22.9
1.0	26.5	27.6	14.0	28.0	18.5
2	14.6	14.9	9.9	15.2	12.2
3	10.4	10.6	8.1	10.8	9.57
5	6.9	7.0	6.3	7.14	7.04
7	5.3	5.4	5.3	5.52	5.76
10	4.1	4.2	4.43	4.26	4.7
20	2.55	2.61	3.13	2.67	3.07
30	2.00	2.04	2.56	2.09	2.40
50	1.50	1.54	1.98	1.57	1.77
70	1.27	1.30	1.69	1.31	1.45
100	1.07	1.09	1.40	1.10	1.17
200	0.77	0.81	0.99	0.80	0.77
300	0.65	0.68	0.81	0.68	0.60
500	0.55	0.56	0.63	0.55	0.44
700	0.50	0.50	0.53	0.48	0.36
1,000	0.46	0.44	0.44	0.42	0.29
2,000	0.42	0.35	0.31	0.33	0.19
3,000	0.40	0.30	0.26	0.29	0.15
5,000	0.385	0.26	0.20	0.24	0.11
7,000	0.390	0.23	0.17	0.21	0.09
10,000	0.405	0.20	0.14	0.19	0.07
20,000	0.45	0.16	0.10	0.15	0.05
30,000	0.47	0.14	0.08	0.13	0.04
50,000	0.49	0.12	0.06	0.11	0.03
70,000	0.50	0.11	0.05	0.10	0.02

SOURCE: R. H. Perry and C. H. Chilton, *Chemical Engineers Handbook,* 5th ed., McGraw-Hill, New York, 1973, pp. 5–64.

Example 4.5 A particle of diameter d and density ρ settles under the influence of gravity. What is its terminal settling velocity?

For terminal settling the drag force F_D equals the force due to gravity F_G. Hence

$$F_D = F_G$$

$$mg = C_D A \rho_m \frac{v^2}{2}$$

for spheres

$$\frac{\pi}{6} d^3 (\rho_p - \rho_m) g = C_D \frac{\pi}{4} d^2 \rho_m \frac{v^2}{2}$$

$$v^2 = \frac{4d(\rho_p - \rho_m)g}{3C_D \rho_m}$$

Unfortunately, C_D depends on Re which depends on v.

To circumvent this difficulty, use the relationship

$$C_D \text{Re}^2 = C_D \frac{v^2 d^2 \rho_m^2}{\mu^2}$$

then substituting for v^2 gives

$$C_D \text{Re}^2 = C_D \frac{d^2 \rho_m^2}{\mu^2} \left[\frac{4}{3} \frac{d \rho_m (\rho_p - \rho_m) g}{3 C_D \rho_m} \right]$$

$$C_D \text{Re}^2 = \frac{4}{3} \frac{d^3 \rho_m (\rho_p - \rho_m) g}{\mu^2} \qquad (4.12)$$

And $C_D \text{Re}^2$ can be computed from Eq. 4.12 since the v term has been eliminated; from a plot of $C_D \text{Re}^2$ versus Re (Fig. 4.3) a value of Re can

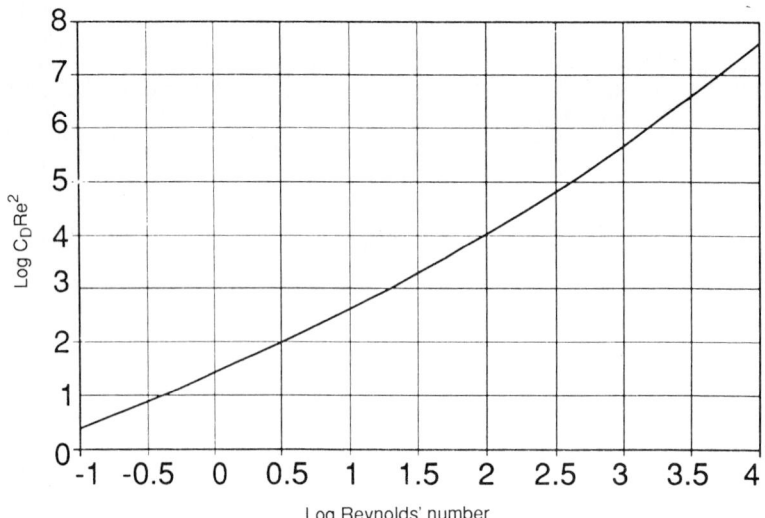

Figure 4.3 Plot of $C_D \text{Re}^2$ versus Re.

be found which yields v. This method, although crude, is valid for determining settling velocities for any size particle in either a gas or a liquid. For most aerosol particles a simpler method is available, as discussed in the next chapter.

Example 4.6 Determine the settling velocity of a 100-μm-diameter gold sphere (ρ = 19.3 g/cm^3) when it settles in air and water.
The density of air is 0.0012 g/cm^3, so using Eq. 4.12 gives

$$C_D \text{Re}^2 = \frac{4}{3} \frac{d^3 \rho_m (\rho_p - \rho_m) g}{\mu^2}$$

$$= \frac{4}{3} \frac{(100 \times 10^{-4})^3 (0.0012)(19.3 - 0.0012)(980)}{(1.82 \times 10^{-4})^2}$$

$$= 918$$

From Fig. 4.3, $C_D \text{Re}^2 = 918$ gives a value of Re = 19.95. Hence

$$\text{Re} = \frac{dv}{\nu}$$

$$v = \frac{\text{Re}\nu}{d} = \frac{19.95(0.151)}{100 \times 10^{-4}} = 301.0 \text{ cm/s}$$

For water, $\rho_m = 1$ and $\mu = 0.01$ so that

$$C_D \text{Re}^2 = \frac{4}{3} \frac{(100 \times 10^{-4})^3 (1)(19.3 - 1)(980)}{(0.01)^2}$$

$$= 239$$

Again, from Fig. 4.3, $C_D \text{Re}^2 = 239$ gives a value of Re = 6.31. Hence

$$v = \frac{\text{Re}\nu}{d} = \frac{(6.31)(0.01)}{(1)(100 \times 10^{-4})} = 6.31 \text{ cm/s}$$

Example 4.7 Given particle density and settling velocity, how can the particle diameter be determined for a settling particle?
Again, equating forces,

$$F_G = F_D$$

$$mg = C_D A \rho_m \frac{v^2}{2}$$

so that for spheres

$$d = \frac{3v^2 C_D \rho_m}{4g(\rho_p - \rho_g)}$$

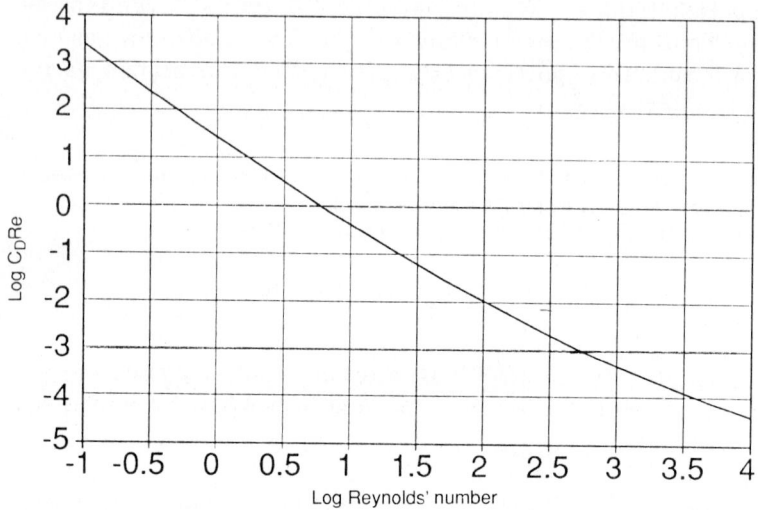

Figure 4.4 Plot of C_D/Re versus Re.

The term C_D can be eliminated by taking the ratio

$$\frac{C_D}{Re} = \frac{C_D\mu}{dv\rho_m} = \frac{C_D\mu 4(\rho_p - \rho_m)g}{3v^3 C_D \rho_m^2}$$

$$= \frac{4\mu g(\rho_p - \rho_m)}{3v^3 \rho_m^2} \quad (4.13)$$

Here C_D/Re can be computed from Eq. 4.13. By using Fig. 4.4 (a plot of C_D/Re versus Re), Re can be found, which then yields d. For small particles settling in the Stokes region, this involved process is not necessary.

Problems

1 Calculate the density of CO_2 at 20°C.

2 Calculate the Reynolds number of a 1-µm spherical sand particle moving in air at a velocity of 10 cm/s (assume NTP).

3 Calculate the Reynolds number of a 10-in ball moving in air at a velocity of 10 cm/s. Is the flow around the ball laminar or turbulent?

4 Calculate the Reynolds number for air flowing through a 10-in-diameter pipe at a velocity of 10 cm/s. Is the flow through the duct laminar or turbulent?

5 A 10-µm-diameter particle settles in air with a velocity of 0.30 cm/s. If the settling of this particle is to be modeled by a 1-in-diameter steel ball moving

in glycerol (viscosity = 1756 cP), what should be the ball's velocity in the glycerol? (Density of glycerol = 1.26 g/cm^3.)

6 Air flows through a 4-in-diameter duct at a rate of 100 ft^3/min. Determine whether this flow is laminar or turbulent within the duct.

7 A 10-μm-diameter particle falls in still air with a velocity of 0.30 cm/s. If the drag coefficient is given by 24/Re, what is the force developed by the falling particle?

8 Using Eq. 4.8, calculate C_D, $C_D\text{Re}^2$, and C_D/Re for Re values of (a) 0.8, (b) 80, (c) 8000.

9 Using a plot of C_D/Re and $C_D\text{Re}^2$ versus Re, find (a) the settling velocity of a 200-μm sand sphere (ρ = 2.65 g/cm^3) and (b) the size of a water droplet that settles at a velocity of 10 cm/s.

Chapter

5

Viscous Motion and Stokes' Law

Introduction

For the case of low-Reynolds-number flow (viscous flow) it is possible to develop an expression for the force resisting the motion of a sphere moving through a fluid based purely on mathematical reasoning. This problem was originally solved by G. G. Stokes, and the expression for force since has become known as *Stokes' law*. For those interested in the mathematics of this problem, Stokes' derivation is given in App. B. Although one doesn't have to understand the derivation to use Stokes' law correctly, it helps to be aware of the assumptions that were made in order to understand how deviations from these assumptions can affect results of calculations made by using Stokes' law.

For Stokes' solution, it was necessary to assume a continuous, incompressible, viscous, and infinite medium with rigid particles and spherical particles. With these assumptions, Stokes found that the resisting force exerted by air on a moving particle, equivalent to the force exerted by moving air on a stationary particle, is

$$F = 3\pi\mu v d \qquad (5.1)$$

where F is the force on the particle, in dynes, μ is the viscosity of the medium in poises, v is the relative velocity between the air and the particle in centimeters per second, and d is the diameter of the sphere in centimeters.

The best proof of the validity of Stokes' law (although indirect) was the Millikan oil drop experiment. Stokes' law has been shown to give a reasonable approximation of the resisting force on spheres in many other situations, and the only stipulation is that the assumptions listed above not be violated.

Example 5.1 A 1-μm unit-density sphere moves through air with a velocity of 100 cm/s. Compute the magnitude of the resisting force offered by the air. Assume $T = 20°C$ and 760-mmHg atmospheric pressure.

$$F = 3\pi\mu vd$$

$$= 3(3.14)(1.82 \times 10^{-4})(100)(1 \times 10^{-4})$$

$$= [g/(cm \cdot s)](cm/s)(cm) = g \cdot cm/s^2$$

$$= 1.72 \times 10^{-5} \text{ dyn}$$

Stokes' law becomes incorrect when assumptions used to derive it cannot be met. It is possible in some cases to develop correction factors broadening the conditions under which Stokes' law is applicable. However, the assumptions may be so broad for the types of problems which are of interest that corrections are not necessary or are impossible to make. In any case, it is useful to examine each assumption in detail to determine when it may or may not be valid.

Continuous Medium

When the diameter of a particle is very small, approaching the mean free path of the molecules in the medium, Cunningham (1910) and also Millikan (1910) showed that because the medium is no longer a "perfect" continuum, the resisting force offered to the particle should be smaller than that predicted by Stokes' law. The difference in the dependence of resistance on particle diameter corresponds to the conditions prevailing at the two extreme particle ranges. For large particles the primary source of resistance is the viscosity of the medium, whereas with small particles or with a highly rarefied medium, viscosity is no longer important and the predominant resisting mechanism is due to the inertia of the gas molecules which the particle encounters. As particle size decreases to near molecular size, the resisting force offered by the medium becomes a function of the cross-sectional area of the particle, consistent with Newton's model for drag (Millikan, 1923).

To correct for this effect, a factor, commonly known as the *Cunningham correction factor*, *slip*, or *Millikan resistance factor*, denoted C_c, must be introduced into the Stokes equation, yielding

$$F = \frac{3\pi\mu vd}{C_c} \tag{5.2}$$

where

$$C_c = 1 + \frac{2\lambda}{d}\left[A + Q \exp\left(-\frac{bd}{2\lambda}\right)\right] \quad (5.3a)$$

or

$$C_c = 1 + \text{Kn}\left[A + Q \exp\left(-\frac{b}{\text{Kn}}\right)\right] \quad (5.3b)$$

The term λ represents the mean free path of the gas molecules. The ratio

$$\text{Kn} = \frac{2\lambda}{d}$$

is known as the *Knudsen number*, being the ratio of the gas mean free path to the particle radius. Values for the constants A, Q, and b have been subject to slight correction over the years so that depending on the age of the reference cited, differences can appear. Table 5.1 lists these constants as presented by several references, and Fig. 5.1 is a plot of C_c versus d using the different values of A, Q, and b. These val-

TABLE 5.1 Definitions for Various Cunningham Correction "Constants"

Constant	Davies (1945)	Fuchs (1964)	Allen and Raabe (1982)	Jennings (1988)
A	1.257	1.246	1.155	1.252
Q	0.400	0.418	0.471	0.399
b	1.100	0.867	0.596	1.100
$A + Q$	1.657	1.664	1.626	1.651

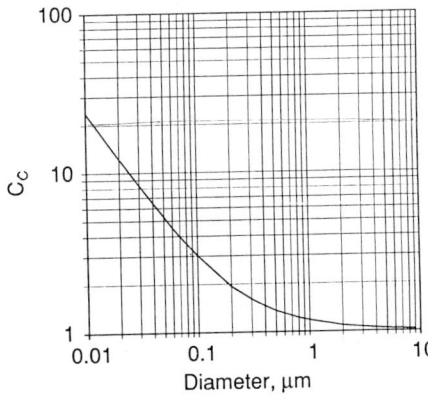

Figure 5.1 Variation of C_c with diameter.

ues are tabulated in Table 5.2. As can be seen, despite seeming dissimilarities, the actual differences in the computed values for C_c are negligible.

The Cunningham correction factor C_c is always equal to or greater than 1. When $d > 2\lambda$, then C_c can be approximated by the expression

$$C_c = 1 + \frac{2\lambda}{d}(A) \tag{5.4}$$

When $d < 2\lambda$, then

$$C_c \approx 1 + \frac{2\lambda}{d}(A + Q) \tag{5.5}$$

The Cunningham correction factor is an important correction to Stokes' law and should always be used when particles are less than 1 μm in diameter.

TABLE 5.2 Computed Values for C_c at $T = 0°C$

d, μm	Davies (1945), $\lambda = 0.065$ μm	Fuchs (1964), $\lambda = 0.065$ μm	Allen and Raabe (1982), $\lambda = 0.066$ μm	Jennings (1988), $\lambda = 0.066$ μm
0.01	22.258	22.161	22.206	22.266
0.02	11.435	11.417	11.472	11.439
0.03	7.838	7.842	7.898	7.841
0.04	6.046	6.060	6.113	6.049
0.05	4.977	4.994	5.045	4.979
0.06	4.268	4.287	4.334	4.269
0.07	3.765	3.784	3.828	3.766
0.08	3.390	3.408	3.449	3.391
0.09	3.100	3.118	3.156	3.101
0.1	2.870	2.887	2.922	2.871
0.2	1.871	1.876	1.889	1.871
0.3	1.562	1.561	1.562	1.562
0.4	1.416	1.412	1.407	1.416
0.5	1.330	1.326	1.318	1.331
0.6	1.275	1.270	1.261	1.275
0.7	1.235	1.231	1.222	1.235
0.8	1.206	1.202	1.193	1.206
0.9	1.183	1.179	1.171	1.183
1	1.164	1.161	1.153	1.164
2	1.082	1.081	1.076	1.082
3	1.055	1.054	1.051	1.055
4	1.041	1.040	1.038	1.041
5	1.033	1.032	1.031	1.033
6	1.027	1.027	1.025	1.027
7	1.023	1.023	1.022	1.023
8	1.021	1.020	1.019	1.021
9	1.018	1.018	1.017	1.018

Example 5.2 Compute the Cunningham correction factor for a silica dust particle (ρ = 2.65 g/cm^3) having a diameter of 0.5 μm. Assume a spherical shape and 20°C.
From Eq. 5.3

$$C_c = 1 + \frac{2\lambda}{d}\left[A + Q \exp\left(-\frac{bd}{2\lambda}\right)\right]$$

The gas mean free path (from Chap. 3) is 0.0687 μm, so

$$C_c = 1 + \frac{(2)(0.687)}{(0.5)}\left\{1.257 + 0.4 \exp\left[-\frac{(1.10)(0.5)}{2(0.687)}\right]\right\}$$

$$= 1.35$$

As mentioned earlier, the slip, or Cunningham correction factor, represents the mechanism for transition from the continuum to the molecular case. For large values of d, the resulting force F is proportional to d whereas for small values of d, F is proportional to d^2.

Incompressible Medium

Air is compressible, but compression is not important for motion in the Stokes region. This assumption can be considered to always be valid.

Viscous Medium

In the derivation of Stokes' law, the assumption of a perfectly viscous medium means that no inertial forces are considered. This was done to linearize the Navier-Stokes equation. If these inertial effects are included in a first-order approximation, it is possible to extend the applicability of Stokes' law up to a Reynolds number of about 5. Then the resisting force can be expressed as

$$F = 3\pi\mu v d(1 + 3/16 \; Re) \tag{5.6}$$

Above a Reynolds number of about 5, Stokes' law, even with this correction, is no longer applicable.

Example 5.3 A 100-μm unit-density sphere moves through air with a velocity of 30 cm/s. Compute the resisting force offered by the air, in dynes. Assume normal temperature and pressure.

$$Re = \frac{dv}{\nu} = \frac{(100 \times 10^{-4})(30)}{0.151} = 1.99$$

$$F = 3\pi\mu v d(1 + 3/16 \; Re)$$

$$= 3(3.14)(1.82 \times 10^{-4})(30)(100 \times 10^{-4})[1 + 3/16 \; (1.99)]$$

$$= (5.15 \times 10^{-4})(1.37) = 7.06 \times 10^{-4} \text{ dyn}$$

Infinite Medium

In viscous flow, perturbations caused by a particle extend large distances into the medium. The presence of other particles moving nearby will have the effect of reducing the resistance of the medium to that particle by setting the medium near the particle in motion. Hence an ensemble of particles will settle faster than they would as isolated entities, and when two equal-sized particles fall along the same axis, the upper of the two will fall faster than the lower, so that they will eventually collide. If the particles are of different diameters, the aerodynamic interaction between the two particles will result in an increase in settling velocity for both particles. When the leading particle is smaller than the trailing particle, its increase in velocity will be greater than the increase for the trailing one. Particle-induced interactions are usually neglected in making estimates of settling rates since with the exception of the most extreme cases particle-particle spacing is relatively large.

Example 5.4 Typical concentrations for condensation nuclei are 30,000 to 50,000 nuclei per cubic centimeter. If each nucleus is 0.01 μm in diameter and particles are present in a concentration of 40,000 per cubic centimeter, estimate average particle spacing, in particle diameters.

$$\text{cm}^3/\text{particle} = \frac{1}{40,000}$$

$$\text{cm}/\text{particle} = \sqrt[3]{\frac{1}{40,000}}$$

$$= 0.029$$

$$= \frac{0.029}{0.01 \times 10^{-4}} = 29,240 \text{ particle diameters}$$

This spacing is sufficiently large that particle-particle interactions can be neglected.

When particles move parallel to a flat surface, resistance is increased due to the drag induced by the surface. This increase is so small and extends such a small distance into the medium (several particle diameters at most) that the effect can be neglected without significant error.

For aerosols in a confined space, other interaction effects are possible. For example, a cloud of sedimenting particles could completely fill a finite volume. Then the downward motion of each single particle creates a downward flow field that tends to pull along neighboring particles. But in a confined space this downward flow is balanced by an upward airflow that tends to lift the entire cloud. The net result is that the downward velocity of the cloud in a finite container will be less than that of a similar cloud in an infinite medium. Figure 5.2 illustrates the two cases of confined and unconfined aerosol sedimentation. For a complete discussion of noninfinite medium effects, see Happel and Brenner (1965).

Rigid Particles

Although rigid particles are assumed, often Stokes' law is applied to nonrigid or liquid droplets. In the case where the drops are large, they are deformed by the motion of the air and will no longer be spherical.

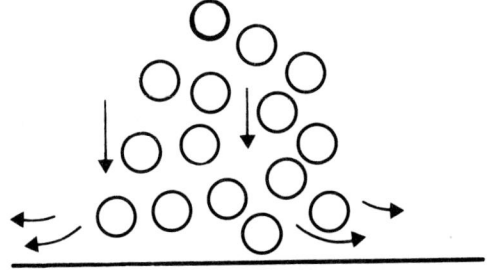

Unconfined Sedimentation

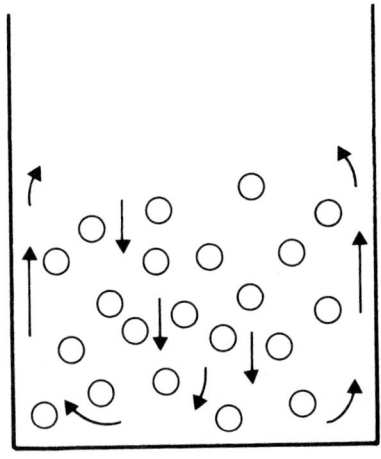

Confined Sedimentation

Figure 5.2 Sketch of generalized flow patterns for cases of unconfined and confined settling.

TABLE 5.3 Terminal Velocities of Water Droplets in Still Air (NTP)

Drop diameter, μm	v_T, cm/s
100	25.6
120	34.5
160	52.5
200	71
400	160
600	246
800	325
1000	403
2000	649
3000	806
4000	883
5000	909
5400	914
5800	917

SOURCE: B. J. Mason, *The Physics of Clouds*, 2d ed., Clarendon Press, Oxford, 1971, p. 594.

Since they tend to flatten out, they offer more resistance to falling and have lower terminal velocities than spherical particles. This effect is not important for freely falling particles having diameters less than a few hundred micrometers. Table 5.3 gives terminal settling velocity data for raindrops of various diameters. Above about 6 mm in diameter the drops fracture and break up while falling.

More important, nonrigid particles can undergo internal circulation as they move through a medium. This circulation reduces the friction at the drop surface so that the resistance offered by the medium to the motion of the drop is reduced. The resisting force then becomes

$$F = 3\pi \mu_m v d \left[\frac{1 + 2\mu_m/(3\mu_p)}{1 + \mu_m/\mu_p} \right] \tag{5.7}$$

where μ_p is the viscosity of the liquid making up the drop and μ_m is the viscosity of the medium. For water droplets in air, the correction factor is for all practical purposes equal to 1, since the viscosity of water is so much greater than that of air. In general, for liquids in air this effect can be neglected.

Example 5.5 Compare the resisting force of air on a 1-μm water droplet falling freely if the liquid nature of the droplet is considered.
Resistance allowing the liquid nature of droplet

$$F = 3\pi \mu_m v d \, \frac{1 + 2\mu_m/(3\mu_p)}{1 + \mu_m/\mu_p}$$

Resistance neglecting the liquid nature of droplet

$$F = 3\pi\mu_m vd$$

Taking the ratio gives

$$\frac{\text{Corrected } F}{\text{Uncorrected } F} = \frac{1 + 2\mu_m/(3\mu_p)}{1 + \mu_m/\mu_p}$$

If $\mu_m = 1.83 \times 10^{-4}$ P and $\mu_p = 0.01$ P, then

$$\frac{\text{Corrected } F}{\text{Uncorrected } F} = \frac{1 + 2 \times 1.82 \times 10^{-4}/(3 \times 10^{-2})}{1 + 1.82 \times 10^{-4}/(1 \times 10^{-2})}$$

$$= \frac{1.0121}{1.0182} = 0.9940$$

It is clear that this effect can be neglected.

If the viscosity of the medium greatly exceeds that of the droplet, the correction factor tends to a limiting value of two-thirds. In this case the resisting force becomes

$$F = 2\pi\mu_m vd \qquad (5.8)$$

which is the resisting force a liquid offers to a bubble rising through it.

Example 5.6 How fast will a 0.1-mm bubble rise in a glass of beer?
If the positive direction is considered to be down, then a negative result would indicate upward motion. Equating the forces gives

$$F_R = F_G = 2\pi\mu_m vd = mg$$

$$v = \frac{(\pi/6)d^3(\rho_p - \rho_m)g}{2\pi\mu_m d} = \frac{1}{12}\frac{d^2(\rho_p - \rho_m)g}{\mu_m} = \frac{10^{-4}(-1)(980)}{12(1)}$$

$$= -0.817 \text{ cm/s}$$

Check:

$$\text{Re} = \frac{dv\rho}{\mu} = \frac{(0.01)(0.817)(1)}{0.01} = 0.817$$

This indicates laminar flow, so assumptions are all right.

When the viscosities of the medium and the particle are the same, then the correction factor has a value of 5/6, and the Stokes resistance is

$$F = \frac{5}{2}\pi\mu_m vd \qquad (5.9)$$

equivalent to the case of a cloud of particles being considered as a single particle having the same viscosity as the air. In this case d is the diameter of the entire cloud.

Example 5.7 Wind blowing on an aerosol can either move it as a cloud or blow through it and dissipate it. Find the concentration of an aerosol made up of 5-μm water droplets which will be just dissipated by the wind.

The force on an individual particle is

$$F_S = 3\pi\mu_m vd$$

The force on an ensemble of particles (assuming spherical cloud of 10-m diameter D) is

$$F_E = \frac{5}{2}\pi\mu_m vD$$

Since resisting forces tend to a minimum value, if the sum of the forces acting on all the particles is greater than the single force acting on the ensemble of particles, the particles will remain as an aerosol cloud. Otherwise, the cloud will dissipate.

Letting c equal the aerosol concentration (particles per cubic centimeter), when the forces are just equal,

$$\frac{5}{2}\pi\mu_m vD = 3\pi\mu_m vdc \frac{\pi}{6}D^3$$

Solving for c gives

$$c = \frac{5}{\pi dD^2}$$

With $d = 5$ μm and $D = 10$ m,

$$c = \frac{5}{\pi \times 10^{-4} \times 10^6} = 3.18 \times 10^{-3} \text{ particles/cm}^3$$

Concentrations greater than this value will result in the cloud's remaining intact. This indicates that it is quite difficult to dissipate a cloud without some external aid other than mere blowing.

Spherical Particle

A final assumption made in the derivation of Stokes' law was that the particles of interest were spheres. In many cases this is not true. Particles may have irregular shapes, depending on how they were formed and the amount of agglomeration which may have taken place. Liquid aerosols are always spherical, so that for liquid aerosols the assumption of sphericity holds. For isometric particles this assumption can also be used with little error. For long chains of particles or flocculated particles, large deviations from Stokes' law are possible.

To use Stokes' law with chains or fibers, several approaches are available. Traditionally a correction factor κ, known as the *dynamic shape factor*, is defined such that

$$F = 3\pi\mu v d_e \kappa \tag{5.10}$$

The term d_e is the diameter of a sphere having the same volume as the chain or fiber, i.e.,

$$\text{Volume (chain or fiber)} = \frac{\pi}{6} d_e^3 \tag{5.11}$$

For a cluster of n spheres of diameter d, $d_e = \sqrt[3]{n}d$. When the aggregate particle size is small, the Cunningham correction factor should be considered.

Quite good estimates of the numerical value of κ have been made experimentally (Stöber and Flachsbart, 1969), and Table 5.4 shows some of these data. For tightly packed clusters, the maximum value for κ is about 1.25.

TABLE 5.4 Values of κ for Different Chain Configurations

n	Configuration	κ
2	oo	1.12
3	ooo	1.27
3	o o / o	1.16
4	oooo	1.32
4	oo with o above and o below	1.25
5	ooooo	1.45
6	oooooo	1.57
4	oo / oo	1.17
7	ooooooo	1.67
5	ooo with o above and o below	1.30
6	oooo with o above and o below	1.43
8	oooooooo	1.73
8	oooooo / o / o	1.64
5	o o / o / o o	1.19
8	oooooo with o above and o below	1.56
6	oo / oo / oo	1.17

SOURCE: Adapted from W. Stöber and H. Flachsbart, *Environmental Science and Technology*, 3, 1280 (1969).

Example 5.8 Determine the aerodynamic diameter of a particle made up of four spheres of 10-µm diameter (unit density) and formed into a tight cluster.

By equating forces, $F_R = F_G$,

$$v_T = \frac{mg}{3\pi\mu d_e \kappa} = \frac{(4)(\pi/6)d^3\rho_p g}{3\pi\mu(\sqrt[3]{4}d)\kappa}$$

The *aerodynamic diameter* d_A can be defined as

$$d_A^2 = \frac{18\mu v_T}{g} \quad (1)$$

Hence

$$d_A^2 = \frac{18(4)(\pi/6)d^3\rho_p g\mu}{3\pi\mu(\sqrt[3]{4}d)\kappa g} = \frac{4d^2\rho_p}{\sqrt[3]{4}\kappa}$$

$$d_A = \left(\frac{\rho_p}{\kappa}\right)^{1/2}\sqrt[3]{4}d$$

For unit-density spheres, $\rho = 1$, so that

$$d_A = \frac{\sqrt[3]{4}d}{\sqrt{\kappa}} = \frac{\sqrt[3]{4}}{\sqrt{1.17}}d = 1.468d = 14.68 \text{ µm}$$

With fibers, measurements are usually in terms of fiber length L and diameter d. Writing the aerodynamic diameter as

$$d_A^2 = \frac{18 v_T \mu}{g} \quad (5.12)$$

and replacing v_T with an expression derived from equating gravitational and resisting forces

$$v_T = \frac{(\pi/4)d^2 L \rho_p g}{3\pi\mu\left(\frac{3}{2}d^2 L\right)^{1/3}\kappa} \quad (5.13)$$

yield

$$d_A = \left(\frac{3}{2}\right)^{1/3}\left(\frac{\rho_p}{\kappa}\right)^{1/2}\left(\frac{L}{d}\right)^{1/3}d \quad (5.14)$$

Stöber (1972) noted that for chainlike aggregates of spheres

$$d_A = 1.077\rho^{1/2}N^{1/6}d \quad (5.15)$$

where N is the number of spheres in the aggregate and d is the diameter of a single sphere. For a fiber the term N could be considered to be proportional to the aspect ratio L/d_f and the term d to the fiber diameter d_f. This implies that fiber length has very little influence on the fiber aerodynamic diameter.

Example 5.9 Using Eq. 5.15, estimate the aerodynamic diameter of an asbestos fiber having a length of 15 µm, a diameter of 0.4 µm, and a density of 2.65 g/cm³.

From Eq. 5.15

$$d_A = 1.077 \rho^{1/2} \left(\frac{L}{d_f}\right)^{1/6} d$$

$$= (1.077)(2.65)^{1/2} \left(\frac{15}{0.4}\right)^{1/6} (0.4)$$

$$= 1.283 \text{ µm}$$

As mentioned above, Eq. 5.15 implies that the aerodynamic diameter of a rod or fiber will be influenced very little by its length, being much more dependent on its cross-sectional diameter. Hence fibers of different lengths but similar cross-sections will have similar aerodynamic properties, despite large differences in mass.

A relatively new way to consider the shape factor correction starts from a more fundamental point of view. According to Stokes' law, the pressure on the surface of a sphere *(form drag)* amounts to about one-third of the total drag with the remainder coming from the tangential shear stress on the surface of the sphere, the so-called friction drag.

Leith (1987) pointed out that for a moving nonspherical object, form drag should be associated with the projected cross-sectional area of the object normal to its motion. He considered friction drag as being associated with the object's surface area. Thus according to Leith, Stokes' law can be written as

$$F_D = 3\pi\mu v (\tfrac{1}{3}d_n + \tfrac{2}{3}d_s) \qquad (5.16)$$

where d_n is the diameter of a sphere whose projected area is the same as the normal projected area of a moving object and d_s is the diameter of a sphere whose effective surface equals that of the object. Then writing Stokes' law in terms of d_n gives

$$F_D = 3\pi\mu v d_n \kappa_n \qquad (5.17)$$

where the theoretical value for κ_n is

$$\kappa_n = \frac{1}{3} + \frac{2}{3}\frac{d_s}{d_n} \qquad (5.18)$$

Using experimental settling measurements of a number of irregular particle shapes, Johnson (1985) developed the following empirical equation for the factor κ_n:

$$\kappa_n = 0.357 + 0.684\frac{d_s}{d_n} + 0.00154\Psi + 0.0104A \tag{5.19}$$

where Ψ is the length ratio, defined as

$$\Psi = \frac{(\text{axis parallel to direction of motion})^2}{\text{projected area normal to direction of motion}}$$

and A = the ratio of the longest axis to the shortest axis in the projected area normal to the direction of motion. In terms of the "volume equivalent" κ, that is, the classical case is

$$\kappa_n = \frac{d_e}{d_n}\kappa \tag{5.20}$$

For the case of a sphere, Johnson's equation gives a value of $\kappa_n = 1.053$; that is, there is about a 5 percent error in the empirical estimate. This error appears to persist for many other shapes as well but for most problems is not particularly significant.

In many practical cases Leith's approach to the definition of the aerosol shape factor has greatly simplified the understanding of this correction to Stokes' law. For example, consider again the aerodynamic diameter of a fiber having a cross-sectional diameter d_f, length L, and density ρ_f. This can be approximated by using Eqs. 5.17 and 5.18 for the case of long axis motion parallel to the flow as

$$d_A \approx \sqrt{\frac{3}{2}}\, d_f \left(\frac{L}{d_f}\right)^{1/4} \rho_f^{1/2} \tag{5.21}$$

or for the case of long axis motion perpendicular to the flow

$$d_A \approx 1.199 d_f \left(\frac{L}{d_f}\right)^{1/4} \rho_f^{1/2} \tag{5.22}$$

Example 5.10 Estimate the aerodynamic diameter of the asbestos fiber in Example 5.9, using Eq. 5.21.

$$d_A \approx \frac{3}{2}(0.4)\left(\frac{15}{0.4}\right)^{1/4}(2.65)^{1/2}$$

$$\approx 2.42\ \mu\text{m}$$

Both of these approximations differ from Eq. 5.15 in the value of the coefficient and in the value of the exponent of the aspect ratio ($1/6$ versus $1/4$). Spurny et al. (1978) reported experimental measurements of asbestos fiber aerodynamic diameters which indicate a range of exponential values of 0.116 to 0.171 with a coefficient of about 1.34. However, even if the details are still not clear, it is clear that for fibers the effect of fiber length on aerodynamic diameter is of much less importance than fiber diameter. This means, e.g., that fibrous aerosols will persist in air much longer than isometric ones for equal fiber and particle mass.

Problems

1 Compare the force resisting the movement of a 100-µm-diameter sphere as it moves through air at a velocity of 1 cm/s to a sphere moving in water at the same velocity.

2 Compute the force on a 10-µm unit-density sphere as it settles at a velocity of 0.3 cm/s.

3 At a Re value of 4, the measured value of C_D for a sphere is 8.472. Determine the error in using Stokes' law with and without the appropriate correction factor.

4 Determine an expression for the force resisting the movement of an air bubble in water (assume Stokes' law holds). Find the value of C_D for such a system when Re = 1.

5 Compute the value of C_c for a 0.5-µm-diameter sphere (*a*) in air at 20°C and 760 = mmHg pressure and (*b*) in air at 0°C and 0.25-atm pressure.

6 Show, using Stokes' law and the slip correction factor, that for very small particles the resisting force is proportional to d^2.

7 Given a particle made up of a two-sphere cluster, each sphere having a density of 2 g/cm³ and a diameter of 1 µm, find the aerodynamic and Stokes' diameter of the cluster.

8 What length of 1-µm-diameter fiber will have the same aerodynamic diameter as a 10-µm unit-density sphere? Assume the fiber density is 2.65 g/cm³.

9 Show that Eqs. 5.20 and 5.21 can be derived from Eq. 5.17 and the definition of aerodynamic diameter. In this derivation, assume that the aspect ratio is sufficiently large that

$$\left(\frac{d_f^2}{2} + d_f L\right)^{1/2} \approx (d_f L)^{1/2}$$

Chapter 6

Particle Kinetics

Settling, Acceleration, and Deceleration

Kinetics is the study of changes in particle motion due to various forces acting on the particle. Particle motion can be rectilinear, i.e., along a straight line, with the particle perhaps accelerating or decelerating as it moves, or the motion can be curvilinear, caused by forces acting to make a particle change its direction of motion. Rectilinear motion is covered in this chapter, curvilinear motion in the next.

When a force is applied to a particle at rest, the particle begins to accelerate. If the particle is in air, the force of the air resisting the motion is zero when the particle is at rest but increases as the motion of the particle increases. As discussed in Chap. 5, at low Reynolds numbers this resisting force is given by Stokes' law. If the accelerating force is constant, eventually a point will be reached where the resisting force and accelerating forces are equal, and the particle will then move at a constant velocity. Often it is important to know how long it will take before a particle starting from rest will reach this constant velocity or, if it has some initial velocity, how long it will take to reach its final velocity. This is important in determining the velocity necessary for capture of particles by a ventilation system and is of interest in determining how quickly particles attain a constant or terminal settling velocity after they are dropped.

Example 6.1   Carbon particles in the exhaust of a diesel truck traveling at 55 mi/h are discharged into the atmosphere. Assuming that as these particles leave the exhaust they have the same velocity as the truck, how far will they travel in air before their motion is essentially that of the air in which they were discharged?

For all practical purposes, they lose their initial velocity immediately on discharge. This will become apparent later in the chapter.

Equation of Motion of an Aerosol Particle

By determining the path of a single particle when it is acted upon by a variety of forces, it is possible to predict particle position and behavior. This can be done by solving force balance equations which then give acceleration, velocity, and position of the particle.

The net difference of the forces acting on the particle is equal to the rate of change of particle momentum. Thus

$$m \frac{d\vec{v}}{dt} = \vec{F}_1 + \vec{F}_2 + \vec{F}_3 + \vec{F}_4 + \cdots \quad (6.1)$$

where m is the mass of the particle. The forces $\vec{F}_1, \vec{F}_2$, etc., may include those which are generally functions of time and the position of the particle, such as electric or magnetic forces, or they can be forces which are constant, such as gravity. These forces are generally balanced against the drag force, which depends on the properties of the medium, field of flow, particle shape, and instantaneous particle velocity. For many aerosol problems, this force is taken to be equal to the Stokes resistance $3\pi\mu vd$, with the appropriate corrections, and it always acts in a direction opposite to the instantaneous particle motion.

If all forces are balanced, that is, $m \, dv/dt = 0$, the particle is not accelerating and moves with a uniform velocity if it moves at all. When the Stokes resistance is equal to zero. the particle velocity with respect to the airstream is zero.

Equation 6.1 represents a system of three differential equations for the coordinates x, y, and z (or for some curvilinear coordinates q_1, q_2, q_3) expressed as functions of time t. Solution of these equations defines a trajectory of the particle for certain initial conditions of position and velocity. Several examples will be examined.

Example 6.2 Write a force balance equation for a particle which is acted upon by gravity in an electric field.

Since the direction of motion of the particle is unspecified, our equation must be flexible in terms of direction. This is done by writing the equation in vector notation (the arrow over the variable indicates that the variable is a vector, it has both magnitude and direction).

Electric force is given by

$$\vec{F}_E = q\vec{E}$$

where q is the charge on the particle and $\vec{E}$ is the field strength (a vector quantity).

The gravitational force is given by

$$\vec{F}_G = mg\vec{G}$$

Here m is the mass of the particle, g is the acceleration due to gravity (both non-vector quantities), and $\vec{G}$ is a unit vector which establishes the direction in which gravity acts.

Then (since velocity is also a vector quantity)

$$m \frac{d\vec{v}_1}{dt} = mg\vec{G} + q\vec{E} + 3\pi\mu d\vec{v}_2$$

Notice that each set of terms in the equation contains one vector quantity (i.e., each term specifies a direction as well as a magnitude). Also notice that $\vec{v}_1$ represents the absolute particle velocity whereas $\vec{v}_2$ is the particle velocity relative to the medium velocity. Thus if $\vec{u}$ is the medium velocity, $\vec{v}_2 = \vec{u} - \vec{v}_1$.

Particle Motion in Air in the Absence of External Forces Except Gravity

Consider the case of a spherical aerosol particle in a homogeneous airstream with no forces acting on the particle except gravity. For simplicity the motion will be assumed to occur only in the Stokes region (in most cases this assumption is valid). Then (similar to Example 6.2),

$$m \frac{d\vec{v}}{dt} = 3\pi\mu d(\vec{u} - \vec{v}) + mg\vec{G} \qquad (6.2)$$

where m is the particle mass, $\vec{v}$ the velocity of the center of gravity of the aerosol particle, $\vec{u}$ the velocity of the airstream near the particle, and $\vec{G}$ the unit vector of the force of gravity. Dividing by $3\pi\mu d$ and rearranging terms give

$$\tau \frac{d\vec{v}}{dt} + \vec{v} = \vec{u} + \tau g \vec{G} \qquad (6.3)$$

where

$$\tau = \frac{m}{3\pi\mu d} = \frac{m}{F_R/v} \qquad (6.4)$$

The factor τ is an extremely important parameter in aerosol studies, as will be shown later. Properties of the particle (diameter and density) and of the medium (viscosity and density) are incorporated in this parameter, which has units of seconds. It represents a relaxation time for the aerosol particle.

For spherical particles of mass m, with $m = (\pi/6) d^3(\rho_p - \rho_m)$, τ becomes

$$\tau = \frac{1}{18} \frac{d^2(\rho_p - \rho_m)}{\mu}$$

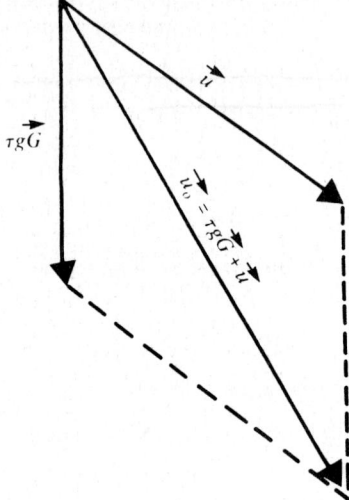

Figure 6.1 Vector diagram showing definition of $\vec{u}_0$.

Since for air $\rho_p \gg \rho_m$, τ is usually written as

$$\tau = \frac{1}{18}\frac{d^2}{\mu}\rho_p \tag{6.5}$$

The terms $\vec{u}$ and $\tau g \vec{G}$ in Eq. 6.3 represent two constant vectors which can be added to form a single constant vector $\vec{u}_0$. This addition is shown schematically in Fig. 6.1.

Example 6.3 Air flows in a horizontal duct with a velocity of 4 cm/s. If the acceleration due to gravity is 980 cm/s², determine the numerical value of the constant vector $\vec{u}_0$ for a 30-μm-diameter particle ($\tau = 2.75 \times 10^{-3}$ s).

$$(u_0)^2 = u^2 + (\tau g)^2$$

$$= (4)^2 + (2.77 \times 10^{-3} \times 980)^2 = 16 + 7.25 = 23.25$$

$$u_0 = 4.82 \text{ cm/s}$$

Expressing the equation of motion in terms of τ and $\vec{u}_0$ gives

$$\tau \frac{d\vec{v}}{dt} + \vec{v} = \vec{u}_0 \tag{6.6}$$

Suppose the cartesian coordinates are aligned such that at $t = 0$ the particle is at the origin. In addition, the coordinates are rotated so that the x axis is parallel to $\vec{u}_0$. Finally, the initial velocity vector of the particle is oriented such that it lies in the xy plane. Then this initial velocity vector can be broken down into x and y velocity components $\vec{v}_{x_i}$ and $\vec{v}_{y_i}$; that is, $\vec{v}_{x_i} + \vec{v}_{y_i} = \vec{v}_i$. Figure 6.2 illustrates the general orientation for solution of Eq. 6.6.

It should be realized that this coordinate system can be rotated at will. Although the orientation chosen is for convenience in solving the equation, it does not necessarily reflect the actual physical orientation of the problem (gravity may not be down, e.g.). Hence in using this development, it is helpful to keep actual particle orientation in mind.

Equation 6.6 in its vector form represents two scalar differential equations, one representing motion in the x direction and the other representing motion in the y direction:

$$\tau \frac{dv_x}{dt} + v_x = u_0 \quad (6.7a)$$

$$\tau \frac{dv_y}{dt} + v_y = 0 \quad (6.7b)$$

Integration of these equations with the initial conditions that $x = 0$ and $y = 0$ and $v_x = v_{x_i}$ and $v_y = v_{y_i}$ at $t = 0$ gives two equations for the velocity of the aerosol particle at any time

$$v_x = u_0 + (v_{x_i} - u_0)e^{-t/\tau} \quad (6.8a)$$

$$v_y = v_{y_i} e^{-t/\tau} \quad (6.8b)$$

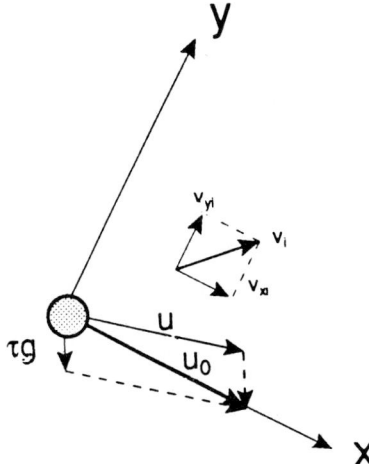

Figure 6.2 General orientation of solution of Eq. 6.6. Vector $\vec{u}_0$ is aligned so it is parallel to the x axis.

and two equations for the particle's position

$$x = u_0 t + \tau(v_{x_i} - u_0)(1 - e^{-t/\tau}) \qquad (6.9a)$$

$$y = v_{y_i}\tau(1 - e^{-t/\tau}) \qquad (6.9b)$$

These four equations completely describe the position and trajectory of the particle at any time, provided particle motion is in the laminar flow (Stokes') region.

Example 6.4 A 30-μm-diameter unit-density sphere ($\tau = 2.75 \times 10^{-3}$ s) falling at a terminal settling velocity of 2.7 cm/s is captured by a horizontal airflow of 100 ft/min which is flowing into a hood. Find its velocity 1 ms later, relative to the point at which it was captured.

First it is necessary to rotate axes so that vector $\vec{u}_0$ lies along x axis. Then u_0 will have a value of

$$u_0 = \sqrt{(50.83)^2 + (2.7)^2} = 50.90 \text{ cm/s}$$

The rotation required is arcsin $(2.7/50.90) = 3°$. Then

$$v_{x_i} = 2.7 \sin 3° = 0.14$$

$$v_{y_i} = 2.7 \cos 3° = 2.70$$

$$e^{-t/\tau} = \exp\left(\frac{-10^{-3}}{2.75 \times 10^{-3}}\right) = 0.695$$

and the velocity in the x and y directions can be calculated as follows:

$$v_x = u_0 + (v_{x_i} - u_0)^{-t\tau}$$

$$= 50.90 + (0.14 - 50.90)(0.695)$$

$$= 15.63 \text{ cm/s}$$

$$v_y = v_{y_i}e^{-t/\tau}$$

$$= (2.70)(0.695) = 1.87 \text{ cm/s}$$

It is now necessary to switch back from the artificial coordinate system to the real one. This can be done as follows:

$$v_V = 15.63 \sin 3° + 1.87 \cos 3° = 2.70 \text{ cm/s}$$

$$v_H = 15.63 \cos 3° - 1.87 \sin 3° = 15.51 \text{ cm/s}$$

To get the particle position, a similar approach would be taken.

Notice that for typical aerosol particle sizes, the exponential terms rapidly disappear. Note also that the particle is rapidly acquiring the velocity of the horizontal airflow.

Terminal Settling Velocity

Equations 6.8 and 6.9 can now be applied to the case of a particle falling under the influence of gravity in still air ($u = 0$). Since the direction of the gravitational force is along the x axis, Eq. 6.8a shows that even with an initial velocity component in some other direction, eventually the only velocity the particle will have will be in the direction of the gravitational force. The velocity in the direction of gravity is given by

$$v_t = \tau g + (v_{x_i} - \tau g)e^{-t/\tau}$$

As time progresses, the particle will attain a constant velocity given by τg. If a particle is initially given a velocity greater than this, it will decelerate until it has reached τg. If the particle's initial velocity is less, it will increase to a value of τg. If a particle falls from rest, it will accelerate until τg is attained. Thus τg represents the *terminal settling velocity* of the particle v_t

$$v_t = \tau g \tag{6.10}$$

Example 6.5 An asbestos fiber is reported to have an aerodynamic diameter of 1.79 μm. Determine its terminal settling velocity.

$$\tau = \frac{1}{18}\frac{d_a^2}{\mu}(1) = \frac{(1.79 \times 10^{-4})^2}{(18)(1.82 \times 10^{-4})}$$

$$= 9.78 \times 10^{-6}$$

$$v_t = \tau g = 9.78 \times 10^{-6} \times 980$$

$$= 9.58 \times 10^{-3} \text{ cm/s}$$

For particles with diameters smaller than about 10 μm (actually, 1 μm is often taken as the cutoff point), it is necessary to include the Cunningham correction factor in calculating the terminal settling velocity. Then

$$v_t = \tau g C_c \tag{6.11}$$

The practice of neglecting C_c is only for convenience in calculations. With the advent of programmable calculators or personal computers, it is now best to always include C_c in the computation of τ wherever possible. In the case of nonisometric particles it is sometimes difficult to determine the appropriate value of C_c to apply. Dahneke (1973a,b,c) has proposed a method for computing C_c for some nonspherical particles, and this approach has been confirmed by Cheng et al. (1988).

Example 6.6 Determine the terminal settling velocity of a 0.5-μm-diameter silica sphere ($\rho = 2.65$ g/cm^3). Include the Cunningham correction factor in the estimate.

From Eq. 5.3,

$$C_c = 1 + \frac{2\lambda}{d}(1.257) \cong 1 + \frac{(2)(7 \times 10^{-6})}{0.5 \times 10^{-4}}(1.257)$$

$$= 1.35$$

$$\tau = \frac{1}{18}\frac{d^2}{\mu}\rho_p = \frac{1}{18}\frac{(5 \times 10^{-5})^2(2.65)}{1.82 \times 10^{-4}} = 2.02 \times 10^{-6} \text{ s}$$

$$v_t = \tau g C_c = (2.01 \times 10^{-6})(980)(1.35) = 2.67 \times 10^{-3} \text{ cm/s}$$

Equation 6.11 is often derived by merely equating Stokes' resistance with the gravitational force. Although conceptually simpler, it does not provide the insights into the time-dependent cases of acceleration or deceleration of the particle to terminal velocity. The rapidity with which the terminal settling velocity is reached is given by the factor $e^{-t/\tau}$. Thus the smaller the value of τ, the more quickly an aerosol particle will reach equilibrium or steady-state conditions. For example, for a 2-μm-diameter unit-density sphere τ has a value of 1.305×10^{-5} s. Since e^{-7} is about 0.001, equilibrium values are essentially reached when $t/\tau = 7$ or for the 2-μm sphere within about 100 μs. Table 6.1 gives values of τ and 7τ for unit-density spheres of other diameters. It is clear that particles smaller than several micrometers in diameter will rapidly accelerate or decelerate to equilibrium conditions, so that generally for these sizes of particles it is possible to neglect the inertial term in Eq. 6.1.

TABLE 6.1 Relaxation Times τ and Equilibrium Time 7τ for Unit-Density Spheres at Atmospheric Pressure and 20°C

Diameter	Relaxation time, s	Equilibrium time, s
0.01	6.77×10^{-9}	4.74×10^{-8}
0.02	1.39×10^{-8}	9.73×10^{-8}
0.04	2.94×10^{-8}	2.06×10^{-7}
0.06	4.67×10^{-8}	3.27×10^{-7}
0.08	6.60×10^{-8}	4.62×10^{-7}
0.1	8.73×10^{-8}	6.11×10^{-7}
0.2	2.28×10^{-7}	1.59×10^{-6}
0.4	6.90×10^{-7}	4.83×10^{-6}
0.6	1.40×10^{-6}	9.78×10^{-6}
0.8	2.35×10^{-6}	1.65×10^{-5}
1	3.55×10^{-6}	2.48×10^{-5}
2	1.32×10^{-5}	9.23×10^{-5}
4	5.08×10^{-5}	3.55×10^{-4}
6	1.13×10^{-4}	7.89×10^{-4}
8	1.99×10^{-4}	1.39×10^{-3}
10	3.10×10^{-4}	2.17×10^{-3}
20	1.23×10^{-3}	8.60×10^{-3}

Stop Distance

Consider the case of a particle having an initial velocity v_{y_i} in the y direction when u_0 is zero. This is equivalent to a particle being projected into still air. If gravity is neglected, it can be seen from Eq. 6.8b that the particle rapidly decelerates to zero velocity. While decelerating, the particle traverses a distance which can be found in Eq. 6.9b when t goes to infinity. This distance

$$y = v_{y_i}\tau \qquad (6.12)$$

is known as the *stop distance* or *horizontal range* of the particle. Equation 6.12 indicates that small particles move very short distances before coming to rest; a 1-μm-diameter particle projected into air at an initial velocity of 1000 cm/s, for example, moves a distance of only 0.0036 cm before stopping.

Example 6.7 Determine the stop distance of a 1.5-μm-diameter unit-density sphere which is projected into still air with an initial velocity of 1000 cm/s. (Neglect gravity.)

$$\tau = \frac{1}{18}\frac{d^2}{\mu}\rho_p C_c$$

$$= \frac{1}{18}\frac{(1.5 \times 10^{-4})^2}{1.82 \times 10^{-4}}(1)\left[1 + \frac{2 \times 7 \times 10^{-6}}{1.5 \times 10^{-4}}(1.257)\right]$$

$$= (6.87 \times 10^{-6})(1.12) = 7.66 \times 10^{-6}\text{ s}$$

$$y = (1000)(7.66 \times 10^{-6}) = 7.66 \times 10^{-3}\text{ cm}$$

Particle Acceleration or Deceleration

For particles injected into a moving airstream (similar to acceleration under the influence of gravity and similar to problems of particle deceleration), it can be seen that the difference between particle velocity and stream velocity decreases by a factor of e for each time period $t = \tau$. Thus within 7τ steady-state conditions are reached.

Example 6.8 A 40-μm-diameter unit-density sphere falls across a slot opening for a ventilation system into which air is being drawn. How long will it take the particle to achieve the velocity of the in-rushing air?

$$\tau = \frac{1}{18}\frac{d^2}{\mu}\rho_p = \frac{1}{18}\frac{(40 \times 10^{-4})^2}{1.82 \times 10^{-4}}(1) = 4.88 \times 10^{-3}\text{ s}$$

$$t = 7(4.88 \times 10^{-3})\text{ s} = 0.034\text{ s}$$

This indicates that within 0.034 s the particle will be caught up and transported by the moving airstream. Although a fairly short time, it may be too long to ensure capture of the particle. It should be apparent from this analysis that small particles ought to be easier to capture with a ventilation system than large particles.

Limitations

Equations of motion presented here were developed for cases of uniform medium velocity and are oversimplified for many other cases regarding aerosols. In addition, evaluation of the equations for the trajectories of aerosol particles is sometimes impossible because of the difficulty in accurately describing the field of flow. Although for laminar flow Eq. 6.6 can be separated into x and y components, with increasing Reynolds number the nonlinearity of the resisting force prevents separation of the vector equation. Fortunately, most aerosol problems can be treated in the low-Reynolds-number regime.

Example 6.9 Determine the diameter of a unit-density sphere that has a Reynolds number equal to 1 at terminal settling velocity.

$$\text{Re} = \frac{v_t d}{0.151}$$

$$v_t = \tau g = \frac{1}{18}\frac{d^2}{\mu}\rho_p g$$

Substituting and rearranging give

$$d^3 = \frac{(0.151)(18)(\text{Re})(\mu)}{\rho_p g} = \frac{(0.151)(18)(1)(1.82 \times 10^{-4})}{(1)(980)}$$

$$= 5.04 \times 10^{-7}$$

$$d = 79.60 \times 10^{-4} \cong 80 \ \mu\text{m}$$

This represents a rough guide for the upper size of particles for which Stokes' law applies. This size will be different for particles of different densities.

One-Dimensional Motion at High Reynolds Numbers

There are occasions when particle motion is so great or particle diameter so large that Stokes' law is no longer applicable. Then some other simplifying approach must be taken. In Chap. 4 this problem was treated for the case of sedimenting particles through the use of plots of $C_D \text{Re}^2$ versus Re and C_D/Re versus Re.

For the generalized case of one-dimensional particle motion, recall that

$$F_D = AC_D\rho_m\frac{v^2}{2} \tag{6.13}$$

$$\text{Re} = \frac{vd\rho_m}{\mu} \tag{6.14}$$

$$\frac{dv}{d\text{Re}} = \frac{\mu}{\rho_m d} \tag{6.15}$$

From Eq. 6.1

$$m\frac{dv}{dt} = F \tag{6.16}$$

so, for spheres, using Eq. 6.13

$$\frac{\pi}{6}d^3\rho_p\frac{dv}{dt} = \frac{\pi}{4}d^2 C_D\rho_m\frac{v^2}{2} \tag{6.17}$$

$$\frac{dv}{dt} = \frac{3C_D\rho_m v^2}{4d\rho_p}$$

Expressing Eq. 6.17 in terms of the Reynolds number

$$\frac{d\text{Re}}{dt} = \frac{3C_D\mu}{4\rho_p d^2}\text{Re}^2 \tag{6.18}$$

gives, on inverting,

$$dt = \frac{4\rho_p d^2}{3\mu}\frac{d\text{Re}}{C_D\text{Re}^2} \tag{6.19}$$

On integration from Re_i to Re_f, Eq. 6.19 yields

$$t = \frac{4\rho_p d^2}{3\mu}\int_{\text{Re}_i}^{\text{Re}_f}\frac{d\text{Re}}{C_D\text{Re}^2} \tag{6.20}$$

Similarly, for displacement, since $s = vt$,

$$s = \frac{4}{3}\frac{\rho_p d^2}{\mu}\int\frac{d\text{Re}}{C_D\text{Re}^2}\frac{\mu\text{Re}}{\rho_m d} \tag{6.21}$$

$$s = \frac{4}{3}\frac{\rho_p}{\rho_m}d\int_{\text{Re}_i}^{\text{Re}_f}\frac{d\text{Re}}{C_D\text{Re}} \tag{6.22}$$

The utility of Eqs. 6.20 and 6.22 lies in their generality. All that is required is an expression for C_D for the range of Reynolds numbers over which the particle is moving. Keep in mind that this solution is applicable *only* to one-dimensional flow or to those cases where one-dimensional flow can be approximated.

Example 6.10 A 6-in-diameter grinding wheel is operated at 1750 revolutions per minute (r/min). How far will a 0.1-mm-diameter particle be thrown from the wheel if gravity is neglected? Assume a spherical particle having a density of 3 g/cm³.

$$\text{Initial velocity} = \omega r = 1750 \times \frac{1}{60} \times 2\pi \times 6 \times 2.54 \times \frac{1}{2} = 1396 \text{ cm/s}$$

$$\text{Re}_i = \frac{dv}{0.151} = \frac{(0.01)(1396)}{0.151} = 92.55$$

From Eq. 4.7

$$C_D = \frac{14}{\text{Re}^{0.5}}$$

This equation is applicable over the range of Reynolds numbers $2 < \text{Re} < 800$.

$$s = \frac{4}{3}\frac{\rho_p}{\rho_m} d \int_{\text{Re}_i}^{\text{Re}_f} \text{Re}^{0.5} \frac{d\text{Re}}{14\text{Re}} = \frac{4}{42}\frac{\rho_p}{\rho_m} d \int_{\text{Re}_i}^{\text{Re}_f} \text{Re}^{-0.5} d\text{Re}$$

Integrating gives

$$\frac{4}{21} \frac{3}{0.0012} (0.01) \text{Re}^{0.5}\big|_{91.9}^{2} = -38.87 \text{ cm}$$

To go from $\text{Re} = 2$ to $\text{Re} = 0$, use $C_D = 24/\text{Re}$. Then

$$s = 33.33 \int_2^0 \frac{\text{Re}}{24} \frac{d\text{Re}}{\text{Re}} = \frac{33.33}{24} \text{Re}\big|_2^0$$

$$= \frac{33.33}{24}(-2) = -2.76$$

Total stop distance $= 38.87 + 2.76 = 41.63$ cm

Notice that the particle travels the greatest distance at the higher Reynolds numbers. The laminar flow contribution to the stop distance is small compared to the intermediate or turbulent flow contribution.

Ideal Stirred Settling

Although with an aerosol each particle will settle at its own terminal settling velocity, settling rarely takes place in absolutely still air since there is always some circulation and mixing. This mixing has the effect of producing a uniform aerosol concentration which decreases with time because of sedimentation.

Consider a cylinder of uniform cross-sectional area A and height H, filled with a monodisperse aerosol having an initial concentration of n_0 particles per cubic centimeter. With mixing, but in the absence of any other external forces, the average number of particles per square centimeter moving in an upward direction and crossing a horizontal plane L which cuts the cylinder at some arbitrary height will equal the average number moving downward. That is,

$$AH \, dn = \tfrac{1}{2}nAv_u \, dt - \tfrac{1}{2}nAv_d \, dt = 0 \qquad (6.23)$$

where v_u and v_d are the particle velocities up and down, respectively.

Now suppose a negative force field is included (i.e., include the force due to gravity which imparts an additional downward velocity v_t on the particles; this is the terminal settling velocity). Then

$$AH \, dn = \tfrac{1}{2}nA(v_u - v_t) \, dt - \tfrac{1}{2}nA(v_d + v_t) \, dt \qquad (6.24)$$

since $v_u = v_d$

$$AH \, dn = -nAv_t \, dt \qquad (6.25)$$

and then

$$\frac{dn}{n} = -\frac{v_t}{H} dt \qquad (6.26)$$

Assuming that v_t is independent of time, position, and concentration, Eq. 6.26 can be integrated to give

$$n = n_0 \exp\left(-\frac{v_t t}{H}\right) \qquad (6.27)$$

Equation 6.27 implies exponential decay of an aerosol concentration in a closed chamber. This is observed in practice.

Since n and n_0 have the same units, they can have *any* units—particles per cubic centimeter, milligrams per cubic meter, etc.

Example 6.11 The smoke concentration in a room is found to be 50 mg/m³. If this aerosol is made up of 0.75-μm-diameter spherical particles (unit density), estimate the concentration in the room 3 h later. The ceiling height is 8 ft.

For 0.75-μm spheres,

$$v_t = \frac{1}{18} \frac{d^2}{\mu} \rho_p C_c g$$

$$= \frac{1}{18} \frac{(0.75 \times 10^{-4})^2}{1.82 \times 10^{-4}} (1)(1.23)(980)$$

$$= 2.07 \times 10^{-3} \text{ cm/s}$$

$$H = 8 \text{ ft} \times 30.5 \text{ cm/ft} = 244 \text{ cm}$$

$$n = n_0 \exp\left(-\frac{v_t t}{H}\right) = 50 \exp\left(-\frac{2.07 \times 10^{-3} \times 3 \times 60 \times 60}{244}\right)$$

$$= 45.62 \text{ mg/m}^3$$

For a polydisperse aerosol Eq. 6.27 can be solved for each size increment and the results summed to get the total aerosol concentration. This approach assumes that each size of particles is independent of every other size. This assumption is valid with moderate mixing and moderate to low aerosol concentrations.

Problems

1 Show that if $v_y = v_{y_i}$ at $t = 0$, the solution to

$$\tau \frac{dv_y}{dt} + v_y = 0$$

is

$$v_y = v_{y_i}\tau(1 - e^{-t/\tau})$$

if $y = 0$ at $t = 0$.

2 Compute the value of τ for a 15-μm-diameter sand particle ($\rho = 2.65$ g/cm^3). Then compute (a) its terminal settling velocity, (b) its Reynolds number at this velocity, and (c) its stop distance if its initial velocity equals its terminal settling velocity.

3 a. What are the diameter and the terminal settling velocity of a unit-density sphere having Re = 0.5 at its terminal settling velocity?
   b. Assuming the particle initially started from rest, how long will it take to reach one-half its terminal settling velocity?

4 A 10-μm gold sphere ($\rho = 19.3$ g/cm^3) is dropped from a 10-ft-high platform. Estimate the time it takes to strike the ground (a) neglecting air resistance and (b) including air resistance.

5 A 200-μm-diameter raindrop falls freely in the atmosphere. Determine its terminal settling velocity. Compare this to the measured value given in Table 5.3.

6 What is the diameter of a unit-density particle which falls with a terminal settling velocity of 200 cm/s?

7 Using Stokes' law, show that

$$C_D = \frac{24}{\text{Re}}$$

8 An approximate formula sometimes used to estimate the settling velocity in feet per minute of airborne dust particles is

$$v = \frac{d^2}{100}$$

where d is the particle diameter in micrometers. Compare the estimate given by this equation to the terminal velocity given by Stokes' law for 10-, 1.0-, and 0.1-μm particles with a density of 2.3.

9 Given three unit-density spherical particles of 2-, 0.2-, and 0.02-μm diameter, compute the sedimentation velocity for each (a) neglecting C_c and (b) correcting for C_c.

10 Examine the settling velocity of a 100-μm unit-density sphere with and without the correction

$$1 + \tfrac{3}{16}Re$$

applied. How much error is introduced by neglecting this correction?

11 Determine the position of the particle described in Example 6.4 at 1 ms after it is captured.

Chapter 7

Particle Kinetics

Impaction

Curvilinear Motion

When particles are transported by air currents, changes in the direction of these currents give rise to accelerating forces on the aerosol particles. Thus spinning of air tends to move the aerosol away from the axis of rotation, or rapid changes of airflow around an obstacle can result in aerosol particles being deposited on that body. This may be one of the principal mechanisms by which particles are removed by nature from the atmosphere. Sampling and collection devices such as impactors or impingers are based on the use of centrifugal forces, as are such other devices as "cyclones" and aerosol centrifuges.

The magnitude of the accelerating force that acts on a particle in curvilinear motion depends on the particle inertia. The greater the inertia of the particle, the greater will be the displacement. Inertia depends on particle mass and velocity. Heavy particles will be displaced more from the streamlines in which they are traveling than light ones, and increases in velocity will increase displacement for a particle of given mass.

When a particle is moving in a circular path around a point a distance r away with an angular velocity of ω, it experiences a radial acceleration of

$$a_r = \omega^2 r \qquad (7.1)$$

and a tangential velocity of

$$v_\omega = \omega r \qquad (7.2)$$

Consider a particle being carried by a volume of air (or other gaseous medium) which is moving in a circular orbit. The particle will have a constant tangential velocity—that of the air mass in which it is contained—and will also experience an outward, accelerating force which at low radial velocities can be approximated by the Stokes resistance. Equating the radial accelerating force with this resisting force gives

$$F_r = F_R \qquad (7.3)$$

$$ma_r = \frac{3\pi\mu d v_r}{C_c}$$

Solving for v_r assuming a spherical particle yields

$$v_r = \frac{(\pi/6)d^3 \rho_p \omega^2 r C_c}{3\pi\mu d} = \tau\omega^2 r \qquad (7.4)$$

$$v_r = \frac{v_\omega^2}{r}\tau \qquad (7.5)$$

an expression for the radial velocity of a particle.

Example 7.1 A 10-μm-diameter unit-density sphere is held in a circular orbit by an electric field. The orbit is 25 cm in radius, and the particle moves around the center of the circle at a rate of 100 r/min.

a. What is the radial velocity of the particle at the instant the electric field is removed? From Eq. 7.4

$$v_r = \tau\omega^2 r = \frac{d^2}{18\mu}\rho C_c \omega^2 r$$

$$= \frac{1}{18}\frac{(10 \times 10^{-4})^2}{1.82 \times 10^{-4}}(1)(1)\left(\frac{100}{60} \times 2\pi\right)^2 (25)$$

$$= 0.851 \text{ cm/s}$$

b. How far will the particle move until its radial velocity is dissipated? The distance the particle will move is just the stop distance s:

$$s = \tau v_r = (3.11 \times 10^{-4})(0.851)$$

$$= 2.64 \times 10^{-4} \text{ cm}$$

Impaction of Particles

When air carrying particles suddenly changes direction, the particles, because of their inertia, tend to continue along their original paths. If the change in air direction is caused by an object placed in the airstream, particles with sufficient inertia will strike the object. This process is known as *impaction*. It is the mechanism by which many large particles are removed from the atmosphere, it is one of the important mechanisms for removal of particles by the lungs, and it is important in air cleaning as well as aerosol sampling. The process of impaction can be modeled by using the equation of motion of an aerosol particle, if an appropriate choice is made of the velocity field. Often compromises have to be made in this choice. In any theoretical development, however, certain factors seem to be important.

Consider a simple model of impaction. Air issues from a long slot of width W at a velocity u. A surface is placed normal to the discharging flow a distance S away. With this configuration, air leaving the slot must make a 90° turn before it escapes. Particles that fail to make this turn strike or "impact" on the surface and are assumed to be retained by that surface.

As a crude first approach (Fuchs, 1964), it can be assumed that the streamlines of the air issuing from the slot are quarter-circles with their centers at C (see Fig. 7.1a) and that $S = W/2$. At point B a particle has a tangential velocity given by $v_\omega = u$ and a radial velocity given by

$$v_r = \frac{dr}{dt} = \frac{v_\omega^2}{r}\tau \tag{7.6}$$

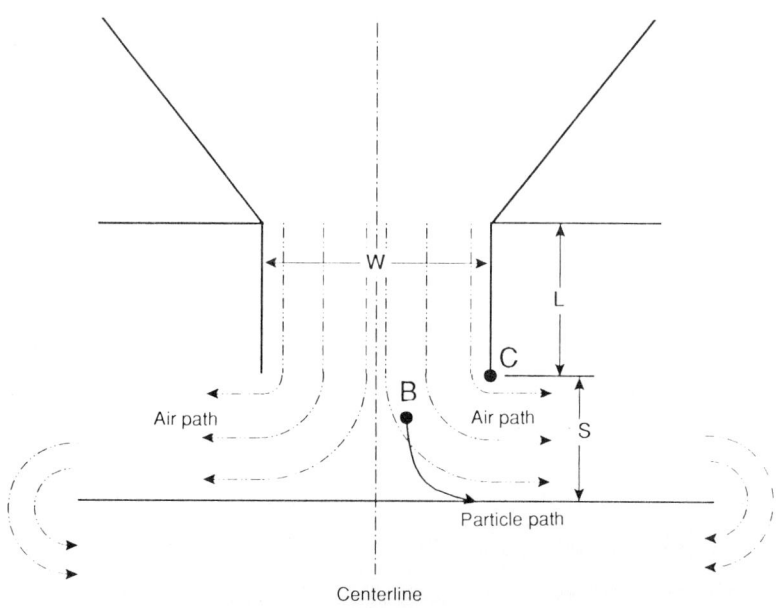

Figure 7.1a Sketch of simple "ideal" impactor.

In a time dt the particle will be displaced toward the surface a distance

$$d\delta = \frac{v_\omega^2}{r}\tau \sin\phi \, dt \tag{7.7}$$

where ϕ is the angle formed by the line connecting points B and C and the plane normal to the airflow which passes through point C (Fig. 7.1b). As the streamlines turn from the slot to be parallel with the surface ϕ goes from 0° to 90°. This change in angle ϕ can be expressed as

$$d\phi = \frac{v_\omega}{r} dt \tag{7.8}$$

In traversing the full 90°, the particle will be displaced a distance δ

$$\delta = \int_0^{\pi/2} v_\omega \tau \sin\phi \, d\phi = v_\omega \tau = u\tau \tag{7.9}$$

That is, the particle will move one stop distance out of its initial streamline while losing all its original velocity parallel to the slot.

Since all particles that lie within a distance δ of the slot centerline are considered to be removed, the overall removal efficiency ϵ of the impactor will be

$$\epsilon = \frac{\delta}{W/2} = \frac{2u\tau}{W} \tag{7.10}$$

This is, of course, only a very crude approximation for impactor efficiency, since the actual flow field is much more complex, varying in

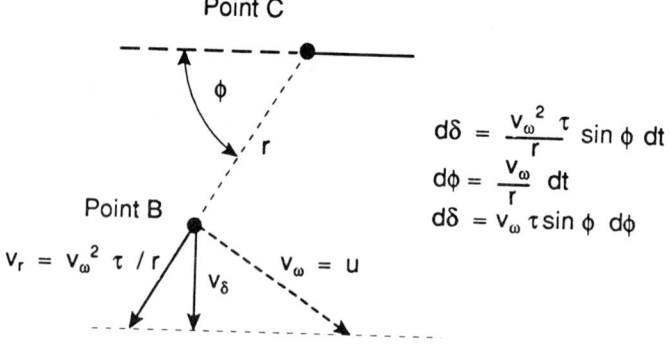

Figure 7.1b Diagram of velocities at point B.

configuration depending on the slot Reynolds number. In general, $S \neq W/2$. This sample model does give some idea of how effective a particular impactor configuration can be, the estimate being reasonably good when $\epsilon \gtrsim 1$ but rapidly losing accuracy for $\epsilon < 0.7$.

Example 7.2 A rectangular slot impactor has slot dimensions of 2.08-cm length and 0.358-cm width.
  Estimate the flow in liters per minute required for this impactor so that 15-μm-diameter unit-density spheres can be collected with near 100 percent efficiency.

$$\tau = \frac{1}{18} \frac{d^2}{\mu} \rho_p C_c$$

$$= \frac{1}{18} \frac{(15 \times 10^{-4})^2}{1.82 \times 10^{-4}} (1)(1.01)$$

$$= 6.95 \times 10^{-4} \text{ s}$$

$$\epsilon = \frac{u\tau}{W/2}$$

$$u = \frac{\epsilon W/2}{\tau} = \frac{(1)(0.358/2)}{6.95 \times 10^{-4}}$$

$$= 258 \text{ cm/s}$$

$$Q = Au = (2.08)(0.358)(258) = 192 \text{ cm}^3/\text{s}$$

$$= 11.51 \text{ L/min}$$

The quantity $2u\tau/W$ is an important dimensionless parameter in impactor studies, known as the *Stokes number*

$$\text{Stk} = \frac{2u\tau}{W} \qquad (7.11)$$

This dimensionless parameter is used to describe impactor behavior. For impactors with rectangular openings, W is the slit width; for circular openings W represents the diameter of the impactor opening. Thus the Stokes number is the ratio of the stop distance to the impactor opening half-width.

Some authors prefer to use the impaction parameter ψ, rather than the Stokes number, to describe impactor properties (e.g., Green and Lane, 1964; Ranz and Wong, 1952). The *impaction parameter* is defined as

$$\Psi = \frac{u\tau}{W} \qquad (7.12)$$

a factor which is one-half the Stokes number.

Example 7.3 Compute Stk, $\sqrt{Stk}$, ψ, and $\sqrt{\psi}$ for a circular jet impactor when $\tau u = 0.004$ cm and the jet diameter is 0.1 mm.

$$Stk = \frac{\tau u}{W/2} = \frac{4 \times 10^{-3}}{0.005} = 0.8$$

$$\sqrt{Stk} = 0.89$$

$$\psi = \frac{Stk}{2} = 0.4$$

$$\sqrt{\psi} = 0.63$$

It is common practice to plot impactor efficiency as a function of either $\sqrt{Stk}$ or $\sqrt{\psi}$ (e.g., Rao and Whitby, 1978). This is done because the particle diameter is present in either term as d^2, making the square root of the term proportional to the particle diameter.

Impactor operation

The characteristic behavior of impactors depends on factors such as nozzle-to-plate distance, nozzle shape, flow direction, and Reynolds numbers for both the jet and the particle. Other factors of importance include the probability that the particles will stick to the impaction surface and particle loss to the walls of the impactor. It is not surprising that with such a variety of possible variables it is quite difficult, if not impossible, to accurately predict impactor characteristics on purely theoretical grounds.

Calculation of the jet or nozzle Reynolds number is straightforward for a round jet; thus $Re = Wv/\nu$, where W is the jet diameter, v the velocity in the jet, and ν the kinematic viscosity. For a flow of Q cm³/s,

$$Re = \frac{4Q}{\pi \nu W} \qquad (7.13)$$

For a rectangular jet the "wetted perimeter" concept must be used (Marple, 1970). That is, the opening width Ω to be used in computing the Reynolds number is defined as $\Omega = 4$ (area/perimeter). For a rectangle of length L and width W, $\Omega = 2WL/(W + L)$. When $L \gg W$,

$$Re = \frac{2Q}{L\nu} \qquad (7.14)$$

For a well-designed impactor, a typical plot of impactor efficiency versus $\sqrt{Stk}$ is shown in Fig. 7.2. It can be seen that the efficiency curve may deviate from the ideal case. In the ideal case, for all efficiencies there would be a single value of $\sqrt{Stk}$ and hence a sharp size cut of the impactor. All particles larger than this size would be collected, and all smaller sizes would be passed. In actuality, this is not

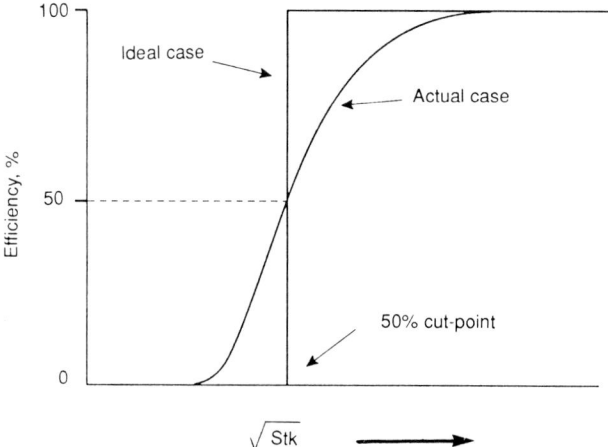
Figure 7.2 Typical impactor stage efficiency curve.

the case, and a range of particle sizes are collected with varying efficiencies. To represent the impactor stage collection characteristic, it is often the practice to choose the 50 percent efficiency point as the representative cut point. The maximum slope at this point most nearly represents the ideal case. In Fig. 7.2 both the actual and ideal cases would be considered to have the same characteristic cut size.

What constitutes a well-designed impactor? According to Marple and Rubow (1986), the minimum value of S/W should be no greater than 1 for a round impactor and 1.5 for a rectangular impactor. As an upper limit, S/W ratios several times greater than these minimums are possible, but design values as close to the minimum as possible are preferred. However, some commercially available impactor designs use S/W ratios of about 0.5 [e.g., the Hering design (Hering and Marple, 1986)].

Figure 7.3 shows theoretical impactor performance when a number of parameters are varied, including jet-to-plate spacing, jet Reynolds number, and throat length-to-width ratio. These curves indicate that impactor efficiency is fairly insensitive to Reynolds number in the range $500 < Re < 3000$ and that impactor efficiency is also relatively independent of S/W and T/W ratios, except for small values of S/W.

The calculations that produced the curves in Fig. 7.3 were repeated in more detail by Rader and Marple (1985), who also included the effect of the physical size of the particle (interception distance) as it approaches the collection plate. Figure 7.4 shows similar curves from this more recent work. Although differences in the results of the two sets of calculations are small, the newer curves are steeper and show a

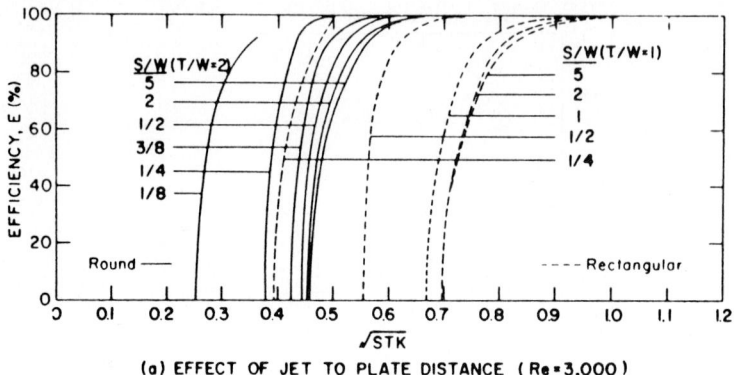

(a) EFFECT OF JET TO PLATE DISTANCE (Re = 3,000)

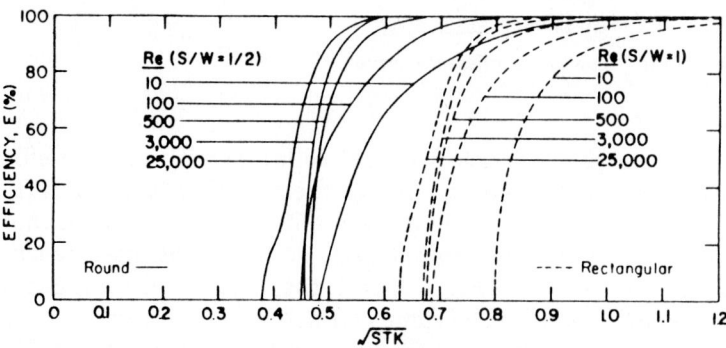

(b) EFFECT OF JET REYNOLDS NUMBER (T/W = 1)

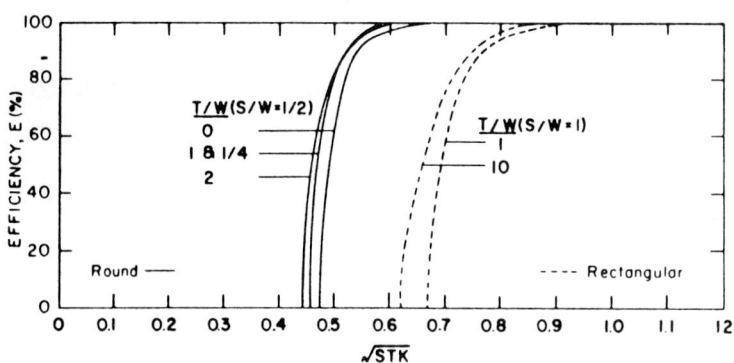

(c) EFFECT OF THROAT LENGTH (Re = 3,000)

Figure 7.3 Impactor efficiency curves for rectangular and round impactors showing effects of jet-to-plate distances in Reynolds number Re and throat length T. W is impactor width or impactor jet diameter. (*From Marple and Willeke, 1979.*) (a) Effect of jet-to-plate distance (Re = 3000). (b) Effect of jet Reynolds number ($T/W = 1$). (c) Effect of throat length (Re = 3000).

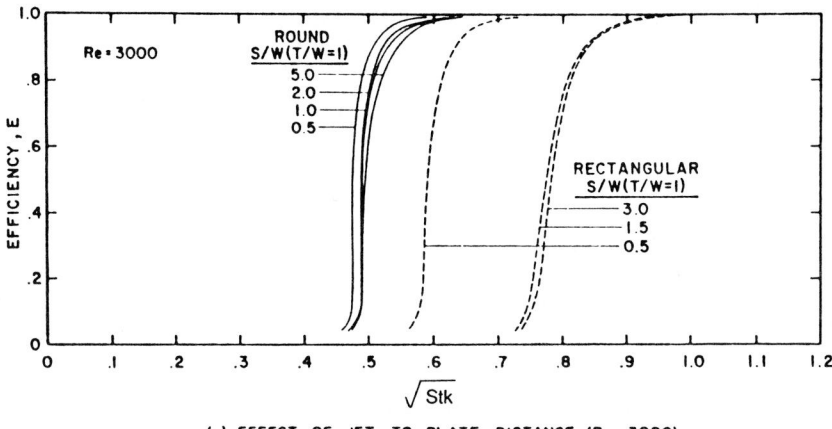

(a) EFFECT OF JET TO PLATE DISTANCE (Re = 3000)

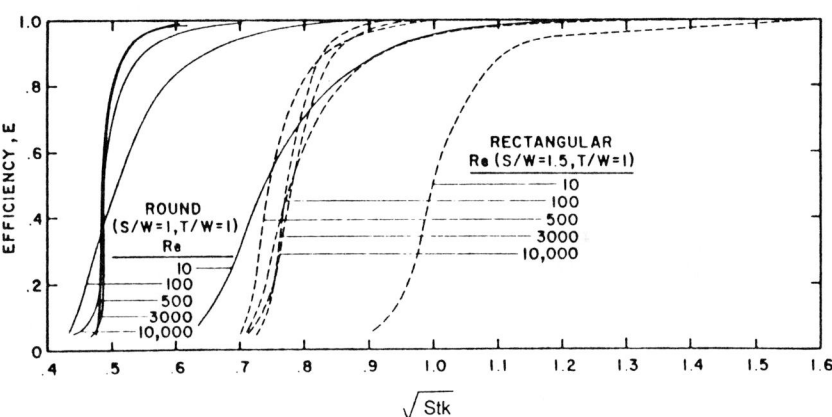

(b) EFFECT OF JET REYNOLDS NUMBER

Figure 7.4 Revised impactor efficiency curves. (*From Rader and Marple, 1985.*)

characteristic S shape that is found in experiment and is most likely due to the inclusion of the interception distance in the calculations.

Example 7.4 A round jet impactor is operated such that the jet Reynolds number is 3000. Using Fig. 7.3b, find the particle diameter (unit-density sphere) that will be collected with 50 percent efficiency if the jet diameter is 0.3 cm.

$$\text{Re} = \frac{av}{0.151} = \frac{(0.3)(v)}{0.151} = 3000$$

$$v = 1509 \text{ cm/s}$$

From Fig. 7.3b, at efficiency = 50 percent and $\sqrt{Stk} = 0.46$,

$$Stk = 0.21 = \frac{\tau v}{0.3/2}$$

$$\tau = \frac{(0.21)(0.15)}{1509} = 2.10 \times 10^{-5} \text{ s}$$

$$= \frac{1}{18}\frac{d^2}{\mu}(\rho) \quad \text{(neglect } C_c\text{)}$$

$$d^2 = \frac{18\tau\mu}{\rho} = \frac{(18)(2.10 \times 10^{-5})(1.82 \times 10^{-4})}{1}$$

$$d = 2.63 \times 10^{-4} \text{ cm} = 2.61 \text{ μm}$$

The theoretical impactor performance data can also be expressed in terms of the 50 percent cut size. Figure 7.5a shows $\sqrt{Stk_{50}}$ plotted as a function of the S/W ratio, and Fig. 7.5b shows the same ordinate plotted as a function of Re. These curves again illustrate that Stk_{50} is quite insensitive to changes in either S/W or Re, except in the extremes.

$$d_{50} = \sqrt{\frac{9 Stk_{50} \mu L W^2}{C_c \rho_p Q}} \quad (7.15)$$

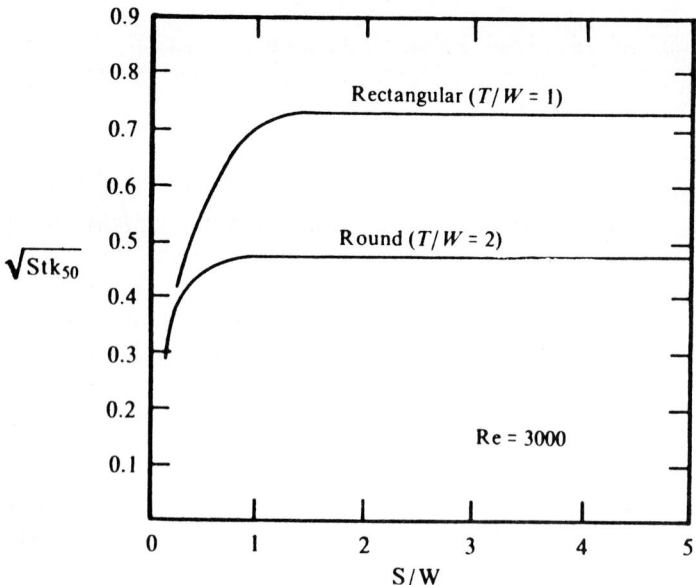

Figure 7.5a The 50 percent cutoff Stokes number as a function of jet-to-plate distance. (*From Marple, 1970.*)

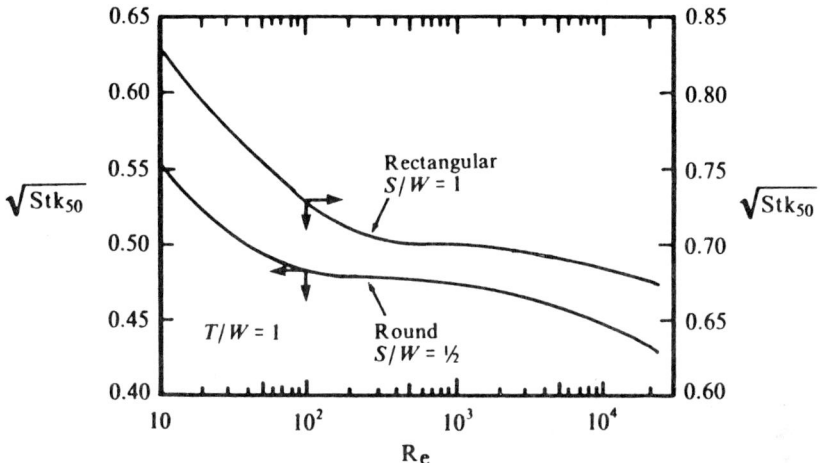

Figure 7.5b The 50 percent cutoff Stokes number as a function of Reynolds number. (*From Marple, 1970.*)

Impactor 50 percent cut points can be estimated from the equation for rectangular jets of length L and width W or

$$d_{50} = \sqrt{\frac{9\text{Stk}_{50}\mu\pi W^3}{4C_c\rho_p Q}} \tag{7.16}$$

for round jets of diameter W. These are rearrangements of Eq. 7.11. As shown in Fig. 7.5a, theoretical estimates of $\sqrt{\text{Stk}_{50}}$ are 0.71 for rectangular jet impactors and 0.46 for round jet impactors.

Particle bounce

The surface on which particles impact is also an important factor in determining impactor efficiency. Particles which bounce off the impaction surface can be carried through the impactor and can distort measurement data. Particle bounce will lower the collection efficiency of a given impactor stage and will lower the apparent mean diameter of the aerosol measured.

Another way of considering the effect of particle bounce is shown in Fig. 7.6. A plot of collection efficiency versus substrate loading indicates efficiencies which never reach 100 percent. Particle bounce can be minimized by using collection media coated with such materials as Vaseline, L and H high-vacuum greases, stopcock grease, oil, or Apiezon (Moss and Kenoyer, 1986).

Internal deposition of material may take place within the impactor and not on the impactor stage. Consider the following experiment:

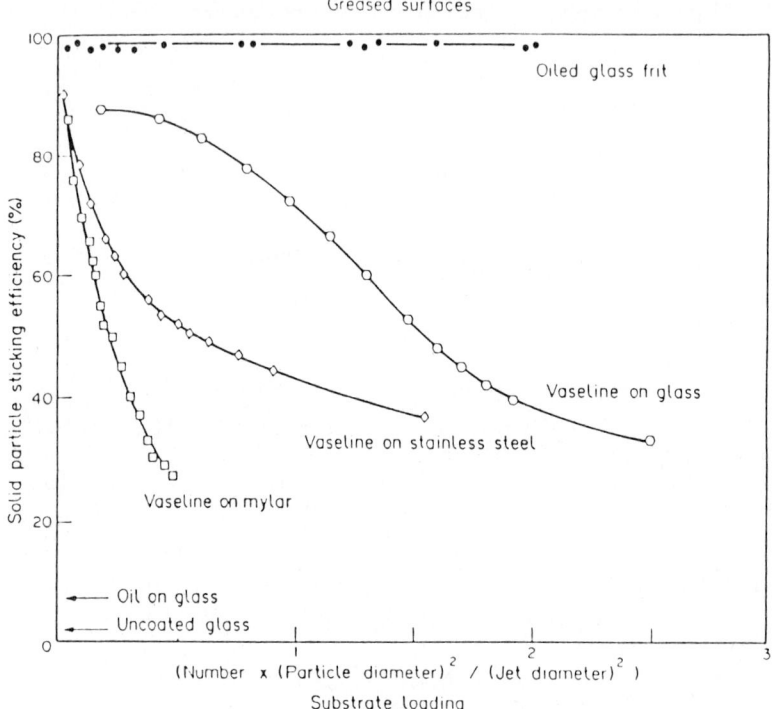

Figure 7.6 Dependence of solid particle sticking efficiency for various surface treatments and substrate loading. (*From Turner and Hering, 1987.*)

An impactor is operated such that there is no rebound and internal deposition can be halted by the presence of an electric field. If, when the impactor is operated with the field on, a certain slot velocity gives a downstream concentration of a monodisperse aerosol that is 50 percent of the upstream concentration, what happens when the field is shut off so wall deposition can take place?

Since deposition is occurring, the downstream aerosol concentration will drop. To increase this concentration so that C/C_0 will again be 50 percent, the impactor flow must be reduced. Thus wall deposition has the effect of raising the value of $\sqrt{Stk_{50}}$ compared to the case of no wall deposition.

Impactors for very small particle sizes

By their very nature, impactors are high-pressure drop sampling devices. Air is drawn at high velocity through a fairly small nozzle to remove small aerosol particles. The particle diameter which can be collected with a 50 percent cut efficiency for a specific set of operating conditions can be determined by recalling that

$$\text{Stk} = \frac{\rho_p d^2 v C_c}{9\mu W} \tag{7.17}$$

Rearranging in terms of d gives

$$d^2 = \frac{9\mu W \text{Stk}}{\rho_p C_c v} \tag{7.18}$$

Then using the 50 percent Stokes number for the 50 percent cut size produces

$$d_{50}^2 = \frac{9\mu W \text{Stk}_{50}}{\rho_p C_c v} \tag{7.19}$$

Since the value of Stk_{50} is nearly constant for similar impactor designs and since the viscosity and particle density are constant, the only way to change the 50 percent cut size for an impactor is to vary the jet velocity, the jet width, or C_c.

The traditional cascade impactor is constructed with a series of jets of decreasing diameters, so that both W and v are varied. Air flows through one jet (or group of jets of the same diameter), removing particles with a certain 50 percent cut size, and then proceeds to a smaller size jet (or series of jets of the same diameter) where particles with a smaller cut size are removed. This process can be repeated until the velocity in the jet approaches sonic velocity, at which point a backup filter catches the remaining small particles which have penetrated the impactor. The practical lower limit for an impactor of this type is about 0.4 μm.

Within the past few years there has been growing interest to measure the diameters of very small particles, i.e., particles with diameters less than about 0.5 μm. As discussed by Hering and Marple (1986), although traditional impactors are not adequate for this task, either low-pressure or microorifice impactors can collect particles with substantially smaller particle diameters than 0.4 μm.

With microorifice impactors W is made very small, on the order of 50 to 150 μm. Velocities are still kept somewhere below 100 m/s (Re ≈ 500 to 3000), but the number of orifices is increased to provide a reasonable total flow so that an adequate amount of sample is collected.

Low-pressure impactors utilize the fact that C_c is a function of not only particle diameter but also, through the gas mean free path, pressure. Therefore at low pressures C_c can be substantially larger than for the same size particle at atmospheric pressure.

Example 7.5 A Hering impactor operates at a flow rate of 1 L/min. Both stage 3 and stage 6 of this single-jet impactor have jet diameters of 0.99 mm and an S/W ratio of 0.5. Air enters stage 6 at 0.185 times atmospheric pressure and leaves with a p/p_0 ratio of 0.146 (Hering and Marple, 1986). Assuming $Stk_{50} = 0.22$ and a unit-density particle, calculate the appropriate d_{50}.

The average velocity in the jet is found by using Q/A adjusted for expansion and the lower or downstream pressure ratio. Then

$$v = \text{flow} \times \frac{p_0/p}{\text{area}} = \frac{1000}{60} \times \frac{1}{0.146} \div \frac{\pi}{4}(0.099)^2 = 14{,}830 \text{ cm/s}$$

Recalling Eq. 7.19

$$d_{50}^2 = \frac{9\mu W Stk_{50}}{\rho_p C_c v}$$

All factors on the right-hand side of this equation are known except C_c, which depends on d_{50}. Thus C_c is moved to the left-hand side of the equation, and the right-hand side is computed.

$$d_{50}^2 C_c = \frac{9\mu W Stk_{50}}{\rho_p v} = \frac{(9)(1.82 \times 10^{-4})(0.099)(0.22)}{(1)(14{,}830)} = 2.41 \times 10^{-9}$$

Since C_c also depends on d_{50}, to find d_{50} it is necessary to use an iterative procedure. A value for d_{50} is assumed and the associated C_c calculated by using the equation

$$C_c = 1 + \frac{2\lambda}{d_{50}p/p_0}\left[1.252 + 0.399 \exp\left(-\frac{1.1 d_{50}p/p_0}{2\lambda}\right)\right]$$

For λ use a value of 0.0687 μm. The correct pressure ratio to use for computing C_c is the *upstream* pressure ratio since that is where the effect of slippage on impaction will be most pronounced. However, as noted above, the jet velocity is computed by using the *downstream* pressure ratio since this represents the increased volume of air going through the impactor nozzle.

After C_c is computed, $d_{50}^2 C_c$ is calculated and compared to that found by using Eq. 7.19 (2.41×10^{-9}). With several iterations the following values are determined:

$$C_c = 7.45 \qquad d_{50} = 0.180 \text{ μm}$$

Pressure drop in impactors

In evaluating impactor operation it is important to know (or estimate) the pressure loss across each of the impactor stages. A simple approach is to assume that all the velocity pressure in the impactor jet is lost due to turbulence. Then the pressure drop ΔP across an orifice can be written as

$$\Delta p = p_{up} - p_{down} = 2\rho v^2 \frac{p}{p_{down}} \qquad (7.20)$$

Hence ρ, p, and v refer to the density, pressure, and velocity at atmospheric or some reference condition. The subscripts *up* and *down* refer to the pressure just upstream and downstream, respectively, of the impactor nozzle. For the first stage of an impactor, $p_{up} = p$. For subsequent stages, the calculated p_{down} of stage n is set equal to the upstream pressure of stage $n + 1$. If v is in units of centimeters per second and ρ in units of grams per cubic centimeter, then the units of p will be dynes per square centimeter.

Analysis of impactor data

The most common configuration for impactors used for aerosol sampling is to have a series of jets of decreasing size, arranged so that the air passes in series from the largest through the smallest slot. This cascade arrangement permits the aerosol to be fractionated into a number of size intervals, depending on the number of impactor stages used. Aerosol mass collected on the different impactor stages is then analyzed to provide size distribution information.

Particle distribution data can be presented as a bar chart where the mass of material collected in the ith stage, denoted m_i, is taken as the mass of particles in the size range $(d_{50})_{i+1}$ to $(d_{50})_i$. The height of the bar then represents the percentage of mass in that interval, and the size range gives the width of the bar.

Example 7.6 A six-stage impactor yields the following data (a filter is placed downstream of the impactor as a final stage to collect all particles which might otherwise escape):

Stage no.	d_{50}, μm	Particle mass collected, μg	Percentage in interval
1	18	0	0
2	11	15	3.30
3	4.4	35	7.69
4	2.65	110	24.18
5	1.7	190	41.76
6	0.95	80	17.58
Filter	—	25	5.49

Plot a histogram showing the particle size distribution.

By computing the percentage in each interval and assuming that the lower collection limit of the filter is 0 μm, Fig. 7.7 can be plotted. A second method is to plot the data in a form similar to Example 2.3.

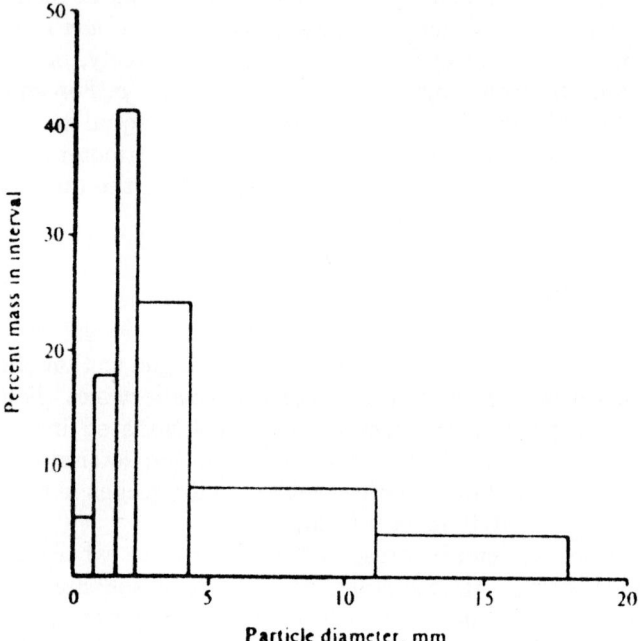

Figure 7.7 Histogram of impactor data.

In many cases a mean (or median) particle size and a standard deviation are desired.

The most common method of data reduction is to plot the stage data on log probability paper. If a straight line can be fitted to the data, a lognormal distribution is assumed. For this plot, the 50 percent cut diameter or median size for a given stage is taken as the characteristic size for that stage. That is, all particles equal to or larger than $(d_{50})_i$ are retained on the ith stage; all smaller particles pass through the stage. Cumulative percentage less than a given size is plotted on log probability paper as a function of particle size, and a line of best fit is drawn through the points. From this line a median diameter and geometric standard deviation can be determined. If particle mass measurements are used as estimates of material deposited on the various stages, and the cut diameters are in terms of aerodynamic diameter, the resultant median value is the *mass median aerodynamic diameter*, sometimes abbreviated as MMAD.

Example 7.7 Given the following impactor data, using a log probability plot, determine MMAD and σ_g.

Stage no.	Aerodynamic $(d_{50})_i$	Mass, μg, collected on stage
0	20	0
1	10	0
2	5	10
3	3	35
4	2	70
5	1	190
6	0.8	60
7	0.5	85
Filter	0	50

Compute the cumulative percentage collected for each stage.

Diameter interval, μm	Percentage in interval	Cumulative percentage less than upper interval size
0–0.5	10	10
0.5–0.8	17	27
0.8–1.0	12	39
1.0–2.0	38	77
2.0–3.0	14	91
3.0–5.0	7	98
5.0–10.0	2	100

From the plot (Fig. 7.8) an MMAD of 1.2 μm is found and a σ_g of 2.0 determined. This is a mass median diameter because the weight or mass of aerosol collected on each stage was used in the analysis.

Errors associated with impactor data

As mentioned earlier, particles can be deposited within the impactor housing or can be reentrained from an impactor stage after collection. Both phenomena can give rise to errors in impactor measurements.

Impactor calibrations must be done carefully to minimize error. In many cases this is not done, and one should be wary of unsubstantiated claims of impactor performance.

Finally, the methods of data analysis presented here are quite crude. Fitting a straight line to data on log probability paper requires the assumption that the data are lognormally distributed, which may not be the case. However, more detailed and sophisticated methods

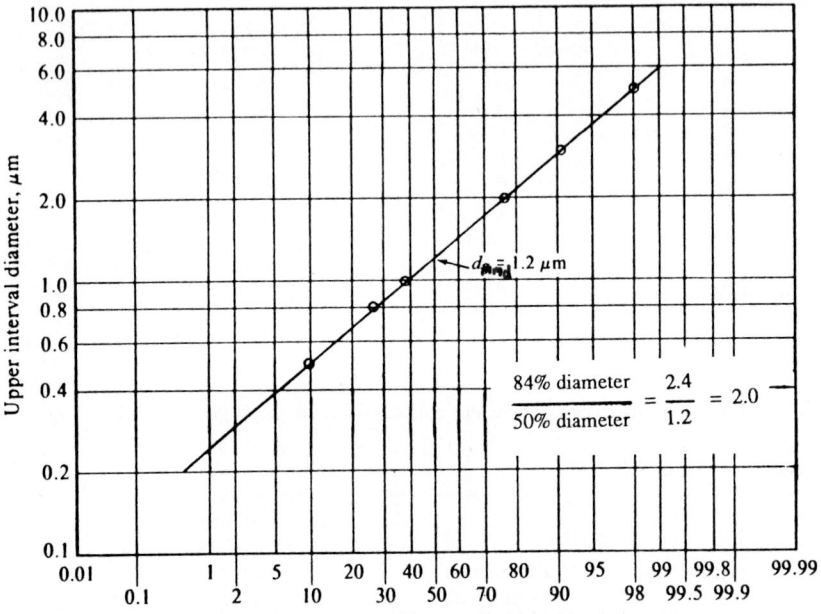

Figure 7.8 Lognormal plot of impactor data given in Example 7.7.

such as the use of iterative calculational schemes seem unwarranted unless very accurate calibration and measurement data are available (Fuchs, 1978).

Impactor Analysis Using Phase Trajectories

The complex flow within an impactor can be studied by using the concept of phase trajectory analysis where the paths of particles with different initial locations and velocities are determined. By analyzing these paths, conclusions can be drawn about a particle's fate as it travels through an impactor. Because in this analysis ideal streamline flow conditions are assumed (which actually may not be the case), phase trajectory analysis helps show how predictions from ideal assumptions may be modified by real-world conditions. A fairly simple case is chosen to illustrate the method.

Suppose an aerosol is flowing from an opening and impacting against an infinite wall with the air moving in a so-called hyperbolic stream. The velocity components of the air will have the form

$$u_{1x} = a_1 x \quad \text{and} \quad u_{1y} = b_1 y \quad \text{where } a_1, b_1 > 0$$

With a planar stream as illustrated in Fig. 7.9, $a_1 = b_1$. The current

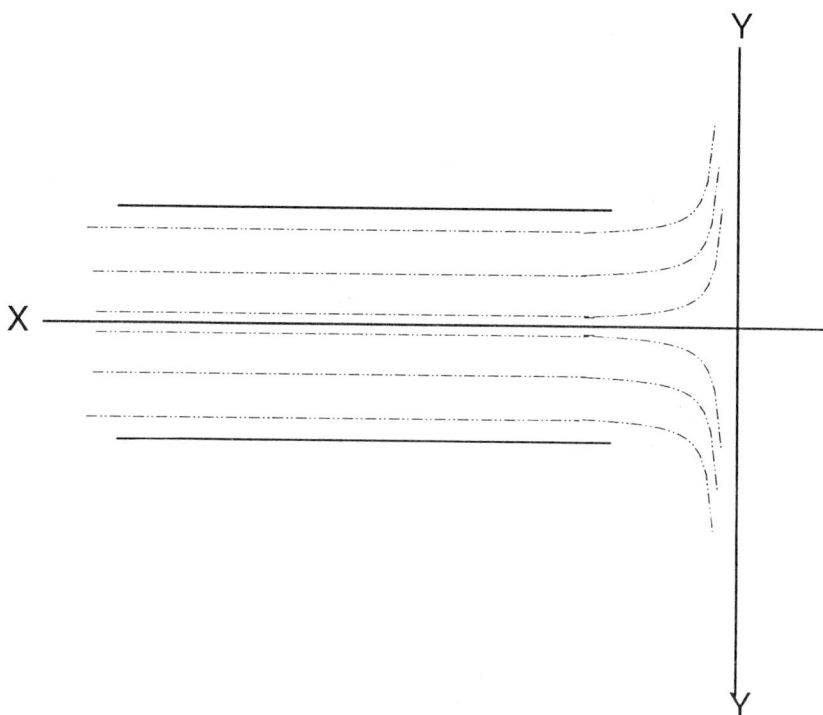

Figure 7.9 Example of a planar stream.

lines are hyperbolas given by

$$x_1 y_1 = \text{constant}$$

If gravity is neglected, the equation of motion for the aerosol particles flowing in the airstream is

$$\tau \frac{d\vec{v}}{dt} + \vec{v} - \vec{u} = 0 \qquad (7.21)$$

where v is the particle velocity. For phase analysis, this equation is broken down into components and put in a dimensionless form by letting u equal a unit velocity and w equal a unit length, so that $\phi = \tau u/w$, $a = a_1 w/u$, and $b = b_1 w/u$. The terms u and w are chosen to be fixed but arbitrary quantities. For example, u could be the average velocity of air leaving the impactor jet and w the jet half-width. Then, recalling that $v = ds/dt$ and $dv/dt = d^2s/dt^2$, the equation of motion, Eq. 6.6, can be expressed in dimensionless quantities as

$$\phi \frac{d^2 x}{dt^2} + \frac{dx}{dt} + ax = 0 \qquad (7.22)$$

$$\phi \frac{d^2y}{dt^2} + \frac{dy}{dt} - bx = 0 \qquad (7.23)$$

Equations 7.22 and 7.23 are linear second-order differential equations having as general solutions

$$x = C_1 \exp \lambda_1 t + C_2 \exp \lambda_2 t \qquad (7.24)$$

and

$$y = C_3 \exp \mu_1 t + C_4 \exp \mu_2 t \qquad (7.25)$$

where the arbitrary constants C_1, C_2, C_3, and C_4 are to be determined from the initial conditions and the eigenvalues $\lambda_{1,2}$ and $\mu_{1,2}$ are

$$\lambda_{1,2} = \frac{-1 \pm \sqrt{1 - 4a\phi}}{2\phi} \qquad (7.26)$$

$$\mu_{1,2} = \frac{-1 \pm \sqrt{1 + b\phi}}{2\phi} \qquad (7.27)$$

So far the method of solution is standard. With a great deal of effort the constants in Eqs. 7.24 and 7.25 can be evaluated from the initial and boundary conditions of the problem, and the flow trajectories of the aerosol particles determined. Phase analysis permits the bypassing of this laborious task, so that something can be learned from these equations without having to evaluate the constants.

Recalling that

$$\frac{dv_x}{dt} = \frac{dv_x}{dx}\frac{dx}{dt} = v_x \frac{dv_x}{dx}$$

the equations of motion, Eqs. 7.22 and 7.23, can be rewritten as

$$\phi \frac{dv_x}{dx} v_x + ax = 0 \qquad (7.28)$$

and

$$\phi \frac{dv_y}{dy} v_y - by = 0 \qquad (7.29)$$

Rearranging Eq. 7.28 gives

$$\frac{dv_x}{dx} = \frac{-ax - v_x}{\phi v_x} \qquad (7.30)$$

and rearranging Eq. 7.29 gives

$$\frac{dv_y}{dy} = \frac{by - v_y}{\phi v_y} \quad (7.31)$$

With these two equations it is possible to plot the fields of flow (velocity as a function of position). For example, by picking an initial y_i and v_{y_i} and plotting these initial points on a graph of v_y versus y, Eq. 7.31 is solved for dv_y/dy. Then by assuming some small increment for dy, a new (v_y, y) can be plotted and the process repeated. Such a plot for v_y as a function of y is shown in Fig. 7.10. Trajectories on this plot can be estimated by knowing that $v_y = \mu_1 y$ and $v_y = \mu_2 y$ will be asymptotes ($\mu_2 < 0$ and $0 < \mu_1 < b$). Also all trajectories must cross the $v_y = by$ line with a slope of zero.

Several observations about particle flow patterns can be made from this plot. First, although particles can enter the impactor with any y velocity at any location, their subsequent paths are quite well determined depending on initial velocity and location. Particles with an absolute value of initial y velocity greater than $\mu_1 y$ or $\mu_2 y$ have the potential for crossing the impactor centerline. Particles inside this range may reverse their direction but will never cross the impactor centerline.

Efficiency calculations for impactors are often based on the assumption that a particle's location relative to other particles never changes so that all particles initially lying within some arbitrary distance from the impactor centerline will be collected, while all those outside

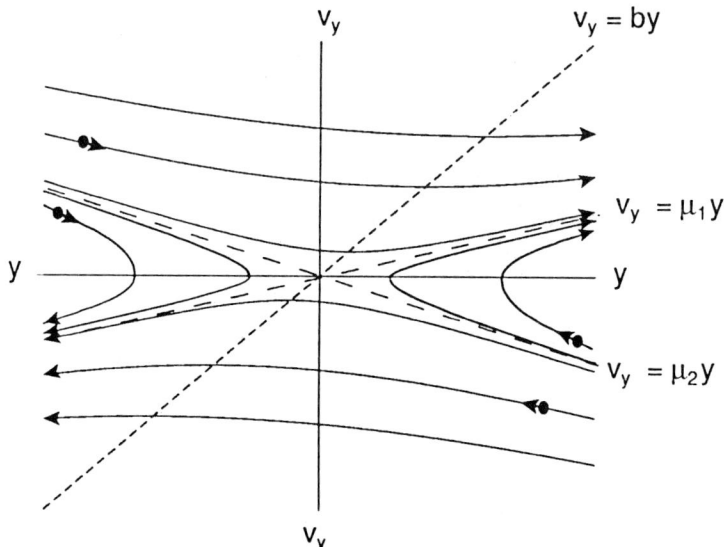

Figure 7.10 Phase diagram, Y axis.

of this boundary will not. Clearly, this is only strictly true if the initial y velocity of all particles is zero.

In all cases, however, particles will eventually be moved away from the impactor centerline with increasing absolute velocity in the y direction as y becomes greater. This is, of course, an obvious conclusion of the analysis.

For motion in the x direction the analysis is more involved because λ can exist as either a real or complex number, depending on whether $\phi > 1/(4a)$ or $\phi \leq 1/(4a)$. For the case $\phi > 1/(4a)$ the plot is of the stable focus type (Fig. 7.11). The right-hand side of Figure 7.11 has been lightly shaded since this is an imaginary zone—this is the space behind the infinite plane.

For any particle starting in the second or third quadrants with any x velocity (either direction), the particle will eventually cross the x axis, implying that sooner or later the particle will deposit on the impaction surface at $x = 0$.

On the other hand, when $\phi \leq 1/(4a)$, the phase diagram is of the singular-node type, Fig. 7.12, similar to free vibration with viscous damping. Under these conditions it is possible that some particles starting in the second or third quadrants will have trajectories that terminate at the origin. This means that the $x = 0$ surface is reached after an infinite time. The practical implication of this observation is that these particles will never be collected.

The conditions for not being collected are if $\phi \leq 1/(4a)$ and $v_{x_i} <$

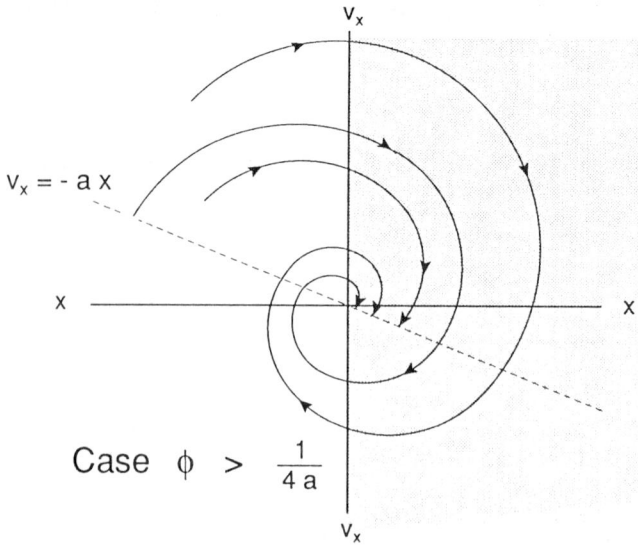

Figure 7.11 Phase diagram, X axis.

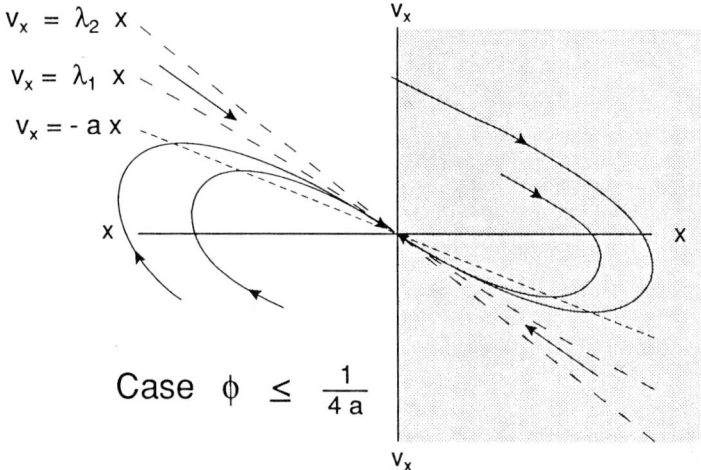

Figure 7.12 Phase diagram, X axis.

$-ax$. But ϕ is just the Stokes number $\phi = \tau u/w$, so that this analysis implies the existence of a *critical Stokes number* Stk_{cr} which describes the conditions under which very small particles will not be collected by impaction. Because this analysis assumes all particles are points and does not take the particle's physical dimensions into account, there is a legitimate question whether in actuality this lower limit does exist. Even so, the method of analysis reveals that the collection efficiency of impactors for small particles may be impaired, illustrating the utility of using phase trajectories for analyzing certain aerosol dynamics problems.

Problems

1 Using simple impaction theory, estimate the minimum particle diameter that will be collected with 100 percent efficiency by a plate placed at a right angle to the flow and 4 in in front of a 4-in-diameter duct out of which air containing an aerosol is flowing at a rate of 100 ft^3/min.

2 In the first stage of the Lundgren impactor, air issues at a flow of 85 L/min through a number of round jets 0.82 cm in diameter at a Reynolds number of 3700.
  a. Determine the number of jets in the first stage of the impactor.
  b. Using the data in Fig. 7.5a, estimate the effective cutoff aerodynamic diameter for this stage.

3 The Andersen sampler is a six-stage circular jet impactor. Each stage has 400 jets of a certain diameter. For stage 3, the jet diameter is 0.028 in. For stage 4, the jet diameter is 0.021 in. Assuming a flow rate of 1 ft^3/min and particles of unit density, determine the range of particle diameters removed by the stage 4 impaction surface. Use the impaction data given in Fig. 7.5b.

4 If we assume that the critical Stokes number for a 1-in-diameter circular orifice is ⅛ and that the pressure drop for an orifice and collecting plate in series in milliatmospheres is $\Delta P = 1.25 \times 10^{-3} \rho_m U$, where ρ_m = fluid density (g/cm³), and U = orifice gas velocity (cm/s), determine the diameter of the smallest size unit-density spherical particle that can be collected at a pressure drop of 2 in of water.

5 Marple and Rubow (1986) state that the jet Reynolds number in an impactor can be expressed in terms of the air mass flow rate $\dot{m}$. Show that for a round jet impactor Re = $\dot{m}W/\mu$ and for a rectangular jet impactor Re = $2\dot{m}W/\mu$. Marple and Rubow point out that this is a useful form for the Reynolds number since $\dot{m}$ is constant for all stages of the impactor.

6 As air flows through an impactor nozzle, it expands since this is a near adiabatic process. According to Hering and Marple (1986), the pressure drop through an orifice with a cross-sectional area A is

$$p/p_0 = \frac{q_0}{Av}\left(1 - v^2\frac{\gamma - 1}{2\gamma RT_0}\right)$$

where p is pressure, T is temperature, R is the gas constant, γ is the ratio of the heat capacities, and q_0 is the volumetric flow through the nozzle. The subscript 0 indicates that the quantity is measured upstream of the nozzle; quantities without a subscript are measured below the nozzle. The core velocity can be computed from

$$v_{core} = \left\{\frac{2\gamma RT_0}{\gamma - 1}\left[1 - \left(\frac{p}{p_0}\right)^{(\gamma-1)/\gamma}\right]\right\}^{1/2}$$

for subsonic flow and

$$v_{core} = \left(\frac{2\gamma RT_0}{\gamma + 1}\right)^{1/2}$$

for sonic flow. Calculate the temperature at the exit of the jet and the core velocity for the Hering impactor stage 6 in Example 7.5.

7 Suppose trajectory analysis yields an equation of the form

$$\frac{dv_z}{dz} = \frac{v_z}{z - v_z}$$

Plot v_z as a function of z.

Chapter 8

Particle Kinetics

Centrifugation, Isokinetic Sampling, and Respirable Sampling

Centrifugation of Particles

As discussed previously, terminal settling velocities of aerosol particles are generally quite small. Under normal circumstances it is unreasonable to expect that simple sedimentation as such will be an effective removal mechanism.

One way to remove these particles from air is to subject them to high centrifugal forces. If an aerosol particle is caused to move in a circular path, it will have a radial acceleration given by Eq. 7.1. This radial acceleration can be likened to the acceleration due to gravity in a gravitational field. By rotating the aerosol, accelerations many times that of gravity can be achieved.

Example 8.1 A centrifuge has a radius of 50 cm and is operated at 500 r/min. Determine the ratio of radial acceleration to gravitational acceleration in this case.

$$\frac{\text{Radial acceleration}}{\text{Gravitational acceleration}} = \frac{[(500/60)(2\pi)]^2(50)}{980} = 1.40 \times 10^2$$

As can be seen from Example 8.1, quite large radial accelerations are possible, indicating that very small particles can be removed in this manner.

In an aerosol centrifuge, particles are made to follow circular paths until they strike the outer wall of the unit. The distance from the inlet that a particle is deposited is indicative of particle size. This can be shown as follows:

Consider a centrifuge having an inner radius R_1 and an outer radius R_2. A particle enters at R_1 and travels across the annulus to be deposited somewhere on the surface at R_2. The radial velocity of the particle is given by Eq. 7.4:

$$v_r = \tau \omega^2 r \tag{8.1}$$

that is,

$$\frac{dr}{dt} = \tau \omega^2 r \tag{8.2}$$

The tangential velocity of the particle will be that of the airstream moving through the centrifuge. With a flow of Q cm³/s and a centrifuge channel depth of h,

$$\bar{u} = \frac{Q}{A} = \frac{Q}{h(R_2 - R_1)} \tag{8.3}$$

This represents an average velocity. In actuality the velocity distribution in the channel at any point r across the radius will be parabolic of a form given by

$$u = \frac{4k\bar{u}}{(R_2 - R_1)^2}(r - R_1)(R_2 - r) \tag{8.4}$$

The factor k can range in value from 1.5 for deep, narrow channels to about 3 for rectangular ones (Tillery, 1979).

If l_D is the tangential distance the particle travels downstream before being deposited, then

$$dl_D = u\, dt \tag{8.5}$$

In terms of r this becomes

$$dl_D = \frac{4k\bar{u}}{(R_2 - R_1)^2} \frac{1}{\omega^2 \tau r}(r - R_1)(R_2 - r)\, dr \tag{8.6}$$

Integrating Eq. 8.6 between the limits R_1 and R_2 gives

$$l_D = \frac{2k\bar{u}}{\omega^2 \tau}\left[\frac{R_2 + R_1}{R_2 - R_1} + \frac{2R_1 R_2}{(R_2 - R_1)^2} \ln\frac{R_1}{R_2}\right] \tag{8.7}$$

Example 8.2 Using Eq. 8.7, find the diameter of unit-density spheres that would be deposited 2 cm from the entrance of a centrifuge that is operated at 300 r/min with an average channel velocity of 100 cm/s. The centrifuge inner radius is 15 cm, and the channel width is 1 cm. Assume $k = 2$.

$$l_D = \frac{2k\bar{u}}{\omega^2 \tau}\left[\frac{R_2 + R_1}{R_2 - R_1} + \frac{2R_1 R_2}{(R_2 - R_1)^2}\ln\frac{R_1}{R_2}\right]$$

$$2 = \frac{2(2)(100)}{[(300/60)(2\pi)]^2 \tau}\left[\frac{16 + 15}{16 - 15} + \frac{2(16)(15)}{(16 - 15)^2}\ln\frac{15}{16}\right]$$

$$\tau = 4.36 \times 10^{-3}\text{ s} = \frac{1}{18}\frac{d^2}{\mu}\rho_p$$

$$d^2 = (4.36 \times 10^{-3})(18)(1.82 \times 10^{-4}) = 1.43 \times 10^{-5}$$

$$d = 37.8\ \mu\text{m}$$

Aerosol centrifuges are useful in the laboratory but have little application in air cleaning. When care is taken so that the aerosol enters the rotating annulus at a single point, the units are often called *aerosol spectrometers*. A discussion of the calibration and operation of the popular Stöber spectrometer design (Stöber and Flachsbart, 1969) is given by Martonen (1989).

Cyclones

An accurate theory to predict cyclone behavior has yet to be achieved. In a cyclone, particle-laden air is introduced radially into the upper portion of a cylinder so that it makes several revolutions inside the cylinder before leaving axially along the cylinder centerline. While making these revolutions, the particles in the air are accelerated outward to the cylinder walls where they either stick and are retained (low particle loading) or are swirled down to a collection port at the bottom of the cylinder (high particle loading). Overall gas motion in the cyclone consists of an inner vortex moving toward the cyclone exit containing the smaller-sized particles and an outer vortex moving in the opposite direction and carrying the larger particles.

Important parameters in cyclone operation can be established by considering simple cyclone theory. Figure 8.1 shows a sketch of a typical cyclone. Air at a flow of Q cm^3/s enters tangentially, revolving N_T times in the cyclone before it is discharged. Dust that is removed from the air spirals down into the dust discharge port.

Assuming that the gas moves through the cyclone as a rigid airstream with a spiral velocity equal to the average velocity at the cyclone inlet, the retention time for an element of gas within the cyclone is

$$t_r = \frac{N_t(2\pi R)}{v_c} \tag{8.8}$$

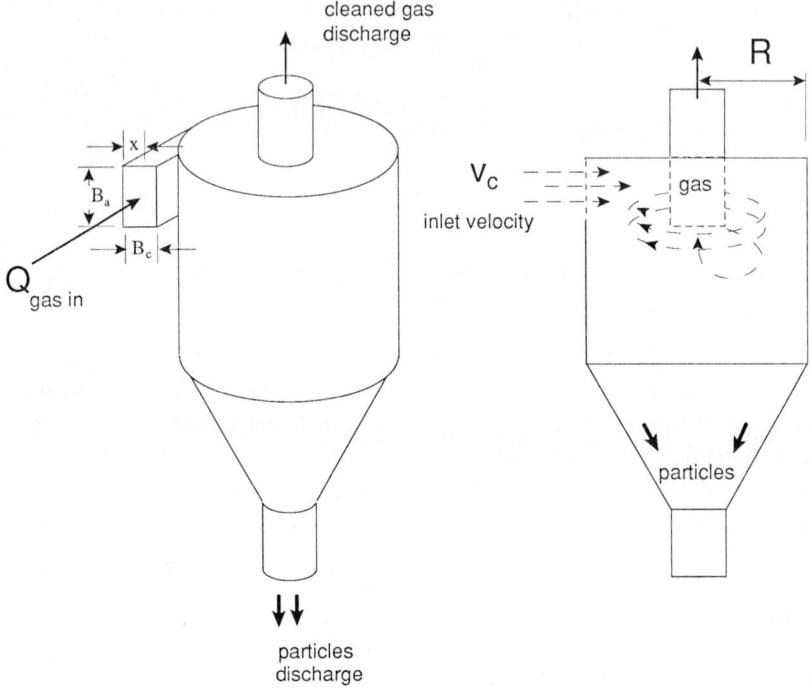

Figure 8.1 Sketch of a cyclone particle collector.

where v_c is the inlet velocity. During this retention time a particle can move a distance x across the width of airstream B_c. Since the particle radial velocity is

$$v_r = \frac{v_c^2}{R}\tau \tag{8.9}$$

the time to go a distance x is

$$t_x = \frac{xR}{v_c^2 \tau} \tag{8.10}$$

and the time to go a distance B_c is

$$t = \frac{B_c R}{v_c^2 \tau} \tag{8.11}$$

Equating this time with the transit time through the cyclone gives

$$\frac{B_c R}{v_c^2 \tau} = \frac{N_t(2\pi R)}{v_c}$$

$$\tau_{\text{crit}} = \frac{B_c}{v_c N_t 2\pi} \tag{8.12}$$

This is the τ of the smallest particles that simple theory says should be collected with 100 percent efficiency. In terms of particle diameter

$$d_{\text{crit}} = \sqrt{\frac{9 B_c \mu}{\rho_p v_c N_t \pi}} \tag{8.13}$$

Cyclone efficiency ϵ is estimated from the ratio x/B_c:

$$\epsilon = \pi N_t \frac{v_c \tau}{B_c/2} \tag{8.14}$$

Since $v_c \tau / (B_c/2)$ = Stk, the cyclone efficiency can be expressed as

$$\epsilon = \pi N_t \text{Stk} \tag{8.15}$$

The factor N_T varies from 0.5 to about 10, depending on cyclone shape and size.

Example 8.3 Estimate the collection efficiency for 5-μm unit-density spheres in a small cyclone having a square entrance of 0.3 cm on a side when operated at a flow rate of 1.7 L/min. Use $N_T = 1$.

$$\frac{B_c}{2} = \frac{0.3}{2} = 0.15 \text{ cm}$$

$$\tau = \frac{1}{18} \frac{d^2 \rho_p}{\mu} = \frac{1}{18} \frac{(5 \times 10^{-4})^2}{1.82 \times 10^{-4}} = 7.63 \times 10^{-5} \text{ s}$$

$$v_c = \frac{Q}{A} = \frac{(1.7 \times 1000)/60}{0.3 \times 0.3} = 314.8 \text{ cm/s}$$

$$\text{Stk} = \frac{v_c \tau}{B_c/2} = 0.160$$

$$\epsilon = \pi N_T \text{Stk} = 0.50 = 50\%$$

The efficiency predicted by Eq. 8.15 is only a rough estimate; the equation estimates a shape in the efficiency-versus-particle-size curve that is different from what is actually observed. There are a number of factors not considered in this elementary derivation. First, laminar flow is assumed, but turbulent flow is often observed in practice. The effect of turbulence will be to move particles away from the cyclone walls or resuspend deposited ones. Hence, turbulence will decrease cyclone efficiency. Second, the width of the cyclone inlet is not as important a parameter as overall cyclone diameter, since it is the width of an element of gas within the cyclone that determines particle deposition, and this width is not strongly controlled by inlet width. Finally, overall cyclone configuration will affect efficiency. This is not taken into account in the simple theory. Equation 8.15 does illustrate the general approach which has been followed in refining cyclone theory, and like the impactor and centrifuge equations, it permits rough estimates of system performance to be made. As can be seen in Fig. 8.2, however, the shape of the efficiency curve as predicted by simple theory is not consistent with experimental observations, the former being concave upward and the latter concave downward. Despite the apparent good fit of some theories to experimental data (e.g., Leith and Licht, 1972), in general, equations developed to predict cyclone efficiency disagree with experimental data (Abrahamson, 1981). In addition, theories that are developed for large, industrial-type cyclones do not give good predictions when applied to small, personal sampling type of cyclones. Chan and Lippmann (1977) present a summary of cyclone collection theories, as applied to small cyclones.

Aerosol concentrations affect cyclone performance with increased concentrations increasing efficiency (Ranz, 1985). Wheeldon and Burnard (1987) showed improved cyclone performance with very high particle concentration (50 g/m^3). Detailed criteria for cyclone design have been summarized by Licht (1980) and include many of the considerations mentioned above.

Isokinetic Sampling

In aerosol sampling, the measured concentration and size distribution should represent as closely as possible the concentration and size distribution of the original aerosol. There are several reasons why the

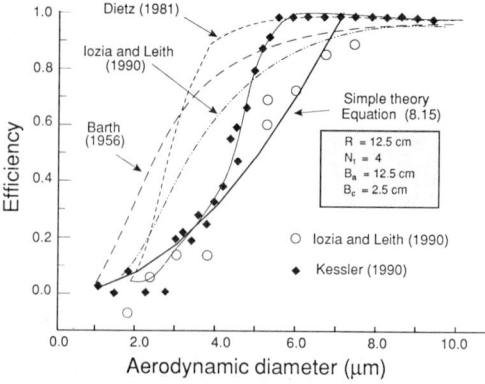

Figure 8.2 Comparison of simple theory with detailed theoretical efficiency and experimental data. (*Adapted from Kessler, 1990.*)

measured concentration can differ from the true concentration. One is that gravitational or inertial deposition of the sample as it flows into the sampling tube can result in loss of larger-sized particles. Also, deposition or selective collection at the mouth of the sampling tube can yield either greater or lesser amounts of larger particles.

Consider the three cases shown in Fig. 8.3. In case a the probe is not aligned with flow. Because of inertia some particles may be lost by impaction, giving a sample concentration which would be less than the actual. In case b, when the collection velocity is greater than stream velocity, some particles, because of their inertia, may fail to follow streamlines and therefore will not be collected, giving a sample concentration that is less than the actual. Finally, in case c, the collection velocity would be less than the stream velocity—the opposite of case b—and the sample concentration might then be greater than the actual.

If the probe is aligned with the flow and the sample velocity is equal to the stream velocity, sampling is said to be *isokinetic*, and the sample as collected should match the actual concentration. If sampling ve-

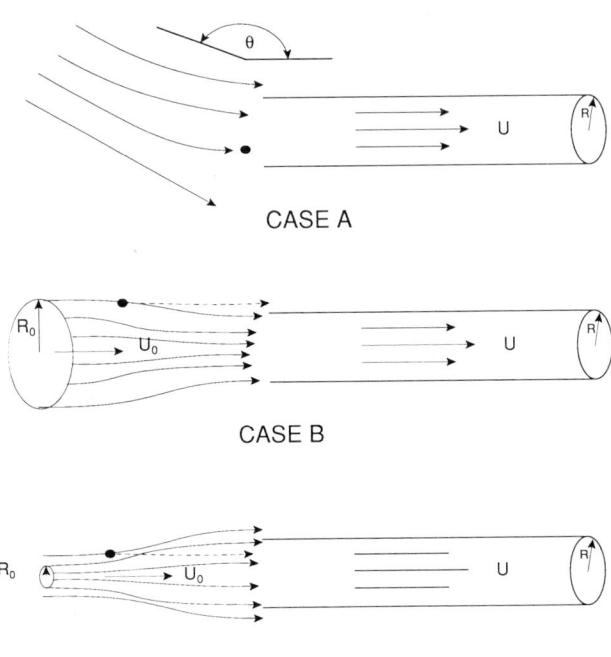

Figure 8.3 Types of anisokinetic sampling.

locity differs from stream velocity, known as *anisokinetic* sampling, particle inertia can give rise to errors in the measured concentrations.

Error from anisokinetic sampling can be investigated theoretically (Davies, 1966). Suppose an aerosol of concentration c_0 flowing with a velocity of u_0 is drawn with a velocity u into a sampling tube of radius R (assume infinitely thin tube walls, sample collected parallel to flow). Gas streamlines entering the tube at some distance away are confined within a cylinder of radius R_0 such that

$$\pi R^2 u = \pi R_0^2 u_0$$

If $u_0 > u$, then $R > R_0$ and the streamlines diverge (Fig. 8.3c). All particles moving within the circular cross-section πR_0^2 enter the tube. The number entering per second = $\pi R_0^2 u_0 c_0$, where c_0 is the stream concentration. Streamlines from the annulus which lie between the cylinders of radii R_0 and R will pass outside the sampling tube. However, a fraction of particles from this space may enter the tube because of their inertia. If α is this fraction, then the number entering per second and from the annulus is equal to $\alpha\pi(R^2 - R_0^2)c_0 u_0$. The sum of the two fluxes, that from the center and that from the annulus, is the total flux entering the tube, equal to $\pi R^2 c u$, where c is the number concentration in the sample:

$$\pi R_0^2 u_0 c_0 + \alpha\pi(R^2 - R_0^2)c_0 u_0 = \pi R^2 c u \tag{8.16}$$

Hence

$$\frac{c}{c_0} = 1 - \alpha + \frac{\alpha u_0}{u} \tag{8.17}$$

When $u > u_0$, then $c < c_0$. When $u < u_0$, then $c > c_0$.

The factor α varies from 0 for small particles to 1 for large ones. The exact value of α is not known. A rough estimate for α has been proposed by Davies (1966):

$$\alpha = \frac{2\text{Stk}}{1 + 2\text{Stk}}$$

where

$$\text{Stk} = \frac{\tau u_0}{R} \tag{8.18}$$

This estimate, although rough, does predict the following experimental observations:

1. $\dfrac{c}{c_0} = 1$ when $u_0 = 0$

2. $\dfrac{c}{c_0} = 1$ when $u_0 > 0$ but $\text{Stk} \ll 1$

3. $\dfrac{c}{c_0} = \dfrac{u_0}{u}$ when $\text{Stk} \gg 1$

4. $\dfrac{c}{c_0} = 1$ when $u = u_0$

The effect of anisokinetic sampling is plotted in Fig. 8.4. This analysis indicates that isokinetic sampling is not necessary when very small particles are sampled. Sometimes, in an effort to be precise,

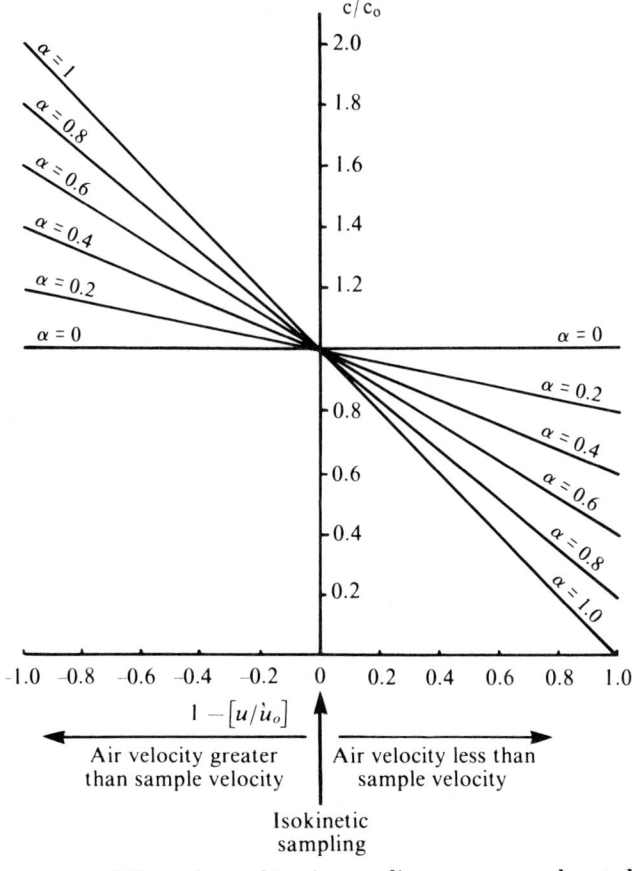

Figure 8.4 Effect of anisokinetic sampling on measured sample concentration.

isokinetic sampling will be required when it is known that all particles to be sampled are well below 1 μm in diameter. Examples would be sampling of fumes or aerosols formed from condensation processes. The result is much added complexity and usually cost, without any increase in the accuracy of the result.

Example 8.4 It is desired to sample fume particles (d = 0.1 μm) that are emitted from a stack at a velocity of 100 cm/s and a temperature of 200°C. (Assume a particle density of 1 g/cm³ and a sample probe diameter of 1 cm.) Determine the sampling error when the sampling velocity is 0.01 of the stream velocity.

$$\text{Stk} = \frac{\tau v_0}{R}$$

At 200°C,

$$\text{Viscosity} = 1.82 \times 10^{-4} \left(\frac{200 + 273}{20 + 273}\right)^{0.5}$$

$$= 2.31 \times 10^{-4} \, P$$

$$\lambda = 0.065 \left(\frac{200 + 273}{273}\right)$$

$$= 0.111 \, \mu m$$

$$\tau = \frac{1}{18} \frac{d^2}{\mu} \rho_p C_c$$

$$= \frac{1}{18} \frac{(0.1 \times 10^{-4})^2}{2.33 \times 10^{-4}} (1) \frac{(4.33)}{0.1}$$

$$= 1.04 \times 10^{-7} \, s$$

$$\text{Stk} = \frac{(1.04 \times 10^{-7})(100)}{0.5} = 2.08 \times 10^{-5}$$

$$\alpha = \frac{2\text{Stk}}{1 + 2\text{Stk}} = 4.16 \times 10^{-5}$$

$$\frac{c}{c_0} = 1 - \alpha + \alpha \frac{u_0}{u} = 1 - (4.16 \times 10^{-5})\left(1 - \frac{100}{1}\right)$$

$$= 1.0041$$

That is, there will be a negligible increase in sample concentration.

Respirable Sampling

For the purpose of estimating the toxic dose of an aerosol, the respiratory system can be divided into a number of functional regions:

1. Alveolar region (for both nose and mouth breathing)
2. Tracheobronchial tree (for both nose and mouth breathing)
3. a. Oral cavity, pharynx, and larynx (for mouth breathing)
   b. Nasopharynx, pharynx, and larynx (for nose breathing)
4. Ciliated nasal passages (for nose breathing)
5. Anterior unciliated nares (for nose breathing)

Regional deposition is dependent on the aerodynamic properties of the particles, usually described in terms of the aerodynamic particle diameter, airway dimensions, and such respiratory characteristics as flow rate, breathing frequency, and tidal volume.

A number of models have been developed to attempt to predict respiratory deposition, especially in the lower airways—the alveolar region. Experimental studies have tended to confirm the validity of the models, recognizing that there is much individual variation and thus a great spread in the results. The results have been clear enough, however, to indicate that all else being equal, deposition of particles in the lungs is greatly influenced by particle size and particle density.

In many cases the dose from airborne toxic materials is dependent on regional deposition in the lungs. A good estimate of this dose is possible if the size distribution of the aerosol is known. For this reason it is important to know mass concentrations within various size fractions. This information can be obtained by (1) carrying out a size distribution analysis of the airborne aerosol or (2) carrying out a size distribution analysis of the collected sample or (3) separating the aerosol into size fractions corresponding to anticipated regional deposition during the process of collection.

With mass respirable sampling, an attempt is made to separate the aerosol into two fractions representing the mass that would be deposited in the alveolar region and the mass that would not be deposited in this region. To do this, it is necessary to define the size distribution of particles deposited in the alveolar region. This material is defined as *respirable dust*.

There are several definitions of respirable dust (Lippmann, 1970). In 1952 the British Medical Research Council (BMRC) defined the respirable fraction in terms of the terminal settling velocity (free-falling speed) by the equation

$$\frac{c}{c_0} = 1 - \frac{v}{v_c} \tag{8.19}$$

where c is the concentration of particles of falling speed v or less, c_0 is the total concentration, and v_c is a constant equal to twice the terminal settling velocity in air of a unit-density sphere having a diameter of 5 µm. This definition was considered to be unsatisfactory in the United States because it was tied to terminal settling velocities.

In 1961 the U.S. Atomic Energy Commission (AEC) established a standard defining respirable dust as that portion of the inhaled dust that penetrates to the nonciliated portions of the lung (the alveolar region). This application of the concepts of respirable dust was intended *only* for "insoluble" particles exhibiting prolonged retention in the lung, and not for soluble particles, nor for those which are primarily chemical intoxicants (Aerosol Technology Committee, 1970).

Respirable dust was defined as follows with sizes in terms of aerodynamic diameters:

Size, µm	10	5	3.5	2.5	2
Respirable, %	0	25	50	75	100

In 1968, the American Conference of Governmental Industrial Hygienists (ACGIH) defined respirable dust as follows:

Aerodynamic diameter, µm	10	5.0	3.5	2.5	2.0
Respirable, %	0	25	50	75	90

Example 8.5 Compare the definitions of respirable dust as given by the BMRC, AEC, and ACGIH.

Neglecting C_c, terminal settling velocities for other sizes of unit-density spheres can be estimated from

$$v_{gd} = v_{g5}\left(\frac{d}{5}\right)^2$$

so Eq. 8.19 can be written as

$$\frac{c}{c_0} = 1 - \frac{v_{g5}(d/5)^2}{2v_{g5}} = 1 - \frac{d^2}{50}$$

when d is expressed in micrometers.

Percentage respirable defined by:	Aerodynamic diameter, µm				
	2	2.5	3.5	5	10
BMRC	92	88	76	50	0
AEC	100	75	50	25	0
ACGIH	90	75	50	25	0

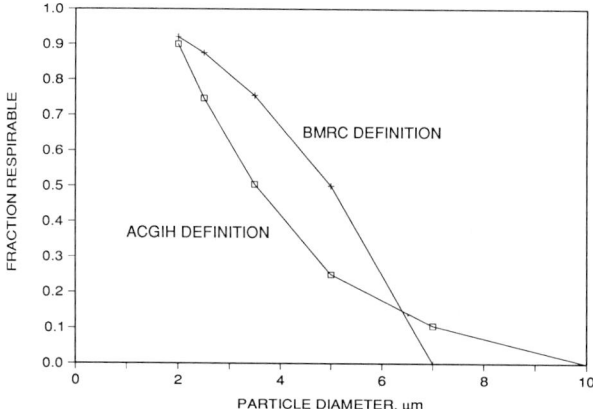

Figure 8.5 Pictorial representation of various definitions of respirable dust.

Figure 8.5 shows a plot of these definitions.

As can be seen from the above table, the ACGIH definition of respirable dust is almost identical with that of the AEC, differing only for a 2-μm-aerodynamic-diameter particle. Lippmann (1989) points out that this difference appears to be a recognition by the ACGIH of the characteristics of real particle separators.

Although there are some differences in the three definitions, under field conditions samples collected using instruments designed to meet any of these three criteria should be comparable. For a more thorough discussion of size-selective sampling, the reader is referred to Lippmann's all-inclusive article (1989) and its 183 references.

Example 8.6 An aerosol made up of unit-density spheres is lognormally distributed with a geometric mean diameter of 2.0 μm and a geometric standard deviation of 2.2. Calculate the respirable fraction of this aerosol as sampled by a sampler which follows the BMRC curve and a sampler which follows the ACGIH curve.

For this comparison the aerosol is broken down into 15 size increments, as shown in Table 8.1. *Respirable fraction* is considered to be the fraction of particles falling into the respirable category as defined above. Hence the respirable fraction in a size interval is the product of the fraction in that size interval and the percentage respirable for that size interval as defined either by the ACGIH or BMRC. The overall respirable fraction is the sum of these products over all size intervals.

The BMRC respirable fraction (RF) is calculated from

$$\text{RF}_{i-\text{BMRC}} = 1 - \frac{d_i^2}{50}$$

TABLE 8.1 Computational Data

Average diameter, μm	Count frequency	BMRC resp. fraction	ACGIH resp. fraction	Estimated deposition, BMRC	Estimated deposition, ACGIH
0.108	0.000	1.00	1.00	0.00	0.00
0.164	0.001	1.00	1.00	0.00	0.00
0.250	0.007	1.00	1.00	0.01	0.01
0.380	0.023	1.00	1.00	0.02	0.02
0.579	0.062	0.99	1.00	0.06	0.06
0.882	0.124	0.98	1.00	0.12	0.12
1.343	0.187	0.96	1.00	0.18	0.19
2.044	0.213	0.92	0.88	0.19	0.19
3.113	0.182	0.81	0.56	0.15	0.10
4.740	0.117	0.55	0.28	0.06	0.03
7.218	0.057	0.00	0.10	0.00	0.01
10.992	0.021	0.00	0.02	0.00	0.00
16.737	0.006	0.00	0.00	0.00	0.00
25.487	0.001	0.00	0.00	0.00	0.00
38.809	0.000	0.00	0.00	0.00	0.00

and the ACGIH respirable fraction is calculated from

$$RF_{i-\text{ACGIH}} = 10^{(0.325 - 0.185 d_i)}$$

The mass respirable results are BMRC, 80.1 percent; ACGIH, 73.3 percent. This result indicates that for the given aerosol the BMRC definition will indicate slightly more respirable mass than the ACGIH definition.

Problems

1 What is the minimum particle diameter collected with 100 percent efficiency by a cyclone precollector of a mass respirable sampler? Assume $R = 0.5$ cm, $B_c = 0.25$ cm, $Q = 1.7$ L/min, $N_t = 5$, $\rho = 1$ g/cm^3, opening is square.

2 A 1-in-diameter tube is used to collect a stack sample from a stack in which air is flowing at a velocity of 30 ft/s. The sampling pump available can pump only at a rate of 1 ft^3/min. Estimate the error in sampling for (a) 10-μm-diameter spheres with $\rho = 2$ g/cm^3 and (b) 0.1 μm-diameter spheres with $\rho = 10$ g/cm^3.

3 A high-volume sampler has an airflow rate of 30 ft^3/min. Design a horizontal elutriator (settling chamber) that could be placed upstream from the sampler to eliminate those particles greater than 10 μm in diameter. Assume a particle specific gravity of 2.3. What diameter particles would be reduced by a factor of 50 percent in this unit?

4 A mass respirable sampler cyclone is designed to operate at a flow rate of 1.7 L/min. Will the mass respirable concentration that is measured be overestimated or underestimated if the sampler is actually operated at a flow rate of 2.0 L/min? Why?

5 Explain why it is not necessary to consider isokinetic sampling when a sample is being collected from still air.

6 Determine the mass respirable sample flow rate to use if a sample is to be collected in a space cabin where the cabin pressure is one-half atmospheric pressure.

7 In the past 20 years size-selective sampling has been applied to ambient sampling as well as personal sampling [John (1984), EPA (1987)]. A method known as the *PM-10 method* samples particles into two size segments, one greater than 10 μm and the other less than 10 μm (the *thoracic fraction*). If a conventional mass respirable sampler operates at a flow rate of 1.7 L/min to collect 3.5-μm particles with a 50 percent efficiency, what flow rate would be necessary to collect 10-μm particles with a 50 percent efficiency? Assume unit-density spheres.

Chapter

9

Brownian Motion and Simple Diffusion

There are two principal ways that extremely small aerosol particles can be removed from an aerosol. The particles can collide with other particles and grow into ones large enough to be removed by gravity or aerodynamic forces (impaction, centrifugal, etc.), or they can migrate to surfaces, stick to those surfaces, and thus be removed.

The process by which these particles migrate, either to a surface or to one another, is called *diffusion,* and their motion is described as *brownian motion*. Diffusion is important in aerosol studies because it represents the major dynamic effect acting on very small particles ($d < 0.1$ μm) and must be considered when the dynamics of these small particles are studied.

Brownian Motion

Small particles suspended in a gas undergo random translational motion because they are being buffeted by collisions with swiftly moving gas molecules. This motion appears almost as a vibration of the ensemble of particles, although there is a net displacement with time of any given particle. Observation of this motion in a liquid was first made in 1828 by the British naturalist Robert Brown (1828), and the phenomenon thus has been called *brownian motion* (also known as brownian movement). Bodaszewski (1883) studied the brownian motion of smoke particles and other suspensions in air and likened these movements to the movements of gas molecules as postulated by the kinetic theory. The principles governing brownian motion are the same, whether the particles are suspended in a gas or in a liquid.

Fick's Laws of Diffusion

When particles are uniformly dispersed in a gas, brownian motion will change the position of the individual particles but will not change the overall particle distribution. When the particles are not uniformly dispersed, brownian motion tends eventually to produce a uniform concentration throughout the gas, the particles moving away from areas of high concentration to regions of low concentration. This process, known as *particle diffusion*, follows the same two general laws that also apply to molecular diffusion, known as *Fick's laws of diffusion*.

Fick's first law of diffusion states that the concentration of particles crossing unit area in unit time J is proportional to the concentration gradient normal to the unit area dc/dx. The constant of proportionality D is known as the *diffusion coefficient*. Symbolically, for the current through a plane set at right angles to the x direction,

$$J = -D \frac{dc}{dx} \tag{9.1}$$

Example 9.1 Particles move by diffusion across a gap 2 cm wide. If the concentration on the left-hand side of the gap is such that it is always 10 times the concentration on the right-hand side (the right-hand side concentration being $10^6 p$ per cubic centimeter) and 100 particles per second per square centimeter crosses the gap, determine the value of the diffusion coefficient D.

$$J = -D \frac{dc}{dx}$$

$$\frac{dc}{dx} = \frac{1 \times 10^6 - 10 \times 10^6}{2} = \frac{(1 - 10) \times 10^6}{2} = -4.5 \times 10^6$$

$$D = -\frac{J}{dc/dx} = \frac{100 \ \frac{p}{\text{cm}^2 \cdot \text{s}}}{4.5 \times 10^6 \ \frac{p}{\text{cm}^3} \cdot \frac{1}{\text{cm}}}$$

$$= 2.22 \times 10^{-5} \ \text{cm}^2/\text{s}$$

Notice that the units of D are centimeters squared per second (in cgs units).

Fick's second law represents the time-dependent case in which the change in concentration of an aerosol, with respect to time at a point in space, is proportional to the divergence of the concentration gradient at that point, the constant of proportionality again being D, the diffusion coefficient (Jost, 1952).

$$\frac{\partial c}{\partial t} = D \nabla^2 c \tag{9.2}$$

The term ∇^2 is the laplacian operator, which in cartesian coordinates (given by x, y, and z) is

$$\nabla^2 = \frac{\partial^2}{\partial x^2} + \frac{\partial^2}{\partial y^2} + \frac{\partial^2}{\partial z^2} \tag{9.3a}$$

in spherical coordinates (given by r, θ, and ϕ) is

$$\nabla^2 = \frac{\partial^2}{\partial r^2} + \frac{2}{r}\frac{\partial}{\partial r} + \frac{1}{r^2 \sin^2\theta}\frac{\partial^2}{\partial \phi} + \frac{1}{r^2}\frac{\partial^2}{\partial \phi^2} + \frac{1}{r^2}\cot\theta\frac{\partial}{\partial \theta} \tag{9.3b}$$

and in cylindrical coordinates (given by r, θ, and z) is

$$\nabla^2 = \frac{\partial^2}{\partial r^2} + \frac{1}{r}\frac{\partial}{\partial r} + \frac{1}{r^2}\frac{\partial^2}{\partial \theta^2} + \frac{\partial^2}{\partial z^2} \tag{9.3c}$$

Thus in cartesian coordinates Fick's second law of diffusion would be written

$$\frac{\partial c}{\partial t} = D\left(\frac{\partial c^2}{\partial x^2} + \frac{\partial^2 c}{\partial y^2} + \frac{\partial^2 c}{\partial x^2}\right) \tag{9.4}$$

These equations, with appropriate boundary conditions, permit in theory the solution of any aerosol problem involving pure diffusion.

Early investigators using a liquid medium found that a particle in brownian motion moves with uniform velocity (Svedberg, 1909), that smaller particles move more rapidly than larger ones (Exner, 1867), that particles travel more rapidly as the viscosity of the medium decreases, and that at constant viscosity the amplitude of the motion is directly proportional to the absolute temperature (Seddig, 1908). These observations are consistent with a theory of brownian motion developed by Albert Einstein (1956) in 1905 and 1906.

Einstein's Theory of Brownian Motion

Consider a cylinder of unit cross-sectional area in which diffusion of particles is taking place along the axis of the cylinder in a single direction. Within the cylinder are two membranes, E and E', a distance of x from one end and a distance dx apart, as shown in the sketch in Fig. 9.1.

Diffusion in the cylinder gives rise to a force from the particles (which could be likened to an osmotic force) acting on the two membranes. The force acting on E is F, whereas the resisting force acting on E' in the opposite direction is F'. The resultant of these forces is

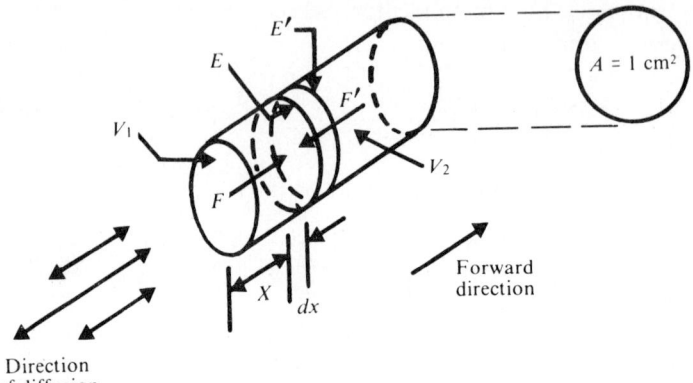

Figure 9.1 Sketch of imaginary cylinder.

$F - F'$. This force acts on the cylinder volume $A\,dx$, but the cross-sectional area of the cylinder A is equal to 1. Thus the force per unit of enclosed volume is $(F - F')/dx$. This is also equal to the osmotic pressure gradient within the enclosed volume dP_e/dx, that is, the pressure gradient between planes E and E'. Letting ΣF_D be the osmotic force per unit volume, we have

$$\Sigma F_D = \frac{F - F'}{A\,dx} = \frac{-dP_e}{dx} \tag{9.5}$$

The osmotic pressure of a solute in a solvent is given by the expression

$$P_e = nRT \tag{9.6}$$

where P_e is the osmotic pressure, R the gas constant, T the absolute temperature, and n the number of particles expressed as gram-molecules per unit volume.

Differentiating with respect to n gives

$$dP_e = RT\,dn \tag{9.7}$$

which, when substituted in Eq. 9.5, gives for the osmotic force

$$\Sigma F_D = -RT \frac{dn}{dx} \tag{9.8}$$

This is the osmotic or diffusional force acting on all the particles per unit volume. If there are n gram-molecules of particles present in the unit volume, then the actual number of particles in the unit volume is determined from the product of n and Avogadro's number N_A. Since $nN_A = c$ is the number of particles per unit volume, the force acting on each particle F_D is

$$F_D = \frac{\Sigma F_D}{nN_A} \qquad (9.9)$$

The resistance offered to the motion of a spherical particle by the medium in which it is moving is given by Stokes' law for small values of the Reynolds number. Equating this resistance with F_D gives

$$F_D = -\frac{RT\, dn/dx}{nN_A} = \frac{3\pi\mu dv}{C_c} \qquad (9.10)$$

Then rearranging terms gives

$$N_A nv = \frac{-RT}{N_A} \frac{C_c}{3\pi\mu d} \frac{d(nN_A)}{dx}$$

or

$$cv = -\frac{RT}{N_A} \frac{C_c}{3\pi\mu d} \frac{dc}{dx} \qquad (9.11)$$

The product cv represents a diffusion current, i.e., the number of particles crossing a unit area in unit time. But Fick's first law of diffusion states that the diffusion current is proportional to the concentration gradient, the constant of proportionality being the diffusion coefficient D. Thus the diffusion coefficient for an aerosol particle is, from Eq. 9.11,

$$D = \frac{RT}{N_A} \frac{C_c}{3\pi\mu d} = kT\frac{C_c}{3\pi\mu d} \qquad (9.12)$$

where k is the Boltzmann constant.

A new term—particle mobility B—can be defined. Mobility represents the velocity given to a particle by a constant unit driving force and is

$$B = \frac{C_c}{3\pi\mu d} \qquad (9.13a)$$

which for a sphere of mass m becomes

$$B = \frac{\tau}{m} \qquad (9.13b)$$

Then the diffusion coefficient is just

$$D = BkT = \frac{\tau}{m}kT \qquad (9.14)$$

Example 9.2 Determine the diffusion coefficient of a cigarette smoke particle (spherical shape, d = 0.25 μm, ρ = 0.9 g/cm³). Assume T = 20°C.

$$C_c = 1 + \frac{2\lambda}{d}\left[1.257 + 0.4 \exp\left(-\frac{1.1d}{2\lambda}\right)\right] = 1.72$$

$$B = \frac{C_c}{3\pi\mu d} = \frac{1.72}{(3)(\pi)(1.82 \times 10^{-4})(0.25 \times 10^{-4})} = 4.0 \times 10^7 \text{ cm/s}$$

$$D = BkT = (4.01 \times 10^7)(1.38 \times 10^{-16})(293)$$

$$= 1.62 \times 10^{-6} \text{ cm}^2/\text{s}$$

The diffusion coefficient has units in the cgs system of cm²/s and is a function only of particle diameter, gas viscosity, and pressure for a given temperature. At normal conditions of temperature and pressure, a 1-μm diameter particle has a diffusion coefficient of 2.76×10^{-7} cm²/s, about 10^6 times smaller than the diffusion coefficient for a typical gas molecule. Diffusion coefficients for other particle sizes are given in Table 9.1.

Brownian Displacement

When a particle moves in brownian motion, the chance that it will ever return to its initial position is negligibly small. Thus, there will be a net displacement with time of any single particle, even though the average displacement for all particles is zero. For example, during a short time interval one particle may move a distance s_1, another a distance s_2, and so on. Some of these displacements will be positive, others negative; some up, others down; but with equilibrium conditions the sum of the displacements will be zero. It is possible to estimate the displacement of any particle in terms of its root-mean-square displacement.

Suppose for simplicity, particles are considered to move only forward or backward along a single axis. Particles moving forward are

TABLE 9.1 Diffusion Coefficients of Spheres of Various Sizes at Normal Temperature and Pressure

Diameter, cm	C_c	B	D, cm/s²
10^{-6}	23.35	1.36×10^{10}	5.50×10^{-4}
10^{-5}	2.97	1.73×10^8	7.01×10^{-6}
10^{-4}	1.17	6.84×10^6	2.76×10^{-7}
10^{-3}	1.02	5.93×10^5	2.40×10^{-8}

those considered to have positive velocities. We can denote the root-mean-square displacement of the forward-moving particles as s, this displacement taking place over the time interval t. During this time interval, on average, only those particles lying a distance of s or less from a plane E will pass through it. The total number of particles displaced per unit area through E is $\frac{1}{2}c_1 s$, where c_1 is the mean concentration of particles within volume V_1, lying to the left of plane E.

By the same arguments, the concentration of particles going the other way from volume V_2 (on the right of plane E) is $\frac{1}{2}c_2 s$. Thus the net flow from left to right is $\frac{1}{2}(c_1 - c_2)s$.

However, for very small values of s we can write

$$\frac{dc}{dx} = \frac{c_2 - c_1}{s} \tag{9.15}$$

the definition of a differential. Then

$$c_1 - c_2 = -s\frac{dc}{dx} \tag{9.16}$$

and the net flow can be written as $\frac{1}{2}s^2\, dc/dx$. In a unit of time the quantity J diffusing through a unit area of plane E is

$$J = -\frac{1}{2}\frac{s^2}{t}\frac{dc}{dx} \tag{9.17}$$

But Eq. 9.17 again resembles Fick's first law of diffusion, giving another expression for D, the diffusion coefficient. In this case

$$D = \frac{s^2}{2t} \tag{9.18}$$

The mean square displacement is then

$$s^2 = 2Dt \tag{9.19}$$

When movement in three dimensions is considered, the displacement over any given time period will be less than in the one-dimensional case, since during part of the elapsed time the particle moves at right angles to the direction of interest. For three-dimensional motion the mean square displacement is

$$s^2 = \frac{4}{\pi}Dt \tag{9.20}$$

Example 9.3 An aerosol made up of 0.25-μm-diameter smoke particles is collected in a spherical flask 5 cm in diameter. How long will it take, on average, for a particle to travel from the center of the flask to its outer edge?

From Example 9.2, for a 0.25-μm spherical particle D is 1.62×10^{-6} cm²/s.

$$s = \sqrt{\frac{4}{\pi}Dt}$$

$$2.5 = \sqrt{\frac{4}{\pi}(1.62 \times 10^{-6})t}$$

$$t = 3.03 \times 10^6 \text{ s}$$

$$= 35.0 \text{ days}$$

This problem illustrates that particles move relatively short distances by diffusion. Thus diffusion is important only when one is considering particles in very small volumes, close to surfaces, or when particle size is so small that the value of D approaches molecular diffusion coefficients.

Brownian Motion of Rotation

Particles comprising an aerosol move randomly in brownian motion because of the gas molecules impacting on them. The random nature of the molecules striking the particles can also cause the particles to rotate, this brownian rotation being described by the equation (Fuchs, 1964)

$$\overline{\theta^2} = 2kTB_\theta t \tag{9.21}$$

The term $\overline{\theta^2}$ is the mean square angle of rotation of the particle about a given axis in time t. For a spherical particle Fuchs (1964) gives the rotational mobility B_θ as

$$B_\theta = \frac{1}{\pi\mu d^3} \tag{9.22}$$

so that Eq. 9.21 becomes

$$\overline{\theta^2} = \frac{2kT}{\pi\mu d^3}t \tag{9.23}$$

Example 9.4 Determine the average number of revolutions a 5-μm spherical particle will make per minute in air at 20°C.

$$\overline{\theta^2} = \frac{2kT}{\pi\mu d^3}t = \frac{2(1.38 \times 10^{-16})(293)}{(\pi)(1.82 \times 10^{-4})(5 \times 10^{-4})^3}(60)$$

$$= 67.89$$

$$\sqrt{\overline{\theta^2}} = 8.24 \text{ rad}$$

$$= 1.31 \text{ r/min}$$

For smooth spherical particles, brownian rotation is of no interest because it produces no observable effect. For particles with some irregularities brownian rotation produces the twinkling effect which is often observed when a beam of light is passed through a cloud of particles. Although Eq. 9.23 was derived for spheres, it can also be applied to isometric particles. For particles smaller than about 1 μm in diameter, the frequency of rotation is faster than the eye can see. On the other hand, with particles having diameters greater than 20 μm or so, brownian rotation is very slow. Thus the twinkling normally observed in a cloud arises only from particles whose diameters lie roughly between 1 and about 20 μm, but since particles smaller than about 10-μm diameter cannot be seen with the unaided eye, the actual range of sizes of twinkling particles is very small.

"Barometric" Distribution of Particles

One consequence of kinetic theory is that particles will have the same average translational energy as molecules when the gas is in equilibrium. Thus it is possible to compute the average velocity of a particle as it moves in brownian motion. Denoting this velocity as v_0,

$$\frac{1}{2} m v_0^2 = \frac{3}{2} kT \tag{9.24}$$

where m is the mass of the particle and $3kT/2$ is the average energy of a particle in the gas. Rearranging terms gives

$$v_0 = \sqrt{\frac{3kT}{m}} \tag{9.25}$$

exactly the same as Eq. 3.10, the equation for the root-mean-square velocity of a gas molecule. Aerosol particles, in their random motion, follow a Maxwell-Boltzmann velocity distribution similar to the molecules. But if they behave similarly to gas molecules, they should also be distributed vertically in equilibrium according to the barometric or Boltzmann equation. This indeed appears to be the case. Monodisperse particles which do not coagulate will be distributed vertically according to the expression

$$c = c_0 \exp\left(\frac{-mgZ}{kT}\right) \tag{9.26}$$

where Z is the height above some reference point at which the concentration c is measured and c_0 is the concentration of particles at the reference height. This effect is of no importance for particles larger than 0.3 μm. For particles with 0.1-μm diameter, at equilibrium essentially all particles will be contained in a band approximately 0.8 mm thick above a given surface. With particles of 0.01-μm diameter, the bandwidth is approximately 50 cm. Thus many very small particles (less than 0.1-μm diameter) will never be removed by sedimentation on their own accord, being constantly buffeted upward by brownian motion. Their removal from air must be carried out by some other mechanism.

Example 9.5 Extremely fine polonium-210 particles (0.01-μm diameter, ρ = 9.4 g/cm^3) are spilled on a laboratory bench. Assuming a barometric distribution is established, at what height will c/c_0 = 0.1?

$$\frac{c}{c_0} = \exp\left(-\frac{mgZ}{kT}\right)$$

$$\ln\frac{c}{c_0} = \frac{-mgZ}{kT}$$

$$Z = \frac{-kT \ln(c/c_0)}{mg}$$

$$= \frac{-(1.38 \times 10^{-16})(293)(\ln 0.1)}{(\pi/6)(10^{-18})(9.4)(980)}$$

$$= 19.30 \text{ cm}$$

Very fine particles will migrate over a surface, possibly as a result of "barometric" resuspension.

Effect of Aerosol Mass on the Diffusion Coefficient

Equation 9.12 indicates that the diffusion coefficient of an aerosol particle is independent of particle density and hence is independent of particle mass. But is this really so? Since particle mass is so much greater than molecular mass and the particles are continually undergoing bombardment by the molecules, one would expect changes in the direction of the particle to be gradual, compared to the rapid changes in direction with molecular diffusion. But if this is true, then particle momentum (mass) should be considered in the particle diffusion coefficient equation.

Two-dimensional trajectories of a typical gas molecule and a typical aerosol particle can be compared in Fig. 9.2. The molecule shows sharp changes in direction, each change occurring when it strikes another molecule. As discussed in Chap. 3, the average distance between hits is defined as the mean free path of the molecule. For the particle, a hit by a single molecule does not appreciably affect its motion. Therefore, its path is not characterized by sharp changes in direction, but by smooth curves representing the combined effect of hits by many molecules.

The problem can be treated by considering the average particle dis-

Gas Molecule   Aerosol particle

Figure 9.2 Gas molecule trajectory compared to aerosol particle trajectory.

placement under the influence of a force whose magnitude and direction vary in a random fashion but whose average magnitude is equal to zero. According to Fuchs (1964), the mean square displacement of an average particle considered in this way is

$$\bar{s}^2 = \frac{2}{3}\bar{v}^2\tau[t - \tau(1 - e^{-t/\tau})] \qquad (9.27)$$

The term $\bar{v}^2$ is equal to $3kT/m$, the mean square Boltzmann velocity. In terms of the diffusion coefficient, Eq. 9.27 becomes

$$\bar{s}^2 = 2D[t - \tau(1 - e^{-t/\tau})] \qquad (9.28)$$

When $t \gg \tau$, Eq. 9.28 reduces to Eq. 9.19, an expression for the displacement of a particle at constant velocity. Since our observation times will generally always be greater than τ, we can conclude that in most instances particle inertia can be neglected in considering particle diffusion.

Example 9.6 Find the ratio of t/τ such that the root-mean-square displacement estimated considering particle inertia (Eq. 9.28) is 10 percent less than the estimate when inertia is not considered (Eq. 9.19).

$$\frac{\text{Eq. 9.28}}{\text{Eq. 9.19}} = \frac{2D[t - \tau(1 - e^{-t/\tau})]}{2Dt} = 0.90$$

$$= 1 - \frac{\tau}{t}(1 - e^{-t/\tau}) = 0.90$$

$$\frac{\tau}{t}(1 - e^{-t/\tau}) = 0.1$$

By trial and error,

$$\frac{\tau}{t} = 0.1$$

i.e., when $t \geq 10\tau$, this correction is unnecessary.

Aerosol Apparent Mean Free Path

Since aerosol particles are continually undergoing molecular bombardment, their paths are smooth curves rather than segments of straight lines. It still is possible to define an *apparent mean free path* for the aerosol particles (Fuchs, 1964). This is the distance traveled by an average particle before it changes its direction of motion by 90°. The apparent mean free path represents the distance traveled by an average particle in a given direction before particle velocity in that direction equals zero. But this is just the stop distance.

At any time, a particle may be considered to be moving in a specific direction with a velocity $v = \sqrt{8kT/\pi m}$. From a definition of the stop distance, the pseudo mean free path l_B is

$$l_B = \tau v = \tau \sqrt{\frac{8kT}{\pi m}} \tag{9.29}$$

At normal pressure and temperature, l_B reaches a minimum at an aerosol particle diameter of 2×10^{-5} cm, but increases only by about a factor of 5 for particles 2 orders of magnitude larger or smaller than this size. Thus, the pseudo mean free path is essentially constant over the size range of interest, having a value of about 10^{-6} cm.

Example 9.7 Compute the apparent mean free paths for unit-density spheres of 0.01-, 0.1-, and 1-μm diameter. Assume $T = 20°C$.

$$l_B = \tau \sqrt{\frac{8kT}{\pi m}}$$

d, μm	C_c	τ	m
0.01	23.35	7.13×10^{-9}	5.24×10^{-19}
0.1	2.97	9.08×10^{-8}	5.24×10^{-16}
1.0	1.17	3.57×10^{-6}	5.24×10^{-13}

$$l_B(0.01 \text{ μm}) = 3.16 \times 10^{-6} \text{ cm}$$

$$l_B(0.1 \text{ μm}) = 1.27 \times 10^{-6} \text{ cm}$$

$$l_B(1.0 \text{ μm}) = 1.59 \times 10^{-6} \text{ cm}$$

Problems

1 Compute the diffusion coefficient in air of a 5-μm unit-density sphere at 20°C.

2 Repeat Prob. 1 for a 0.5-μm unit-density sphere.

3 Repeat Prob. 2 for temperatures of 0 and 100°C.

4 Estimate the root-mean-square displacement for a 2-μm silica dust particle (ρ = 2.65 g/cm^3) over a 10-min period.

5 How long on average will it take a 0.25-μm cigarette smoke particle (assume sphere, ρ = 0.9 g/cm^3) to diffuse (*a*) 1 ft and (*b*) 10 ft?

6 Assuming that a person can distinguish individual flashes of light appearing at a frequency of 5 per second or less, estimate the minimum size of a particle that will appear to twinkle in a beam of sunlight.

7 Using Eq. 9.26, determine the height at which nitrogen molecules, molecular weight = 14, will have a concentration that is one-tenth the concentration at h = 0; hence, estimate the height above the earth where p/p_0 = 0.1 atm.

8 What is the significance (if any) of the observation that the apparent mean free path goes through a minimum value at about 0.2 μm?

Chapter 10

Particle Diffusion

In particle diffusion the migration or movement of aerosol particles down a concentration gradient is considered. Since particles will always tend to move from regions of high concentration to regions of low concentration (Chap. 3), there will always be a tendency for aerosols to migrate to walls or other surfaces where the concentration, because of deposition, is essentially zero. As discussed in Chap. 9, the range over which this migration occurs is quite small. In those cases where the aerosol is in a fairly confined space to begin with, such as in the lung or in a small sampling tube, loss of the aerosol by diffusion can be significant. In this chapter, methods for estimating this loss are described.

Steady-State Diffusion

Consider the case of the diffusion of particles in a gas where the concentration of particles, although varying at different points within a gas, does not change with time. An example is the diffusion of particles from a zone of constant concentration $c = c_0$ to a wall, where the airborne concentration is assumed to be zero. Suppose it is desired to know the deposition rate of particles on the wall due to diffusion. From Fick's first law, the diffusion current J is

$$J = -D \frac{dc}{dx} \qquad (10.1)$$

If δ is the distance from the zone of constant concentration to the wall, then the concentration gradient dc/dx will be

$$\frac{dc}{dx} = \frac{0 - c_0}{\delta} = -\frac{c_0}{\delta} \qquad (10.2)$$

so that number of particles striking a unit area of the wall in unit time is

$$J = \frac{Dc_0}{\delta} \tag{10.3}$$

Example 10.1 An aerosol flowing through a tube is kept at a constant concentration inside the tube to within 1 mm of the tube wall. If the aerosol is made up of 0.5-μm-diameter spheres and the concentration in the tube is 10^3 particles per cubic centimeter, estimate the wall deposition rate, in particles per square centimeter per second. Assume $T = 20°C$.

$$C_C = 1 + \frac{2\lambda}{d}\left[1.257 + 0.4 \exp\left(-\frac{1.1d}{2\lambda}\right)\right] = 1.35$$

$$D = BkT = \frac{C_c}{3\pi\mu d}(kT)$$

$$= \frac{(1.35)(1.38 \times 10^{-16})(293)}{(3\pi)(1.82 \times 10^{-4})(5 \times 10^{-5})}$$

$$= 6.35 \times 10^{-7} \text{ cm}^2/\text{s}$$

$$J = \frac{Dc_0}{\delta} = \frac{6.35 \times 10^{-7} \times 10^3}{0.1} = 6.35 \times 10^{-3} \text{ particles/(cm}^2 \cdot \text{s)}$$

Except for very long tubes, this is a negligible deposition rate. Unfortunately, it is quite difficult to estimate δ, the concentration boundary layer thickness.

Non-Steady-State Diffusion

In Example 10.1 the case where the aerosol concentration does not change with time was considered. In many practical situations, however, the aerosol concentration does change with time, possibly as a result of diffusion and subsequent loss of particles to a wall or other surface. In this event, Fick's second law, Eq. 9.2, must be used. Solution of this equation is possible in many cases, depending on the initial and boundary conditions chosen, although the solutions generally take on very complex forms and the actual mechanics involved to find these solutions can be quite tedious. Fortunately, there are several excellent books available which contain large numbers of solutions to the transient diffusion equation (Barrer, 1941; Jost, 1952). Thus, in most cases it is possible to fit initial and boundary conditions of an aerosol problem to one of the published solutions. Several commonly occurring examples follow.

Infinite volume, plane vertical wall

Consider the case of a plane vertical wall that is in contact with an infinitely large volume of aerosol having the same initial concentration throughout. It is desired to estimate the rate of deposition of aerosol on a unit area of wall, assuming that all particles hitting the wall stick to it. The conditions of the problem make it one-dimensional. By letting x be the distance from the wall, Eq. 9.2 becomes

$$\frac{\partial c}{\partial t} = D \frac{\partial^2 c}{\partial x^2} \qquad (10.4)$$

since the concentration gradients in the y and z directions are zero. With the initial concentration $c(x, 0) = c_0$ and the boundary condition $c(0, t) = 0$ for $t > 0$, the solution to Eq. 10.4 is

$$c(x,t) = \frac{2c_0}{\sqrt{4\pi Dt}} \int_0^x \exp\left(-\frac{\zeta^2}{4Dt}\right) d\zeta \qquad (10.5)$$

or

$$c(x,t) = c_0 \, \mathrm{erf}\left(\frac{x}{\sqrt{4Dt}}\right) \qquad (10.6)$$

where erf represents the probability or error function, a tabulated function (see App. C). When the argument of this function is small, the function is small and erf(0) = 0. For arguments greater than about 2.6, erf(x) = 1. Steep concentration gradients occur initially close to $x = 0$, gradually decreasing with time (Fig. 10.1).

Recalling that the diffusion current J, which is the number of particles crossing unit area in unit time, is equal to the diffusion coefficient times the gradient (Fick's first law), we see that evaluating the gradient at $x = 0$ gives the number of particles deposited in time interval dt

$$J \, dt = -D \frac{\partial c}{\partial x} dt$$

$$x = 0$$

From Eq. 10.6 and App. C,

$$\frac{\partial c}{\partial x} = \frac{2c_0}{\sqrt{\pi}} \frac{1}{\sqrt{4Dt}} \exp\left(-\frac{x^2}{4Dt}\right) \qquad (10.7)$$

which is to be evaluated at $x = 0$. Then

$$J \, dt = dN = c_0 \sqrt{\frac{D}{\pi t}} \, dt \qquad (10.8)$$

at $x = 0$, which on integrating from $t = 0$ to $t = t$ gives N, the number of particles deposited per unit area in the time interval t:

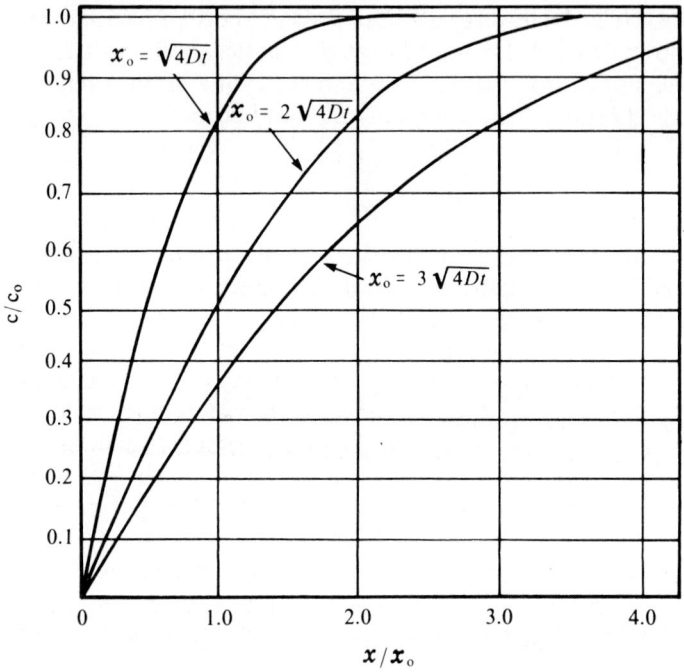

Figure 10.1 Plot of solution of $\partial c/\partial t = D(\partial^2 c/\partial x^2)$ (semi-infinite case). Solution: $c/c_0 = \mathrm{erf}(x/4Dt)$.

$$N = c_0 \sqrt{\frac{4Dt}{\pi}} \qquad (10.9)$$

Example 10.2 Estimate the number of 0.1-μm-diameter particles deposited per square centimeter per hour on a wall placed next to a semi-infinite aerosol containing 100 particles per cubic centimeter.

From Table 9.1, D for 0.1-μm-diameter particles is 6.96×10^{-6} cm²/s;

$$N = c_0 \sqrt{\frac{4Dt}{\pi}} = 100 \left[\frac{(4)(6.96 \times 10^{-6})(60 \times 60)}{\pi} \right]^{1/2}$$

$$= 17.86 \text{ particles} \quad (\text{cm}^2 \cdot \text{h})$$

It is interesting to compare Eqs. 10.3 and 10.9. Equation 10.3 represents steady-state conditions, in which the concentration a distance δ away from the wall is always constant, while Eq. 10.9 relates to the case where the concentration near the wall decreases as particles are lost to the wall.

Two vertical walls a distance *H* apart

Suppose, instead of being semi-infinite, the cloud is contained between two vertical walls spaced a distance H apart. With the same initial conditions as before and the boundary condition that $c = 0$ at both $x = 0$ and $x = H$ when $t = 0$, the solution to Eq. 10.4 becomes

$$\frac{c}{c_0} = \text{erf}\left(\frac{x}{\sqrt{4Dt}}\right) - \left[\text{erf}\left(\frac{H+x}{\sqrt{4Dt}}\right) - \text{erf}\left(\frac{H-x}{\sqrt{4Dt}}\right)\right] \quad (10.10)$$

Figure 10.2 shows a plot of c/c_0 as a function of $4Dt/H^2$ for the cases where $x = H/2$ and $x = H/20$. When H is large compared to x, the solution is equivalent to the single-wall infinite-medium case.

Figure 10.3 is a plot of c/c_0 versus time, measured at $x = H/2$ for monodisperse particles having a diameter of 1 μm when $H = 2$ cm. Note that there is essentially no change in concentration until sometime after 10^5 s, and then a fairly rapid decrease in concentration takes place. It can be concluded that except for very small particles or very small tubes, pure diffusion will have a small to negligible influence on the concentration changes in an aerosol flowing through a tube.

But concentration change, and hence "diffusive" deposition, is observed. Thus, there must be an additional mechanism operating which tends to enhance deposition of small particles by "diffusion."

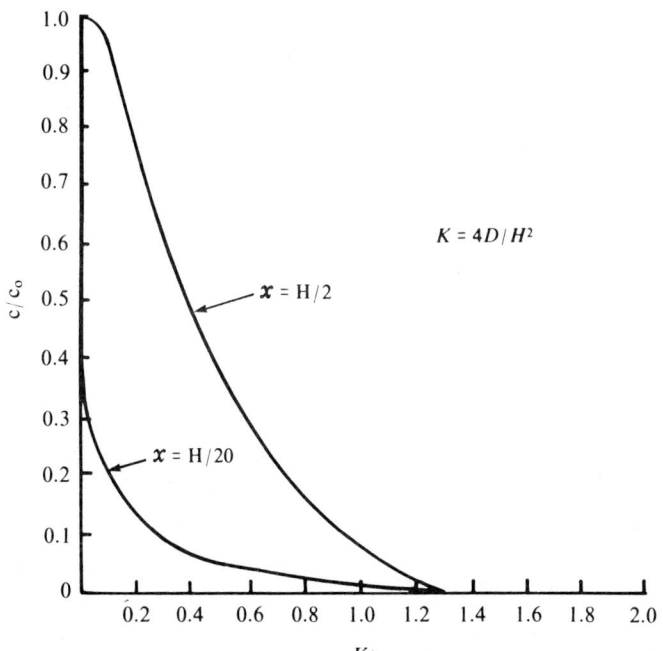

Figure 10.2 Finite case. Change in concentration by diffusion occurring between two walls spaced a distance H apart; Eq. 9.10.

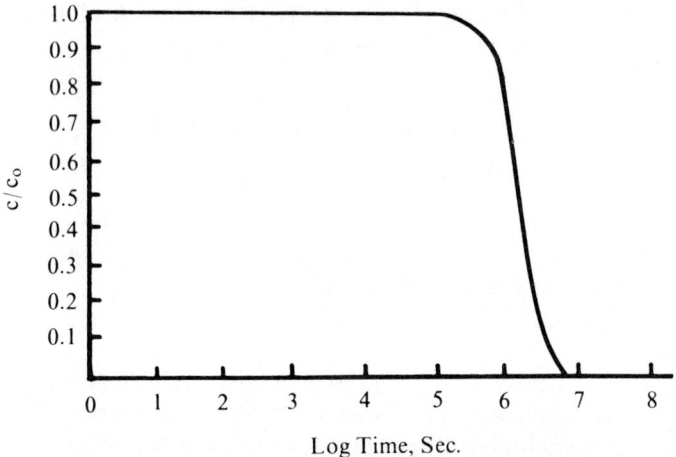

Figure 10.3 Diffusion of monodisperse 1-μm-diameter spheres. Concentration measured at center of two plates spaced 2 cm apart.

Diffusion in Flowing Airstreams—Convective Diffusion

Thus far, only models for diffusion of particles in still or stagnant air have been investigated. Often, however, the air in which the particles are suspended is not stagnant but has some overall motion. As an example, consider the smoker in a room full of people. Although there may be no perceptible air movement, when a cigarette is lighted, the odor of tobacco smoke is quickly detectable throughout the room. Even if molecular diffusion coefficients were used to describe the motion of the tobacco smoke particles, transport rates by diffusion are too small to explain the appearance of smoke so quickly in all parts of the room. What occurs is that particles are entrained and transported by the moving air within the room. Convective diffusion describes this phenomenon.

General Equations of Convective Diffusion

First, consider the flux of particles in a fluid through unit area in unit time. Particles can be transported by molecular diffusion or by the moving fluid. For molecular diffusion, from Fick's first law, $J_D = -D \text{ grad } c$, whereas if particles are entrained in a moving liquid,

$$J_{conv} = c\vec{v} \quad (10.11)$$

The total mass flux is the sum of the two fluxes and is expressed as

$$J = cu - D \operatorname{grad} c \tag{10.12}$$

From Eq. 10.12 it can be shown (Levich, 1962) that the time rate of change of concentration—similar to Fick's second law—is

$$\frac{\partial c}{\partial t} = D\nabla^2 c - u\nabla c \tag{10.13}$$

This is the general convective diffusion equation for particles in an isothermal gas when the particles are not subjected to any forces other than the convective motion of the gas and the molecular motion of the gas molecules.

If a volume source is present, such as a gas-phase reaction which produces Q_0 particles per cubic centimeter throughout the volume, Eq. 10.13 becomes

$$\frac{\partial c}{\partial t} = D\nabla^2 c - u\nabla c + Q_0 \tag{10.14}$$

In rectangular cartesian coordinates, Eq. 10.13 can be written

$$\frac{\partial c}{\partial t} + u_x \frac{\partial c}{\partial x} + u_y \frac{\partial c}{\partial y} + u_z \frac{\partial c}{\partial z} = D\left(\frac{\partial^2 c}{\partial x^2} + \frac{\partial^2 c}{\partial y^2} + \frac{\partial^2 c}{\partial z^2}\right) \tag{10.15}$$

When $u_x = u_y = u_z = 0$, indicating no convective motion of the gas, Eq. 10.15 reverts to the "pure" diffusion case. The terms u_x, u_y, and u_z are not necessarily equal, nor are they usually constant, since convective velocities decrease as a surface is approached. Equation 10.15 thus represents a second-order partial differential equation with variable coefficients. These types of equations are usually quite difficult to solve. However, often it is sufficient to consider only the steady-state solution, i.e., the case where $\partial c/\partial t = 0$, indicating that the concentration at any point within the system is not changing with time. Then Eq. 10.15 becomes

$$u_x \frac{\partial c}{\partial x} + u_y \frac{\partial c}{\partial y} + u_z \frac{\partial c}{\partial z} = D\left(\frac{\partial^2 c}{\partial x^2} + \frac{\partial^2 c}{\partial y^2} + \frac{\partial^2 c}{\partial z^2}\right) \tag{10.16}$$

Convective diffusion defined by the Peclet number

If u_0 is the average velocity in a system where both molecular diffusion and convective diffusion are taking place, L is a characteristic length, and c_0 is a representative concentration, then Eq. 10.16 can be put in dimensionless form by making the following substitutions: $U_x = u_x/u_0$, $X = x/L$, $C = c/c_0$, etc. Equation 10.16 becomes

$$U_x \frac{\partial C}{\partial X} + U_y \frac{\partial C}{\partial Y} + U_z \frac{\partial C}{\partial Z} = \frac{1}{\text{Pe}} \left(\frac{\partial^2 C}{\partial X^2} + \frac{\partial^2 C}{\partial Y^2} + \frac{\partial^2 C}{\partial Z^2} \right) \quad (10.17)$$

where Pe defines the dimensionless ratio

$$\text{Pe} = \frac{u_0 L}{D} \quad (10.18)$$

known as the *Peclet number*. It is clear that one side of Eq. 10.17 represents molecular motion while the other side represents convective motion. Since all the dimensionless terms in Eq. 10.17 are essentially 1 (Levich, 1962), the Peclet number describes the relationship between diffusion and convection in a manner similar to the role played by the Reynolds number in fluid flow. When the Peclet number is small, molecular diffusion predominates. When it is large, convective transport predominates and diffusion can be neglected.

Example 10.3 Determine the Peclet number for 0.25-μm-diameter spheres being mixed in a room 10 ft wide, 20 ft long, and 10 ft high if air is circulating in the room at a rate of 6 air changes per hour.

$$\text{Volumetric flow rate} = \text{volume} \times \frac{\text{changes}}{\text{h}} = 10 \times 20 \times 10 \times 6$$

$$= 12{,}000 \text{ ft}^3/\text{h} = 200 \text{ ft}^3/\text{min}$$

Using room cross-sectional area,

$$u_0 = \frac{200}{10 \times 10} = 2 \text{ ft/min} = 1.01 \text{ cm/s}$$

$$D \text{ for 0.25-μm spheres} = BkT = \frac{C_c}{3\pi \mu d} kT$$

$$= 1.62 \times 10^{-6} \text{ cm}^2/\text{s}$$

$$\text{Pe} = \frac{u_0 L}{D} = \frac{1.01(20 \times 30.5)}{1.62 \times 10^{-6}}$$

$$= 3.81 \times 10^8$$

Pe is very large; hence convection predominates and diffusion can be ignored.
In this solution 20 ft was chosen for L because we were interested in mixing along the entire length of the room. However, if either the width or the height were chosen for L, the resulting conclusion would be exactly the same!

Tube Deposition

For the special case of aerosol deposition in a tube in which both molecular and convective diffusion are important, several mathematical expressions have been derived from the convective diffusion equation,

Eq. 10.13, by assuming laminar flow, steady-state conditions ($\partial c/\partial t = 0$), diffusion in the direction of the aerosol flow to be negligible, and 100 percent sticking of the aerosol particles that reach the tube surface. Then Eq. 10.13 becomes

$$D\frac{\partial^2 c}{\partial R^2} + \frac{\partial c}{\partial R} = u(R)\frac{\partial c}{\partial R} \qquad (10.19)$$

where $u(R)$ is the velocity profile along the z or axial direction and R is a radial distance.

It is possible to make a rough estimate of the diffusional deposition in a tube of radius R by assuming a residence time of $t = L/u$, where u is the velocity in a tube of length L:

$$c_{out} = c_{in} - N\left(\frac{\text{area of tube}}{\text{volume of tube}}\right)$$

$$= c_{in} - 2c_{in}\sqrt{\frac{DL}{\pi\mu}}\left(\frac{2RL\pi}{\pi R^2 L}\right)$$

$$\frac{c_{out}}{c_{in}} = 1 - \frac{4}{\sqrt{\pi}}\sqrt{\frac{DL}{uR^2}} \qquad (10.20)$$

From the simple approach given above, it appears that deposition is controlled by a dimensionless ratio of terms

$$\phi = \frac{DL}{uR^2} = \frac{\pi DL}{Q} \qquad (10.21)$$

Furthermore, the result is interesting because it indicates that for diffusional deposition with a fixed flow rate, deposition is the same whether one uses a large tube with a low velocity or a small tube with a high velocity.

Cheng (1989) has summarized solutions to Eq. 10.19 for various geometries, as shown in Fig. 10.4. The general solution of Eq. 10.19 is expressed as a series of exponential functions

$$\frac{c_{out}}{c_{in}} = \sum_{n=1}^{\infty} A_n \exp(-\beta_n \phi) \qquad (10.22)$$

where A_n and β_n are constants. Figure 10.4 shows solutions to Eq. 10.22 for round and rectangular channels.

For penetration through a series of screens Cheng (1989) points out that the solution can be likened to the filter fan model of Kirsch and Stechkina (1978). In this case

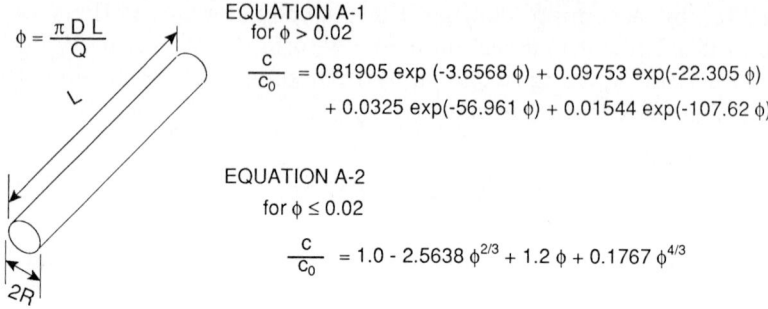

Figure 10.4 Particle deposition equations. (*From Cheng, 1989.*)

$$\frac{c_{out}}{c_{in}} = \exp\left[-\chi^n\left(2.7\text{Pe}^{-2/3} + \frac{1}{\kappa}\mathcal{R}^2 + \frac{1.24}{\kappa^{1/2}}\text{Pe}^{-1/2}\mathcal{R}^{2/3}\right)\right] \quad (10.23)$$

where

$$\chi = \frac{4\alpha h}{\pi(1-\alpha)d_f} \quad (10.23a)$$

$$\kappa = -0.5\ln\frac{2\alpha}{\pi} + \frac{2\alpha}{\pi} - 0.75 - 0.25\left(\frac{2\alpha}{\pi}\right)^2 \quad (10.23b)$$

with n as the number of screens; d_f, the diameter of a single fiber; h, the thickness of a single screen; α, the solid volume fraction of the screen; $\mathcal{R}$, the interception parameter d_p/d_f; and Pe, the Peclet number.

Example 10.4 An aerosol made up of 0.1-μm-diameter particles ($D = 7.01 \times 10^{-6}$ cm^2/s) flows through a 1-cm-diameter tube at 15 L/min. If the tube is 100 ft long, estimate c_{out}/c_{in}.

First ϕ is found by using Eq. 10.21:

$$\phi = \frac{DL}{uR^2} = \frac{\pi DL}{Q}$$

$$\phi = \frac{\pi DL}{Q} = \frac{\pi \times 7.01 \times 10^{-6} \times 100 \times 30.5}{15 \times 1000/60} = 2.69 \times 10^{-4}$$

Then, from Eq. 10.20

$$\frac{c_{out}}{c_{in}} = 1 - \frac{4}{\sqrt{\pi}}\sqrt{\phi} = 0.963$$

Using Eq. A-1 from Fig. 10.4,

$$\frac{c_{out}}{c_{in}} = 0.962$$

and using Eq. A-2 from Fig. 10.4,

$$\frac{c_{out}}{c_{in}} = 0.990$$

As shown earlier, particle diffusion coefficients are fairly small (on the order of 10^{-4} to 10^{-8} cm^2/s), resulting in large Peclet numbers unless u_0 (the average velocity in the system) is quite small or the characteristic length is quite small. In most cases of interest, average convective velocities are 0.01 cm/s or greater, not sufficiently small by themselves to ensure a small Peclet number (and hence a diffusion-controlled problem). Thus, whether diffusion or convection predominates generally depends solely on the definition of the characteristic length L. This has already been defined as the length over which the major change in concentration takes place.

Consider air flowing over a flat surface. If the average concentration of particles in the air is c_0 and the concentration at the surface is 0, it is expected that the concentration change from c_0 to 0 would occur mainly near the surface. This has already been shown to be the case for molecular diffusion alone. The distance over which this concentration change occurs is the characteristic length. In other words, molecular diffusion is important in convective diffusion only in the small region close to surfaces. Here it is extremely important, since not only are concentration gradients decreasing sharply but also velocity gradients are decreasing.

The zones where these gradients occur are often called boundary layers. For example, the aerodynamic boundary layer is the region near a surface where viscous forces predominate. Boundary layers exist with both laminar and turbulent flow and may be either solely laminar or turbulent with a laminar sublayer themselves (Landau and Lifshitz, 1959).

Laminar boundary layer

With air flowing over a surface, the boundary layer thickness δ increases along the surface in the direction of flow (Davies, 1966):

$$\delta = 5\sqrt{\frac{xv}{u_0}} \qquad (10.24)$$

The term δ represents distance from the surface to the point where 99 percent of u_0, the mainstream velocity, is reached (see Fig. 10.5). In Eq. 10.24a, x is the distance measured in the direction of flow from the starting point to the point of interest, and v is the kinematic viscosity. If a linear Reynolds number Re_x is defined as $u_0 x/v$, Eq. 10.24a can be written as

$$\delta = \frac{5x}{\sqrt{\text{Re}_x}} \qquad (10.24b)$$

For laminar airflow in a tube, when δ approaches the tube radius, Poiseuille flow or a parabolic flow profile is fully developed. This is accomplished by the acceleration of the central portion of the flow. However, when Re_x exceeds a value lying somewhere between 10^4 and 10^6, the laminar boundary layer becomes so thick that it is no longer stable, and a turbulent boundary layer develops.

Example 10.5 Twenty liters of air flows per minute into a 2-in-diameter tube of circular cross-section.

   a. Find the boundary layer thickness a distance of 1 cm into the tube.
To determine the boundary layer thickness at 1 cm,

$$\text{Re}_x = \frac{u_0 x}{v} = \frac{4(20{,}000/60)}{\pi(2.54 \times 2)^2} \frac{1}{0.151} = \frac{16.45}{0.151}$$

$$= 109.0$$

Given Re_x, the thickness δ can be determined.

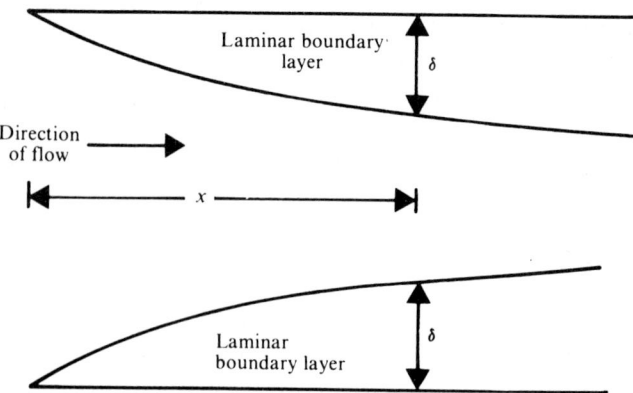

Figure 10.5 Development of laminar boundary layer (not to scale).

$$\delta = \frac{5x}{\sqrt{\mathrm{Re}_x}} = \frac{5(1)}{\sqrt{109}} = 0.48 \text{ cm}$$

b. At what distance into the tube will a parabolic laminar flow profile be fully established?

Let δ = tube radius = 1×2.54 cm. Then

$$\delta = 5\sqrt{\frac{x\nu}{u_0}}$$

$$\delta^2 = 5^2 \frac{x\nu}{u_0}$$

$$x = \frac{\delta^2 u_0}{5^2 \nu} = \frac{(1 \times 2.54)^2 (16.45)}{(25)(0.151)}$$

$$= 28.13 \text{ cm}$$

With fully developed laminar flow, the velocity at any point r in a circular tube of radius R can be expressed by the equation

$$u(r) = u_m \left(1 - \frac{r^2}{R^2}\right) \tag{10.25}$$

where u_m is the maximum centerline flow velocity, or twice the average tube velocity.

Turbulent boundary layer

A turbulent boundary layer is actually made up of three zones, a viscous or laminar sublayer immediately adjoining the wall, a buffer zone, and finally a turbulent zone making up the main boundary layer (Schlicting, 1968). Generally speaking, turbulent boundary layers are thicker than laminar boundary layers.

Concentration boundary layer

Since both laminar and turbulent boundary layers contain laminar or viscous layers, it would seem logical that diffusion would primarily take place across these regions. If the boundary layer thickness were known, assuming a linear decrease in concentration, Eq. 10.3 could be used to estimate diffusion current. Unfortunately, the point of uniform velocity is not necessarily the point of uniform concentration. This is because particles, with their large inertia compared to air, can be carried into laminar boundary regions by mixing as well as by diffusion. The value for δ in Eq. 10.3 will always be less than the equivalent value for the aerodynamic boundary layer thickness, in some cases being only one-tenth or even smaller (Levich, 1962).

Thus, there are actually two boundary layers of interest: the aerodynamic boundary layer which is a result of the velocity gradient established at the boundary and the diffusion or concentration boundary layer resulting from the concentration gradient which exists near the surface.

For turbulence it is convenient to describe particle flux in terms of an eddy diffusion coefficient, similar to a molecular diffusion coefficient. Unlike a molecular diffusion coefficient, however, the eddy diffusion coefficient is not constant for a given temperature and particle mobility, but decreases as the eddy approaches a surface. As particles are moved closer and closer to a surface by turbulence, the magnitude of their fluctuations to and from that surface diminishes, finally reaching a point where molecular diffusion predominates. As a result, in turbulent deposition, turbulence establishes a uniform aerosol concentration that extends to somewhere within the viscous sublayer. Then molecular diffusion or particle inertia transports the particles the rest of the way to the surface.

As particle size increases, particles tend to lag behind the eddy motion of the turbulent air. Particle size may be so large that particles are influenced only slightly or not at all by the turbulence. In this case particles will not be deposited by turbulent motion. Smaller particles that follow turbulence, even though they might lag behind, can be deposited by being projected across the boundary layer if the boundary layer thickness is less than the particle stop distance (Sehmel, 1968). Since increasing turbulence tends to increase particle motion, increases in turbulence will tend to enhance particle deposition for a given size particle. But at a given level of turbulence (Reynolds number), calculations made by Davies (1965) indicate that there exists a particle size having a maximum rate of deposition, as shown in Fig. 10.6. These deposition rates can be expressed in terms of a "deposition" or "diffusion" velocity.

The diffusion velocity

If a concentration boundary layer δ can be defined, then the number of particles deposited per unit area in unit time becomes

$$J = \frac{c_0 D}{\delta} \quad (10.26)$$

By dividing J by c_0, a term having the units of velocity results. This function v_D is called the diffusion velocity, defined as

$$v_D = \frac{J}{c_0} = \frac{D}{\delta} \quad (10.27)$$

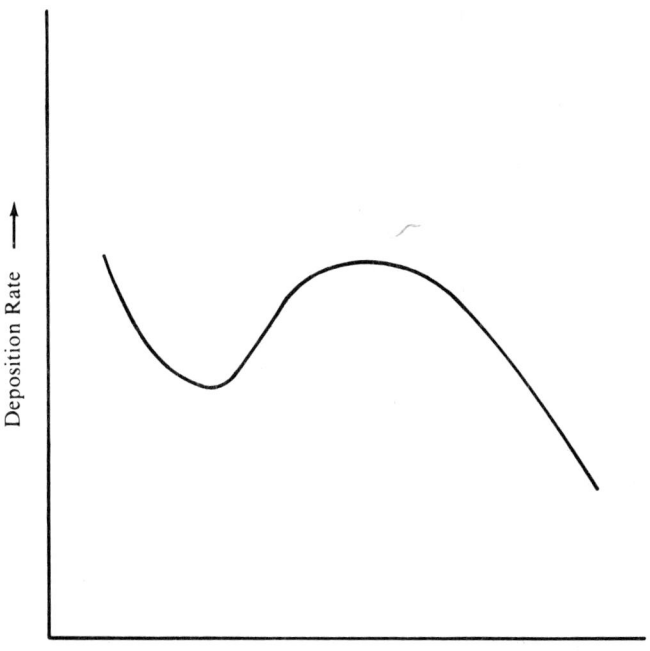

Figure 10.6 Schematic diagram of deposition velocity as a function of particle diameter (Reynolds number fixed).

Following this same logic, a diffusion "force" can be defined as

$$F_{\text{diff}} = \frac{3\pi\mu v_D d}{C_c} = \frac{kT}{\delta} \quad (10.28)$$

which is dependent on particle and medium properties only as δ is dependent on these properties.

Application of diffusion velocity

Consider a well-mixed ensemble of particles flowing through a tube of radius R with no other factors but diffusion tending to remove the particles from the flow. With diffusion velocity considered as a net movement of particles to the tube surface, in an interval of 1 s there will be $J(2\pi R)$ particles deposited per unit length of tube. In a time $dt = dx/u_0$, a 1-cm length of aerosol traverses a distance dx. Thus in this time $J(2\pi R)\, dx/u_0$ particles are removed, and the change in concentration is the number of particles removed divided by the volume from which they are removed, or

$$dc = \frac{J(2\pi R)\,dx}{u_0}\frac{1}{\pi R^2} = cv_D\frac{2\,dx}{u_0 R} \tag{10.29}$$

Integrating, with the initial condition that $c = c_0$ at the entrance of the tube, gives

$$\ln\frac{c}{c_0} = -\frac{2v_D L}{u_0 R} \tag{10.30}$$

where L is the overall length of the tube. If deposition velocity is constant over the length of the tube considered, it is possible to estimate deposition within the tube from Eq. 10.30.

Conversely, this equation can be used to determine deposition velocities from experimental data.

Example 10.6 Unit-density 2-μm spheres are deposited while flowing at a rate of 24 L/min through a 0.21-in-diameter tube. If the concentration downstream of a 100-cm tube length is 87 percent of the initial or upstream concentration, estimate the particle deposition velocity.

$$\ln\frac{c}{c_0} = -\frac{2v_D L}{u_0 R}$$

$$u_0 = \frac{24 \times 10^3}{(60\pi)(0.21 \times 2.54/2)^2} = 1.79 \times 10^3 \text{ cm/s}$$

$$R = \frac{0.21 \times 2.54}{2} = 0.267 \text{ cm}$$

$$v_D = 0.332 \text{ cm/s}$$

Experimental determinations of v_D are complicated by entrance effects as well as by the effect of gravity, which is usually ignored. As a result, only order-of-magnitude accuracy has been achieved from predictions made by using the equations of this section. Even so, it should be clear that deposition is largely determined by the properties of the fluid flowing near the wall. Factors such as surface shape or roughness, since they affect this fluid flow, will have a marked effect on deposition, even at low stream velocities.

Problems

1 Using the barometric equation, compute the height at which 50 percent of 0.05-μm unit-density spheres would be suspended by molecular impacts.

2 Estimate the apparent mean free path of 0.1-μm unit-density spheres in air at 20°C and 760-mmHg pressure.

3 For a 0.01-µm unit-density sphere, how short must a diffusion experiment be to have as much as 1 percent error in distance measurements? How much shorter for a 10 percent error?

4 A cloud of 0.05-µm spheres is held in a large container. The initial concentration is 10^4 particles per cubic centimeter. After 20 min, what is the aerosol concentration (in particles per cubic centimeter) 0.1 mm from the wall? How long will it take the aerosol to decrease to a concentration of 10^3 particles per cubic centimeter 1 cm from the wall?

5 A sheet of glass 4 cm by 4 cm is inserted into a cloud containing 10^5 0.02-µm spherical particles per cubic centimeter. If a microscope is used with a viewing area of 50 µm × 50 µm to view these particles and 100 particles are observed per field, what is the average areal density of particles on the glass? How long must a sample be collected to achieve this density?

6 Compare Eq. A-1 in Fig. 10.4 given by Cheng (1989) to that of Gormley and Kennedy (1949), also given in Fig. 10.4 as Eq. A-2, by determining the value of ϕ for which $c_{out}/c_{in} = 0.5$.

7 Compare the equation given in Fig. 10.4 for a rectangular channel, Eq. B-1, to Eq. A-1 in the same figure for values of c_{out}/c_{in} of 0.2, 0.4, 0.6, and 0.8. Use the same cross-sectional area for the channel and the tube and have $W = 3H$. Does deposition appear to be related to surface area?

8 In the portable diffusion battery of Sinclair (1972), air is flowed through a porous cylinder 1.38-in in length, 1¾-in in diameter containing 14,500 holes, each 0.009 in in diameter. Samples can be drawn out at different points along the length. Show that the equivalent length (i.e., the length of a battery consisting of one tube) is equal to the actual length times the number of holes.

9 For the diffusion battery of Prob. 8, determine the maximum particle diameter which can be collected with 50 percent efficiency with a total flow through the unit of 1 L/min. For penetration use Eq. A-1 in Fig. 10.4.

Chapter 11

Thermophoresis

Introduction

Thermal gradients either within particles or in the supporting medium can be responsible for motion of aerosol particles by creating forces which act on the individual particles. Here the discussion is not about convective motion of the medium set up by thermal gradients which carries particles with it, but with thermal forces which act directly on individual particles to cause motion.

Tyndall (1870) first reported the existence of a dust-free space surrounding a hot body, and other investigators subsequently demonstrated that under the influence of a temperature gradient, aerosol particles move from hot to cold regions, i.e., they move away from the source of heat (Rayleigh, 1882, 1884). The dust-free zone surrounding a hot body is well defined, with particles not crossing the seemingly impenetrable barrier of the dust-free zone. Figure 11.1 illustrates the type of dust-free zones formed by various shapes, and Fig. 11.2 is a photograph of the dust-free zone surrounding a cylindrical rod.

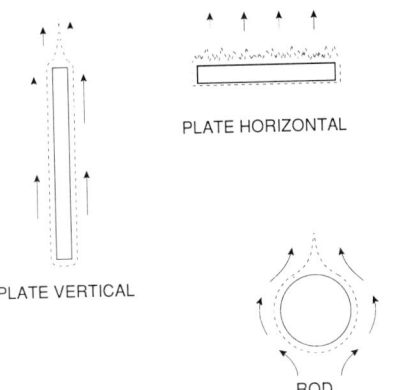

Figure 11.1 Types of dust-free zones formed by thermophoresis.

Figure 11.2 Photograph of the dark space surrounding a heated brass rod.

Early Observations of Thermophoresis

Early observations on the width of the dust-free space were made by Aitken (1884a, b) and Lodge and Clark (1884). Later Watson (1936) and Miyake (1935) developed an empirical formula which fit experimental observations that the dust-free space increases with increasing temperature of the body and decreasing air pressure, and decreases with increasing molecular weight of the surrounding gas. For example, Watson found an empirical relationship between the width in centimeters of the dust-free space σ_{df} and temperature in the form

$$\sigma_{df} = L\Delta T\, H^{-0.38} \qquad (11.1)$$

where ΔT is the temperature difference in degrees Celsius between the surface of the hot body and free air, L is a constant having a value of about 1.56×10^{-4} for vertical plane surfaces or 7.3×10^{-5} for horizontal rods, and H represents the convective heat loss of the body in calories per square centimeter per second. The term H can be roughly approximated by $H \approx 1 \times 10^{-4}\, \Delta T^{1.25}$.

It was also found that the magnitude of the dark space depends on the size and shape of the body as well as inversely on the pressure of the medium.

Example 11.1 A vertical heating element is held at a temperature of 100°C higher than the surrounding air. Estimate the width of the dust-free space around the surface of the element.

$$\sigma_{df} = L\Delta T\, H^{-0.38} = (1.56 \times 10^{-4})(100)\,[1 \times 10^{-4}(100)^{1.25}]^{-0.38}$$

$$= (1.56 \times 10^{-4})(100)(3.72) = 5.80 \times 10^{-2}\ \text{cm}$$

Thermophoretic forces produce very obvious effects near areas of significant temperature gradients. For instance, one can often observe a black deposit on the wall just above a hot-water radiator or pipe. Convection currents conduct the warm gas and particles over the radiator, but since the cooler surfaces nearer the radiator are not protected by a dust-free space, deposition takes place. On a ceiling or on walls of rooms heated by convection, one can often see a replica of the construction behind the plaster formed by deposited particles. Again, the dust is deposited on the cooler portions of the surface: on spaces between the laths if the laths are poor heat conductors and directly opposite the laths if they are good conductors. In a room that is heated by direct radiation, such as by an open fire, the walls and furniture of the room are warmer than the air, so that particles suspended in the air are not deposited by thermal forces (Lodge, 1883; Gibbs, 1924).

Thermal deposition of particles inside boilers or heat exchangers can lead to reduced efficiency of the units (Fuchs, 1964) whereas thermophoresis taking place in ducts or chimneys before equilibrium temperatures are reached can account for an appreciable fraction of the total deposition which occurs. Thermophoresis may protect surfaces against particle deposition (Stratmann et al., 1988).

Thermophoretic forces can be used in sampling aerosols; the particles are passed through the dark space surrounding a hot body and are collected with nearly 100 percent efficiency on a cold surface placed nearby. To date, however, there has been no successful utilization of thermal forces for large-scale air cleaning.

Theory

There have been many attempts to devise a suitable theory which describes thermophoresis, but as yet a complete solution has not been found. In principle the approach is quite simple.

Consider molecular motion in a temperature gradient. The movements will be more vigorous at higher temperatures. When a particle is placed in this gradient, the momentum transferred to one side of the particle exceeds that transferred to the other side, so that a net force results. To determine this net force, it is necessary to know exactly the velocity distribution of the molecules at the particle surface. Among other things this depends on the ratio of particle size and pressure of the medium, the Knudsen number ($Kn = 2\lambda/d$), because, depending on this ratio, the particle itself can have very little or a significantly great influence on the velocity distribution of nearby molecules. A complete theory must take this varying influence into account. At present it is most convenient to consider several size ranges (or Knudsen numbers) when one is attempting to predict the magnitude of the thermal force.

Thermophoresis in the Free Molecule Region ($Kn \gg 1$)

Consider first the case when the particle is much smaller than the mean free path of the gas molecules. This represents the condition where $Kn \gg 1$ and is often called the *free molecule region*.

In the free molecule region, molecules colliding with a particle will travel on average many particle diameters away from the particle before colliding with another molecule. Thus it is extremely unlikely that the molecule and particle will ever meet again or that the molecule will affect other molecules which may collide with the particle. Therefore, the effect of the collision of the molecule with the particle is immediately lost, and the particle itself exerts virtually no influence on the velocities of the surrounding gas molecules.

Simple theories describing the thermal force when $Kn \gg 1$ were first put forth by Einstein (1924) and Cawood (1936), and although they have since been shown to be inexact, they are accurate enough to illustrate the method of approach.

Suppose a small particle of diameter d is placed in the middle of a cylinder having the same diameter as the particle and a length equal to twice the mean free path of the gas, or 2λ. Gas molecules traversing this cylinder from either direction, then, on average, strike the particle without colliding with each other. Now consider only the motion of molecules parallel to the axis of the cylinder. The momentum imparted to the right-hand side of the particle per unit time is

$$\left(\frac{1}{6}n_1v_1\right)\left(\frac{\pi}{4}d^2\right)(mv_1) \qquad (11.2a)$$

and to the left-hand side is

$$\left(\frac{1}{6}n_2v_2\right)\left(\frac{\pi}{4}d^2\right)(mv_2) \qquad (11.2b)$$

where n_1 and n_2 represent the number of molecules per unit volume at the right- and left-hand faces of the cylinder, respectively, v_1 and v_2 the respective mean molecular velocities, and m the weight of the gas molecule. Then the net change of momentum per unit time is

$$-\left(\frac{1}{3}\right)\left(\frac{\pi}{4}d^2\right)\left(\frac{n_1mv_1^2}{2} - \frac{n_2mv_2^2}{2}\right) \qquad (11.3)$$

Assuming that there is little difference in the number of molecules per unit volume on either side of the particle, $n_1 \approx n_2 \approx n$. Since the change in momentum per unit time is the force on the particle,

$$F_T = -\left(\frac{n}{3}\right)\left(\frac{\pi}{4}d^2\right)\left(\frac{1}{2}mv_1^2 - \frac{1}{2}mv_2^2\right) \qquad (11.4)$$

Now if $\frac{1}{2}mv^2$ is replaced by $(3/2)kT$ and the equation is multiplied by $\lambda/\lambda = 1$, then

$$F_T = -\left(\frac{n}{3}\right)\left(\frac{\pi}{4}d^2\right)\left(\frac{3}{2}kT_1 - \frac{3}{2}kT_2\right)\frac{\lambda}{\lambda} = -\frac{n\pi d^2}{4}k\left(\frac{T_1 - T_2}{2\lambda}\right)\lambda \qquad (11.5)$$

Here T_1 and T_2 represent the temperatures at the two faces of the cylinder. Thus the temperature gradient across the cylinder is

$$\frac{\partial T}{\partial x} = \frac{T_1 - T_2}{2\lambda} = \nabla T$$

In addition, $nk = p/T$, the gas pressure divided by the average gas temperature. By making substitutions, the thermal force acting on the particle becomes, for large Kn,

$$F_T = -\frac{\pi}{4}d^2 p\lambda \frac{\nabla T}{T} \tag{11.6}$$

Since the product $(p\lambda)$ is a constant, this equation implies that the thermal force at large Knudsen numbers should be independent of the pressure of the medium.

Thermal force is often given in terms of the thermal conductivity of the gaseous medium. Assuming that the thermal conductivity of a gas κ_g can be given by the expression

$$\kappa_g = {}^{25}\!/_{64} \pi \bar{v} n C_v \lambda$$

and that the relationship $C_v = 5/2k$ holds for diatomic gases (such as O_2 and N_2), the expression for thermal force can be written in terms of heat conductivity as

$$F_T = -\frac{32}{125}d^2 \kappa_g \frac{\nabla T}{\bar{v}} \tag{11.7}$$

where $\bar{v}$ is given by Eq. 3.8.

Using the momentum transfer method, Waldmann (1959) and Derjaguin and Bakanov (1959) found the following expression for the thermal force on a particle when Kn $\gg$ 1:

$$F_T = -\frac{8}{15}d^2 \kappa_g \frac{\nabla T}{\bar{v}} = -\frac{1}{2}\pi\mu v \frac{d^2}{\lambda}\frac{\nabla T}{T} \tag{11.8}$$

This is an equation of the same form as Eq. 11.7 but indicates a force about 2 times larger than the magnitude determined by elementary theory.

Subsequently others have refined Eq. 11.8 to account for the type of molecular reflection from the particle surface (whether diffuse or specular, elastic or inelastic). Mason and Chapman (1962), by assuming all molecules to be reflected elastically, suggest increasing Eq. 11.8 by a factor of $1 + 4\pi/9$. More recently Talbot et al. (1980) derived an expression of the same form as Eq. 11.8 but increased by a factor of

$$1 + \frac{5\pi}{32}(1 - a_t)$$

where a_t is the thermal accommodation coefficient. For most purposes, however, it is sufficient to assume a value of 1 for a_t.

Although strictly valid only for the case Kn = ∞, according to Schmitt

(1959) and Jacobsen and Brock (1965), Eq. 11.8 is in error only by about 5 percent when Kn = 10 and by 10 percent when Kn = 5.

Example 11.2 Compute the ratio of the thermal force to the gravitational force for a 0.05-μm-diameter particle (ρ_p = 1 g/cm^3) in ambient air (T = 20°C) if the temperature gradient across the particle is 1000°C/cm.

$$F_g = \frac{\pi}{6}d^3 g = \frac{\pi}{6}(0.05 \times 10^{-4})^3(980) = 6.41 \times 10^{-14} \text{ dyn}$$

$$F_T = \frac{1}{2}\pi\mu\nu \frac{d^2}{\lambda}\frac{\nabla T}{T}$$

$$= \frac{\pi}{2}(1.82 \times 10^{-4})(0.151)\left[\frac{(0.05 \times 10^{-4})^2}{0.0687 \times 10^{-4}}\right]\left(\frac{1000}{293}\right)$$

$$= 5.36 \times 10^{-10} \text{ dyn}$$

$$\frac{F_T}{F_g} = 8.35 \times 10^3$$

To determine the thermophoretic velocity, Stokes law can be utilized by assuming that the Cunningham or slip correction factor (Eq. 5.3) is applicable for cases where Kn ≫ 1. Thermophoretic velocity will be independent of particle diameter since $C_c \approx \text{Kn}(A + Q)$ when Kn ≫ 1. Then, equating the thermal force (Eq. 11.8) with the resisting force (Stokes law) and solving for the thermophoretic velocity v_T give (Talbot et al., 1980)

$$v_T = \frac{1}{3}\nu(A + Q)\frac{\nabla T}{T} \tag{11.9}$$

where ν is the kinematic viscosity. By using a value of $(A + Q)$ = 1.65, Eq. 11.9 reduces to

$$v_T = 0.55\nu\frac{\nabla T}{T} \tag{11.10}$$

It is also possible to derive an equation for the thermophoretic velocity by considering that the suspended particles are a dilute suspension of giant molecules mixed with a much greater number of smaller molecules. This was done by Mason and Chapman (1962) who found essentially the same form for F_T as that given in Eq. 11.9.

Thermal Forces in the Continuum Regime (Kn ≤ 0.2)

When Kn ≤ 0.2, the particles are described as being in the continuum regime. A theoretical description of particle motion in this flow regime is complicated by several phenomena which affect molecule velocities and hence the particle motion. First, unlike the case of large Knudsen numbers, the presence of the particle does influence molecular velocities near the particle surface. Molecules rebounding from the particle surface are very likely to again strike that surface after one or several molecular collisions. Thus a velocity distribution from "free" space cannot be used for molecules but must be modified to account for the effect of the particle surface nearby.

The particle surface itself can also affect molecular velocities in two ways. First, molecular velocities of the rebounding molecules will depend on the type of rebound (whether specular or diffuse). Since the fraction of molecules rebounding either specularly or diffusively will depend on both particle and gas composition, these two factors should also be of importance in determining the thermal force.

Second, the particle surface may add to or subtract from the velocity of the diffusively reflected fraction of molecules by acting as a heat source or sink. Since the ability of a particle to act as a source or sink of heat depends on its thermal conductivity, the thermal conductivity of a particle should also influence thermal force.

Even with an adequate description of molecular velocities near the particle surface, it is not possible to completely establish all variables influencing thermal force. This is because there also exists a so-called thermal slip flow or creep flow at the particle surface. Reynolds (see Niven, 1965) and others have pointed out that as a consequence of kinetic theory, a gas must slide along the surface of a solid from the colder to the hotter portions. However, if there is a flow of gas at the surface of the particle up the temperature gradient, then the force causing this flow must be countered by an opposite force acting on the particle, so that the particle itself moves in an opposite direction down the temperature gradient. This is indeed the case, known as *thermal creep*. Since the velocity appears to go from zero to some finite value right at the particle surface, this phenomenon is often described as a *velocity jump*. A "temperature jump" also exists at the particle surface.

Epstein's Equation

The first theory of the thermal force acting on a particle which took thermal creep into account was developed by Epstein (1929), using the slip formula proposed by Maxwell (1880). Epstein's equation was of the form

$$F_T = -H_E \frac{9\pi\mu^2 d}{4\rho_m T} \nabla T \tag{11.11}$$

TABLE 11.1 Thermal Conductivity of Several Materials

Material	Thermal conductivity, cal/(cm · s · K)	Material	Thermal conductivity, cal/(cm · s · K)
Air at 20°C	0.000056	Iron	0.16
Aluminum oxide powder	0.08*	Magnesium oxide	0.0003
Asbestos	0.00019	Magnesium oxide powder	0.09*
Carbon	0.01	Mercury	0.02
Castor oil	0.00043	Paraffin oil	0.00030
Clay	0.0017	Platinum fume	0.167*
Fused silica	0.0024	Quartz	0.023
Glass	0.002	Silver fume	0.963*
Glycerol	0.00064	Sodium chloride	0.016
Granite	0.005	Stearic acid	0.0003
		Zinc	0.265*

SOURCE: Values marked with * from Keng and Orr (1966); other values from Hinds (1982).

where the parameter H_E has the value

$$H_E = \frac{1}{1 + C/2} \quad (11.12)$$

and $C = \kappa_p/\kappa_m$ is the ratio of the thermal conductivity of the particle to the thermal conductivity of the medium.

Table 11.1 lists thermal conductivity data for various materials.

Example 11.3 Using Epstein's equation, calculate the thermal force on a 1-μm-diameter glycerol particle in air at standard pressure and temperature when it is placed in a temperature gradient of 1000°C/cm.

From Table 11.1, $C = 0.00064/0.000056 = 11.43$. Then

$$F_T = H_E \frac{9\pi\mu^2 d}{4\rho_m T} \nabla T = \frac{1}{1 + 11.43/2} \frac{9\pi(1.82 \times 10^{-4})^2 (1 \times 10^{-4})}{4(0.0012)(293)} (1000)$$

$$= 9.87 \times 10^{-9} \text{ dyn}$$

It was initially thought that Epstein's theory satisfactorily described the thermal motion of large aerosol particles (Rosenblatt and LaMer, 1946). The theory, however, predicted essentially no thermal force acting on particles of high thermal conductivity (since $H_E \approx 0$). Experiments by Schadt and Cadle (1957, 1961) and others showed that thermal forces do indeed act on highly conductive as well as poorly conducting particles.

Brock's Equation

Brock (1962a) extended and improved Epstein's equation by taking into account convective flow near the particle and by using more complete boundary conditions than Epstein. He then derived an equation for thermal force which can be put in a form similar to Epstein's equation:

$$F_T = -H_B \frac{9\pi\mu^2 d}{4\rho_m T} \nabla T \qquad (11.13)$$

where H_B is defined as

$$H_B = \frac{4}{3} \frac{C_s}{1 + 3C_m \text{Kn}} \frac{1 + C_t \text{Kn} C}{1 + C_t \text{Kn} C + C/2} \qquad (11.14)$$

In his original derivation Brock used a value of ¾ for C_s, the thermal slip coefficient. More recent data by Ivchenko and Yalamov (1971) have shown that $C_s = 1.147$ for complete thermal accommodation. In Brock's equation

$$C_t = \frac{15}{8}\left(\frac{2 - a_t}{a_t}\right) \qquad (11.15)$$

and

$$C_m = \frac{2 - a_m}{a_m} \qquad (11.16)$$

As mentioned earlier, the factor a_t is the thermal accommodation coefficient and a_m the momentum accommodation or "reflection" coefficient. From the data of Rosenblatt and LaMer (1946), Schmitt (1959), and Keng and Orr (1966), as a first approximation a value of 1.25 seems reasonable for C_m, whereas for C_t a value of 2 is a good approximation (Brock, 1962b). These numbers then imply values of 0.89 for a_m and 0.97 for a_t.

On the other hand, Peterson et al. (1989) used values of $C_t = 2.20$ and $C_m = 1.146$ for the coefficients in Brock's equation. These values would imply $a_t = 0.92$ and $a_m = 0.93$.

For poorly conducting particles

$$\frac{H_E}{H_B} \approx \frac{4C_s}{3(1 + 3C_m \text{Kn})}$$

so that $H_B \approx H_E$ at Kn = 0.15. For the case of very small Knudsen numbers, Kn ≪ 1, $H_B \approx H_E$ when C_s is set equal to ¾.

Derjaguin and Yalamov's Equation

Derjaguin and his colleagues approached a theory of thermophoresis for large particles somewhat differently than Brock and Epstein.

First, they considered the aerosol particles to be large molecules immersed in a cloud of smaller molecules. They also pointed out that temperature stresses in a gas can give rise to very slight gas flows which could possibly influence experimental results.

In addition, although Brock considered a temperature jump at the particle-gas surface along with a velocity jump, Derjaguin and colleagues computed that the magnitude of this effect was so small that it could be neglected. Derjaguin and Yalamov (1965) then derived an expression for thermal velocity which can be given in the form

$$F_T = -H_D \frac{9\pi\mu^2 d}{4\rho_m T} \nabla T \tag{11.17}$$

which is the same form as previous equations except that

$$H_D = \frac{2}{3} \frac{4 + C/2 + C_t \text{Kn} C}{1 + C/2 + C_t \text{Kn} C} \tag{11.18}$$

Experimental data have been presented by Derjaguin et al. (1966) which tend to show reasonably good agreement between Eq. 11.17 and experimental data for particles having low heat conductivity (Vaseline oil) and particles having high heat conductivity (sodium chloride). But the variability in all experimental data presented to date indicates that at present there exists no completely adequate theoretical description of the physical factors which give rise to the thermal force acting on particles in the case where Kn < 1. This remains a task for future researchers in the field. [For a further discussion of this point, see Fuchs (1982).]

Thermophoretic Velocity

To determine thermophoretic velocity, the Stokes resisting force is equated with the thermal force. Then

$$v_T = -\frac{3}{4} H \frac{\mu}{\rho_m T} \nabla T\, C_c = -\frac{3}{4} H \nu \frac{\nabla T}{T} C_c \tag{11.19}$$

with the value for H depending on whether one uses the equation of Epstein, Brock, or Derjaguin and Yalamov. If the thermal velocity is by Epstein, then

$$v_T = -\frac{3}{4} \frac{1}{1 + C/2} \frac{\nu C_c}{T} \nabla T \tag{11.20}$$

or by Brock [as modified by Talbot et al. (1980)], then

$$v_T = -\frac{C_s}{1 + 3C_m\text{Kn}} \frac{1 + C_t\text{Kn}C}{1 + C_t\text{Kn}C + C/2} \frac{\nu C_c}{T} \nabla T \qquad (11.21)$$

or by Derjaguin and Yalamov, then

$$v_T = -\frac{1}{2} \frac{4 + C/2 + C_t\text{Kn}C}{1 + C/2 + C_t\text{Kn}C} \frac{\nu C_c}{T} \nabla T \qquad (11.22)$$

Example 11.4 Using Eqs. 11.20, 11.21, and 11.22, compute the thermophoretic velocity for a 1.5-μm-diameter NaCl particle. Use $C_t = 2.20$, $C_s = 1.147$, and $C_m = 1.146$. Assume $T_\infty = 30°C$ and $\nabla T = 1000°C/cm$.

$$\text{Kn} = \frac{2\lambda}{d} = \frac{2(0.0687)}{1.5} = 0.092$$

$$C = \frac{\kappa_p}{\kappa_g} = \frac{0.016}{0.000056} = 285.7$$

$$C_c = 1.12$$

Using Eq. 11.20,

$$v_T = \frac{3}{4} \frac{1}{1 + C/2} \frac{\nu C_c}{T} \nabla T$$

$$= \frac{3}{4} \frac{1}{1 + 285.7/2} \frac{(0.159)(1.12)}{303}(1000)$$

$$= (0.750)(6.95 \times 10^{-3})(0.586) = 0.003 \text{ cm/s}$$

Using Eq. 11.21,

$$v_T = \frac{C_s}{1 + 3C_m\text{Kn}} \frac{1 + C_t\text{Kn}C}{1 + C_t\text{Kn}C + C/2} \frac{\nu C_c}{T} \nabla T$$

$$= \left(\frac{1.147}{1 + 3(1.146)(0.095)}\right)$$

$$\left(\frac{1 + (2.20)(0.095)(285.7)}{1 + (2.20)(0.095)(285.7) + 285.7/2}\right) \frac{(0.159)(1.12)}{303}(1000)$$

$$= (0.865)(0.298)(0.586) = 0.151 \text{ cm/s}$$

Using Eq. 11.22,

$$v_T = -\frac{1}{2} \frac{4 + C/2 + C_t\text{Kn}C}{1 + C/2 + C_t\text{Kn}C} \frac{\nu C_c}{T} \nabla T$$

$$= -\frac{1}{2} \frac{4 + 285.7/2 + (2.20)(0.092)(285.7)}{1 + 285.7/2 + (2.20)(0.092)(285.7)} \frac{(0.159)(1.12)}{303}(1000)$$

$$= (0.500)(1.015)(0.586) = 0.297 \text{ cm/s}$$

Thermophoretic Velocity for All Particle Sizes

Talbot et al. (1980) have shown that thermophoretic velocity determined from Brock's equation (Eq. 11.14) degenerates into thermophoretic velocity determined from Waldmann's equation (Eq. 11.8) when the limit of $\lambda/d \to \infty$ (except for the multiplication factor $C_s/C_m \approx 1$). They point out that there appears to be no theoretical justification for this result except that it appears to fit the available experimental data quite well.

Figure 11.3 is a plot of "reduced" thermophoretic velocity as a function of Knudsen number showing some experimental data along with curves for Brock's and Derjaguin and Yalamov's equations. It can be seen that although these equations all predict the *form* of the data set, there appears to be still much room for improvement in both data analysis and theory.

A schematic plot of thermophoretic velocity (Eq. 11.21) as a function of particle diameter for air at normal temperature and pressure is shown in Fig. 11.4. It can be seen that the thermophoretic velocity decreases from a high value at small particle sizes to a somewhat lower constant value for large particle sizes. The range of the region of changing v_T is approximately $0.01 < d\ \mu\text{m} < 40$, and the thermal conductivity effect of a particle begins to become apparent above about $d \approx 0.2\ \mu\text{m}$.

The Dust-free Space

As mentioned earlier, thermal forces give rise to a dust-free space around bodies that are warmer than their immediate environment. Formulation of an equation which describes the width of this dust-free space appears to be quite difficult, generally involving numerical so-

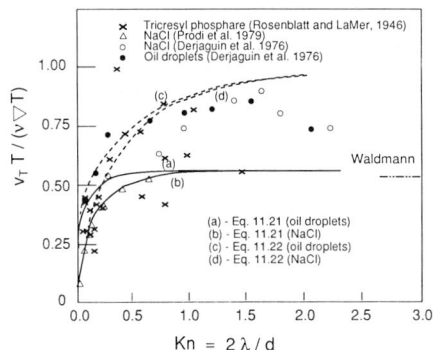

Figure 11.3 (*From Talbot et al., 1980.*)

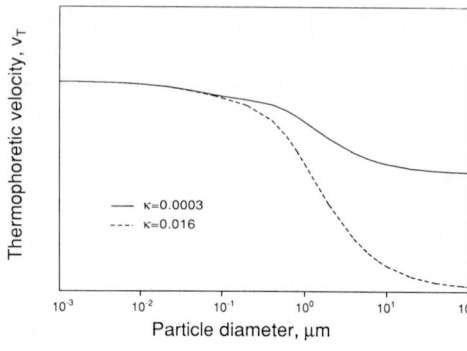

Figure 11.4 Thermophoretic velocity for particles of two different thermal conductivities.

lution of one or several second-order differential equations (e.g., see Goren, 1977).

Stratmann et al. (1988) give an "approximate" equation for the particular case of flow normal to the surface of a heated flat plate.

$$\sigma_{\text{df}} = 0.9610\left(\frac{\nu}{a}\right)^{1/2} H^{1/2} \left(\frac{T_w - T_\infty}{T_w}\right)^{1/2} \text{Pr}^{0.189} \qquad (11.23)$$

where Pr is the Prandtl number (Pr $\equiv c_p \nu \rho_g / \kappa_g$) and a is a constant. The term c_p represents the heat capacity of the gas. For typical temperatures in air the Prandtl number can be considered to be a constant with a value of 0.7 (see Perry and Chilton, 1973). It is interesting that in both this equation and Eq. 11.1 the thickness of the dust-free space appears to be a function of the square root of the temperature difference between the surface and the region away from the surface.

Example 11.5 Using H_B for H in Eq. 11.23, compare the thickness of the dust-free space predicted by Eqs. 11.1 and 11.23. Use $C_t = 2.20$, $C_s = 1.147$, and $C_m = 1.146$. Assume $T_\infty = 30°C$, $T_w - T_\infty = 100°C$, Kn = 0.095, $C = 285.7$, and $a = 10$.

Using Eq. 11.23,

$$H = H_B = \frac{4}{3} \frac{C_s}{1 + 3C_m \text{Kn}} \frac{1 + C_t \text{Kn}C}{1 + C_t \text{Kn}C + C/2}$$

$$= \frac{4}{3} \frac{1.147}{1 + 3(1.146)(0.095)} \frac{1 + (2.20)(0.095)(285.7)}{1 + (2.20)(0.095)(285.7) + (285.7)/2}$$

$$= (4/3)(0.865)(0.298) = 0.343$$

$$\sigma_{\text{df}} = 0.9610\left(\frac{\nu}{a}\right)^{1/2} H^{1/2} \left(\frac{T_w - T_\infty}{T_w}\right)^{1/2} \text{Pr}^{0.189}$$

$$= 0.9610\left(\frac{0.159}{10}\right)^{1/2} (0.343)^{1/2} (0.248)^{1/2} (0.7)^{0.189}$$

$$= 0.033 \text{ cm}$$

By using Eq. 11.1

$$\sigma_{df} = L \, \Delta T \, H^{-0.38} = (1.56 \times 10^{-4})(100)[1 \times 10^{-4}(100)^{1.25}]^{-0.38}$$
$$= (1.56 \times 10^{-4})(100)(3.72) = 5.80 \times 10^{-2} \text{ cm}$$

Problems

1 Friedlander (1977) gives for thermophoretic velocity at large Knudsen numbers the equation

$$v_T = \frac{3v \, \nabla T}{4(1 + \pi a_t/8)T}$$

and Derjaguin and Yalamov (1972) give for the same Kn range the equation

$$v_T = -0.37 \frac{\lambda}{T} \bar{v} \, \nabla T$$

Compare these two equations with Eq. 11.10 given above. By how much do they differ?

2 Using Brock's equation, determine the thermophoretic force on a 1-μm-diameter glycerol particle. For this calculation use $C_s = 1.147$, $C_t = 2.20$, and $C_m = 1.146$. How does this estimate of thermal force compare with the estimate made by using Epstein's equation (Eq. 11.11)?

3 Using Derjaguin and Yalamov's equation (Eq. 11.17), determine the thermophoretic force on a 1-μm-diameter sodium chloride particle. For this calculation use $C_s = 1.147$, $C_t = 2.20$, and $C_m = 1.146$. How does this estimate of thermal force compare with the estimate made by using Epstein's equation (Eq. 11.11) and Brock's equation (Eq. 11.14)?

4 Show that Eq. 11.10 becomes similar in form to Eq. 11.21 when $\lambda/d \to \infty$.

5 How much will the calculation of H_B be changed if the constants $C_s = 0.75$, $C_t = 2$, and $C_m = 1.25$ are used instead of $C_s = 1.147$, $C_t = 2.20$, and $C_m = 1.146$?

6 Infrared lights are used to raise the surface temperature 20°C over ambient temperature of semiconductor chips during manufacture. Estimate the width of the dust-free space above these chips.

Chapter

12

Aerosol-Charging Mechanisms

Introduction

Up to this point, aerosol particles have been considered to be uncharged; i.e., electric forces acting on or between particles were neglected. Most aerosols carry some electric charge which may be continually transferred between particles or gained or lost, depending on a number of external factors. The role of electricity in aerosol behavior is not completely understood, even though there is great interest in this particular phenomenon for such diverse reasons as the prevention of dust explosions or better prediction of particle behavior. It was measurement of charge on aerosol particles that gave the first accurate measure of the unit charge of an electron. Electric forces offer a highly efficient air-cleaning method, and the study of very small particles is most conveniently carried out by analyzing their mobility or movement in an electric field. The possibility of electrostatic propulsion for space vehicles has also generated interest in electrical phenomena of aerosols.

Several electrical properties may be of interest in aerosol studies. These could include the distribution of charges carried by aerosol particles and the velocity of a charged particle in an electric field. This latter property, e.g., is important in determining such things as deposition rates or charge transfer rates.

Definition of Force

Suppose a charged, dilute, monodisperse aerosol made up of spherical particles is placed in a uniform electric field, and the movements of the particles making up the aerosol are observed. Some particles will rise, others will fall, and still others will remain suspended. From this observation, the conclusion can be drawn that the electric field acts as a "field of force" that is superimposed on other forces already present, in this case gravity. However, as seen from the different motions of the particles, their trajectories (whether up or down) are determined by an additional factor as well, in this case the charge carried by each individual particle. Since the magnitude and direction of the electric force acting on each particle appear to depend on not only the direction and strength of the field but also the charged state of the particle (including the sign of the charge), a force vector $\vec{F}_E$ is defined which is equal to the product of the field strength vector $\vec{E}$ (independent of the charged state of the particle) and some scalar quantity called the *charge q,* on the particle that is,

$$\vec{F}_E = q\vec{E} \qquad (12.1)$$

If e is the elementary unit of charge [in cgs units = 4.8×10^{-10} electrostatic units (esu)], then

$$q = ne \qquad (12.2)$$

where n is the number of elementary units of charge on the particle. The algebraic sign of the charge is conventionally determined in such a way that the particle is repelled by a charge of a similar sign.

Example 12.1 A 10-μm-diameter unit-density sphere carries a negative charge equal to 100 electrons. If it is placed in an electric field having a strength of 10 statvolts/cm, determine the force in dynes acting on the particle.

$$\vec{F} = q\vec{E} = ne\vec{E}$$
$$= (100)(4.8 \times 10^{-10})(10)$$
$$= 4.8 \times 10^{-7} \text{ dyn}$$

This is a very small force, less than one hundred-millionth of the force required to lift a fly.

The cgs electrical units are such that when charge is given in electrostatic units and field strength is in statvolts per centimeter, the resulting force is in dynes. The direction of the force is the same as the field except that negatively charged particles will be attracted toward the positive end of the field, and vice versa.

It is customary that electrical parameters be given in terms of "practical" units (volts, amperes, coulombs, etc.), so conversion factors are required. Practical units are used so that the numbers usually encountered will have values which are not extremely large or small. See App. D for a more complete discussion of electrical units.

Particle Mobility

The motion of a particle in an electric field depends on two electrical factors: field strength and particle charge. The motion of particles having varying charges and sizes can be compared by considering what their velocities would be in an electric field of unit strength. This velocity, called the *particle mobility* Z_p, is defined by setting qE equal to $3\pi\mu vd$ and solving for v. Then when E equals unity, v becomes the particle mobility Z_p, or

$$Z_p = \frac{qC_c}{3\pi\mu d} \qquad (12.3)$$

Example 12.2 Determine the mobility of a 10-μm-diameter unit-density sphere when it carries 100 unit charges. Remember, $E = 1$ statvolt/cm is included in the definition of Z_p.

$$Z_p = \frac{qC_c}{3\pi\mu d} = \frac{neC_c}{3\pi\mu d} = qB$$

$$= \frac{(100)(4.8 \times 10^{-10})(1)}{(3)(3.14)(1.83 \times 10^{-4})(10^{-3})}$$

$$= 0.028 \text{ cm/s}$$

This represents the velocity the particle would attain when placed in an electric field having a strength of 1 statvolt/cm.

If the particle mobility is known, it is easy to determine the electric force acting on the particle, provided the field strength is also known. However, the field strength may not be constant but may have some spatial or temporal distribution, that is, $\vec{E} = f(x,y,z,t)$. In addition, q may vary from particle to particle and may vary on a single particle with time in a discontinuous, stochastic manner. Thus, except for quite simple cases, it is exceedingly difficult to predict particle motion in an electric field with accuracy.

Some appreciation of the electrical behavior of aerosols can be gained, however, by considering separately the two factors in Eq. 12.1, q and $\vec{E}$.

Particle Charge q

Particles can be electrified by a number of different sources acting singly or in combination. The basic processes which give rise to a charge on a particle are direct ionization, static electrification, collisions with ions or ion clusters (either with or without an external electric field present), or ionization of the particle by electromagnetic radiation such as ultraviolet light, visible light, or gamma radiation. These processes can be considered separately.

Direct ionization of the particle

Little is known of this electrification mechanism. For one thing, aerosol densities are generally so small that even though one would expect more ionization taking place in a particle than in an equal volume of air, there are generally at least several orders of magnitude more air mass than particle mass per unit volume of space. Since ionization is primarily a mass-dependent phenomenon, there will be at least several orders of magnitude more ionization taking place in the air than in the suspended particles. Thus particle charging should result more from attachment of air ions than by direct ionization. Direct ionization of the particle is not an important particle-charging mechanism.

Static electrification

A second particle-charging mechanism is static electrification. This mechanism arises from one or a combination of several other mechanisms, making theoretical interpretation in terms of a single mechanism very difficult, if not impossible (and most experimenters have attempted to interpret their results in terms of a single mechanism). Five basic mechanisms can result in static electrification. These are examined for their importance in aerosol physics.

Electrolyte effects. In this case, solutions of liquids of high dielectric constant exchange ions with metals or solid surfaces. For example, a drop of a high-dielectric liquid swept from a metal surface will develop and can carry away a high charge. For a given surface and liquid, droplets will all have a net charge of the same sign, so that the droplets will repel each other. This is probably an important mechanism in aerosol charging, although its importance is not well established. Table 12.1 lists dielectric constants for various materials.

Contact electrification. A second static electrification process is contact electrification. Here electrons migrate from clean, dry surfaces of dis-

TABLE 12.1 Dielectric Constants of Liquids at Normal Temperature 20°C, esu

Oil	2–2.2
Turpentine	2.2–2.3
Methyl alcohol	31
Ethyl alcohol	24.3
Sodium chloride	5.9
Water	78
Magnesium oxide	9.65
Glass	5–10
Polyethylene	2.25
Air	1
CCl_4	2.2
PVC	3.3–4.5

similar metals to metals with lower work functions. This process requires that there be no impurities between surfaces and is strictly electronic in nature. Because of this requirement, contact electrification is probably not an important charging mechanism for aerosols.

Spray electrification. A third static electrification process is spray electrification. Surface forces in liquids of high dielectric constants increase the concentration of electrons or negative ions in the outer liquid surface (Lenard, 1915). The disruption of these surfaces by atomization or bubbling imparts a predominantly negative charge to the smaller droplets, while the larger ones will be neutral, positive, or negative in approximately equal proportions. The size of all droplets produced may be altered by subsequent evaporation or condensation. Dissolved salts generally reduce the magnitude of the charge compared to charges produced in pure liquids, and the effect is usually reduced as the dielectric constant of the liquid is reduced, until a point is reached as in the case of pure hydrocarbons where little charging is observed. The charged droplets produced by spray electrification generally have only several units of charge per drop. Spray electrification is important in aerosol charging and very often operates in conjunction with electrolytic effects. This tends to confuse and complicate any attempt at analysis.

Tribo electrification. The fourth static electrification method is frictional electrification or triboelectrification. In this mechanism charge is imparted to dry nonmetallic particles when they come in contact with metals or with other particles. Although triboelectrification is a very common charging mechanism, reasons for its occurrence remain fairly obscure. Some points are well known. For example, it is possible to estimate the sign of each charge when two different materials come into contact. This is shown in Table 12.2. Materials high on this table will be most likely to develop a positive charge on contact, while those on the lower end are most likely to develop a negative charge. These charges can be produced by particle-particle interaction or by particle-surface interaction, although particle-particle interaction seems to produce more highly charged particles (Miller and Heinemann, 1948).

Example 12.3 Quartz particles flow through a glass tube. Estimate the sign of the charge produced by static electricity on the particles and the tube. From Table 12.2,

Charge on particles	−
Charge on tube	+

If the tube were made of copper instead of glass, the signs would be

Charge on particles   +

Charge on tube   −

Many aerosol experiments have suffered because this relationship has not been clearly understood.

TABLE 12.2 Charge Preference in Frictional Charging

+ End
Asbestos
Mica
Glass
Calcite
Quartz
Magnesium
Lead
Gypsum
Zinc
Pyrite
Copper
Silver
Silicon
Sulfur
Rubber
− End

In the case of high concentrations of explosive dusts flowing through an ungrounded duct, sufficient charge may accumulate on the duct to produce a spark discharge and resulting explosion. This electrification is inhibited when relative humidities exceed 50 or 60 percent, thought to be due to the formation of a thin moisture layer on the particles. If the moisture contains sufficient dissolved material to make this layer conductive, the charge will not accumulate. This explanation is consistent with the observation that relative humidity, not absolute humidity, is important in dust explosions, since deposition of water on the particles, not the presence of water vapor, prevents charging by triboelectrification.

Flame ionization. A final static electrification method is the ionization of particles in a flame. This effect was first observed as early as 1600, and it has recently become the subject of much interest because of potential application in such diverse areas as direct generation of electricity, control of combustion processes by applied electric fields, and the like (Lawton and Weinberg, 1969). In the reaction zones of hydrocarbon/air or hydrocarbon/oxygen flames, ion concentrations of 10^9 to 10^{12} ions per cubic centimeter have been measured. Positive ions are definitely present, but there is some controversy as to whether negative ions or free electrons predominate. The presence of particulate material in the flame (e.g., soot particles) greatly enhances the concentration of free charge (Einbinder, 1957). Also, it appears that the smaller the particle size, the more free charge that is developed. For example, carbon particles of about 0.02-μm diameter produced in an oxyacetylene flame carried, on average, about 10 unit charges per particle, representing an overall charge of 1×10^{18} charges per gram.

Collisions with ions or ion clusters

The best understood of the three main charging mechanisms for aerosols is that involving the collision of ions or ion clusters with aerosol particles. Air ions or ion clusters arise from a number of processes. They can be formed by attachment of either positive or negative charges produced by alpha, beta, or gamma photons as they lose energy following emission from a radioactive source (Cooper and Reist, 1973) or from various types of electric discharges.

Two distinct processes are involved in charging that can act either singly or in combination. In the first process, *diffusion charging,* particles are charged in the absence of an external electric field by collisions with diffusing ions. With the second method, *field charging,* particles are charged by ions moving in an orderly direction in an external electric field. The two processes can be considered analogous to molecular diffusion and convective diffusion. Charging rates are faster for field charging than for diffusion charging. For very small particles, diffusion charging is important even in the presence of an external field.

To study charging mechanisms theoretically for either diffusion charging or field charging, it is necessary to make several assumptions regarding the aerosol. First, the particles are assumed to be spherical. This assumption is reasonable for isometric particles. Second, it is also assumed that the particles are monodisperse. The effect of polydispersity complicates but does not invalidate theory. Third, there are no interactions between individual particles. Finally, the ion concentration and electric field near each particle are assumed to be uniform. These last two assumptions are essentially true for all natural and industrial aerosols. Thus except in the most extreme cases, theory should be adequate without other modification.

Diffusion charging—unipolar ions

In diffusion charging, particles are charged by unipolar ions (ions having the same sign) in the absence of an applied electric field. Collisions of ions and particles occur as a result of random thermal motion of the ions, the brownian motion of the particles being generally neglected.

A simple theory for diffusion charging was first proposed by White (1963). He considered that ions diffuse in a gas in accordance with the postulates of kinetic theory except that when an ion strikes a particle, it stays, thus accumulating charge. However, this accumulation of charge on the particle produces an electric field which tends to prevent additional ions from reaching the particle. Thus in White's theory the rate of accumulation of charge on a particle decreases as the charge on the particle increases.

The number of ions striking and attaching to a spherical particle of diameter d per unit time is

$$\frac{dn}{dt} = \frac{\pi}{4} d^2 N v_{avg} \tag{12.4}$$

where N is the number of ions near the particle and v_{rms} is the root-mean-square velocity of the ions. From kinetic theory, the density of ions in a potential field varies according to

$$N = \overline{N} \exp \frac{V}{kT} \tag{12.5}$$

in which $\overline{N}$ is the average ion concentration and V is the potential energy per ion. For a particle accumulating charge, the potential energy of an ion of the same charge a distance R from the center of the particle with n charges is $V = -ne^2/R$.

Very close to the particle surface, the ion concentration is given by

$$N = \overline{N} \exp \left(\frac{-2ne^2}{dkT} \right) \tag{12.6}$$

From Eq. 12.4, the rate of change of ions per unit time dn/dt becomes

$$\frac{dn}{dt} = \frac{\pi}{4} d^2 v_{avg} \overline{N} \exp \left(\frac{-2ne^2}{dkT} \right) \tag{12.7}$$

For an initially uncharged particle, integration of Eq. 12.7 gives

$$n = \frac{dkT}{2e^2} \ln \left(1 + \frac{\pi d v_{avg} \overline{N} e^2 t}{2kT} \right) \tag{12.8}$$

A characteristic charging time t' can be defined as

$$t' = \frac{2kT}{\pi d u_{avg} \overline{N} e^2} \tag{12.9}$$

such that Eq. 12.8 can be written as

$$n = \frac{dkT}{2e^2} \ln\left(1 + \frac{t}{t'}\right) \tag{12.10}$$

Furthermore, a characteristic charge n' can be defined as

$$n' = \frac{dkT}{2e^2} \tag{12.11}$$

so that Eq. 12.8 can be expressed in the dimensionless form

$$\frac{n}{n'} = \ln\left(1 + \frac{t}{t'}\right) \tag{12.12}$$

Figure 12.1 is a plot of Eq. 12.12 showing charge accumulation by diffusion charging as a function of time. It can be seen from Fig. 12.1 that there is a fairly rapid increase in particle charge initially, followed by a much slower increase later on. No ultimate charge is inherent in the diffusion charging process, however, since the particle is able to charge indefinitely. In actuality the charge on the particle is limited by emission of charge from the particle. It is clear, though, that the numerical value of the charge is relatively insensitive to ion concentration and time, whereas it is quite dependent on particle size.

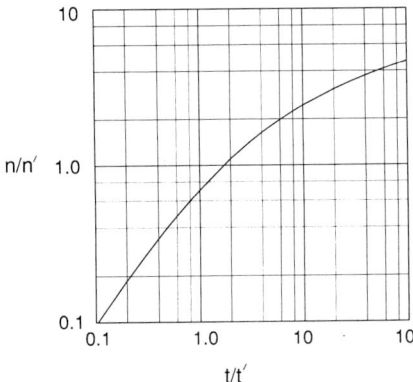

Figure 12.1 Plot of dimensionless diffusion charging.

Example 12.4 Estimate the charge which would develop in 10 s by diffusion charging if a 0.5-μm-diameter spherical particle were placed in an ion field containing 5×10^8 ions per cubic centimeter. Assume 20°C temperature and use $v_{rms} = 2 \times 10^4$ cm/s.

$$n = \frac{dkT}{2e^2} \ln\left(1 + \frac{\pi d v_{avg} \overline{N} e^2 t}{2kT}\right)$$

$$= \frac{(5 \times 10^{-5})(1.38 \times 10^{-16})(293)}{(2)(4.8 \times 10^{-10})^2} \times$$

$$\ln\left[1 + \frac{(\pi)(5 \times 10^{-5})(2 \times 10^4)(5 \times 10^8)(4.8 \times 10^{-10})^2(10)}{(2)(1.38 \times 10^{-16})(293)}\right]$$

$$= 4.39 \ln 44{,}754 = (4.39)(10.71)$$

$$n = 46.98 = 47 \text{ charges}$$

There has been criticism of White's derivation because there are two charging mechanisms at work during diffusion charging and White ignores one of them. There is diffusion of ions directly onto the particle (sometimes called the *Coulomb effect*), and this is the only effect considered by White. However, ions can also be attracted to the particle by an image force (the *image effect*). Thus an ion passing near a particle which otherwise would not hit the particle may be attracted to it by this image force.

The theory for diffusion charging given in Eq. 12.12 represents a simple approximation for diffusion charging in which the image force is neglected and only the Coulomb force is considered. According to Fuchs (1971), this approximation is valid when $2e^2/(dkT) \ll 1$, that is, when $d \geq 0.2$ μm. More accurate models for estimating diffusion charging which include the image force have been given by Fuchs (1963), Gentry (1972), and Hoppel (1977) among others, but these models are quite cumbersome and difficult to use. Figure 12.2 shows a comparison of the various predictions for a 0.018-μm-diameter sphere. When $2e^2/(dkT) \ll 1$, the models should all give the same result.

Figure 12.3 shows a plot of the Fuchs (1963) theory for larger particle sizes along with the experimental points of Liu and Pui (1977). Also shown on this figure is a plot of White's equation, Eq. 12.12, with $v_{avg} = 2 \times 10^4$ cm/s.

In the examples given a value of $v_{avg} = 2 \times 10^4$ cm/s was used for the mean thermal speed of the ions, as opposed to a value of 5×10^4 cm/s for air molecules as used by White. This is because the ions charging the aerosol particle are considered to be associated with molecular clusters, rather than with single molecules.

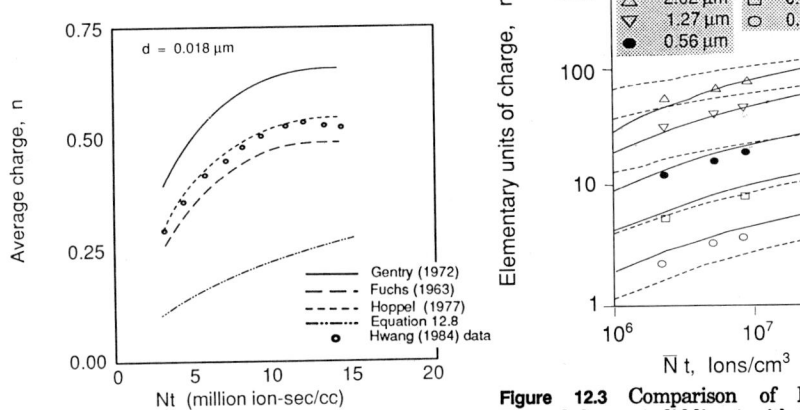

Figure 12.2 Unipolar diffusion charging. (*After Davison and Gentry, 1985.*)

Figure 12.3 Comparison of Fuchs-Bricard theory (solid lines) with that of White (dashed lines). Experimental points of Liu and Pui are also shown.

Example 12.5 Estimate the ionic mean thermal speed which corresponds to ions of the hydrated proton $H^+(H_2O)_6$. Use a temperature of 20° C.

From Eq. 3.8

$$\bar{v}_6 = \sqrt{\frac{8kT}{\pi m}}$$

The term m is the ion mass = $6 \times 18 + 1 = 109$ amu.

$$m = \frac{109}{6.02 \times 10^{23}} \, g$$

$$= 1.81 \times 10^{-22} \, g$$

$$\bar{v}_6 = 23{,}846 \text{ cm/s}$$

What would this speed be if the hydrated proton were of the form $H^+(H_2O)_{24}$? Ion mass = $24 \times 18 + 1 = 433$ amu.

$$\bar{v}_{24} = \bar{v}_6 \sqrt{\frac{109}{433}} = 1.20 \times 10^4 \text{ cm/s}$$

A more serious fault in White's derivation is the lack of appreciation of the stochastic nature of the charge acquisition process. For example, Eq. 12.12 indicates that for small particles and short charging times, fractions of charges are possible. This is clearly an impossibility. Thus results computed from these equations should be considered to represent average rather than specific values (Boisdron and Brock, 1969; Natanson, 1960).

Field charging

Unlike diffusion charging, field charging takes place in an ordered field of unipolar ions, i.e., in a region where the ions are in an electric field and hence have ordered motion (Rohmann, 1923; Pauthenier and Moreau-Hanot, 1932). Suppose an uncharged spherical conducting aerosol particle were suddenly placed in a uniform electric field. The field near the particle would be distorted, as illustrated in Fig. 12.4, so that gas ions, following the field lines, would immediately begin to charge the particle. The dashed lines in the illustration indicate the limits of the field which passes through the sphere. All ions traveling within these limits are considered to strike the particle and charge it.

However, as the particle becomes charged, it will start to repel some of the incoming ions. This repulsion results in an alteration of the field configuration which accordingly reduces the charging rate. A point will eventually be reached where no further charging of the particle takes place. This point is known as the *saturation charge* of the particle. When one-half the saturation charge on the particle is reached, the electric field surrounding the particle is similar to that shown in Fig. 12.5. Notice that both the ions available to make contact with the sphere and the particle area available for contact have been reduced.

The ion current to the particle at any time is a function of the ions which are available to reach the particle and the particle area available to accept the ions. Symbolically this is written as

$$i = \frac{dq}{dt} = \frac{d(en)}{dt} = jA(n) \tag{12.13}$$

where j is the ion current density in the undistorted field just away from the particle and $A(n)$ is the cross-sectional area of the undisturbed ion stream entering the particle when it is charged with n ions.

The value $A(n)$ is computed from the total electric flux which enters the particle when n ions are present, i.e.,

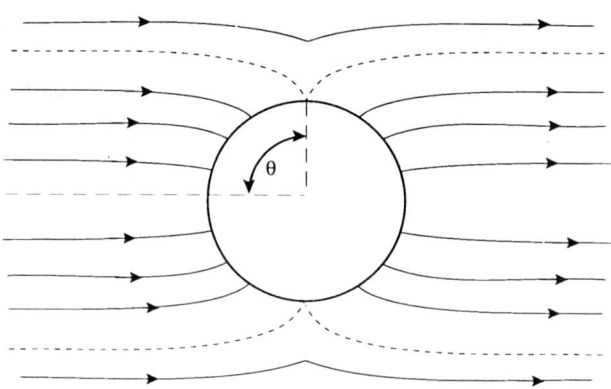

Figure 12.4 Electric field around an aerosol particle, particle uncharged.

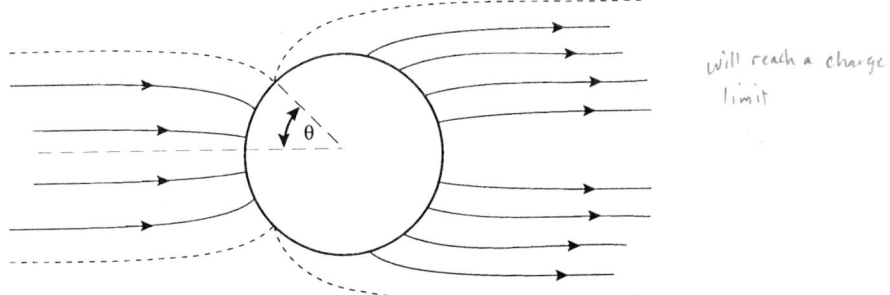

Figure 12.5 Electric field around a partially charged aerosol particle.

will reach a charge limit

$$A(n) = \frac{\psi(n)}{E_0} \tag{12.14}$$

Here $\psi(n)$ is the electric flux entering the particle and E_0 the undistorted electric field strength in the vicinity of the particle.

The electric flux entering the particle is equal to the product of the field at the surface of the particle and the area perpendicular to it, or

$$\psi(n) = \oint E_0 \, dA \tag{12.15}$$

The electric field E_1 at any point on the surface of a sphere that is placed in an initially uniform electric field can be shown to be

$$E_1 = \chi E_0 \cos \theta \tag{12.16a}$$

where $\chi = 3\epsilon_0/(\epsilon_0 + 2)$ and ϵ_0 is the dielectric constant of the sphere. At the same time, however, charges which have collected on the sphere produce a repelling field which acts to prevent the arrival of additional ions. This repelling field E_2 can be given by

$$E_2 = -\frac{4ne}{d^2} \tag{12.16b}$$

The net electric field is

$$E = E_1 + E_2 = \chi E_0 \cos \theta - \frac{4ne}{d^2} \tag{12.16c}$$

When $\theta = \theta_0$, $E = 0$.

Thus the total electric flux entering the particle is

$$\psi(n) = 2 \int_0^{\theta_0} \left(\chi E_0 \cos \theta - \frac{4ne}{d^2} \right) \left(\frac{\pi}{2} d^2 \sin \theta \right) d\theta \tag{12.17}$$

which on integration becomes

$$\psi(n) = \chi \frac{\pi d^2}{4} E_0 \left(1 - \frac{4ne}{\chi E_0 d^2}\right)^2 \quad (12.18)$$

The assumption made here is that the particle is much larger than the ion mean free paths, so that the ions can be considered to follow the lines of force. For large particles (in the continuum region) this assumption is valid. Also, since ion mobility is much greater than particle mobility, particle velocity can be ignored at this point.

The limiting or saturation charge occurs when $\psi(n) = 0$. Setting $\psi(n) = 0$, replacing n with n_s, and solving Eq. 12.18 for n_s gives

$$n_s = \frac{\chi E_0 d^2}{4e} \quad (12.19)$$

This is the maximum number of charges which can be placed on a particle of diameter d by a field of strength E_0.

Example 12.6 Earth's electric field is 1.28 V/cm over the ocean. What is the maximum electric charge which can exist on a 10-μm spherical particle over the ocean due to the earth's electric field? Assume $\chi = 3$.

$$E_0 = 1.28 \text{ V/cm} = \frac{1.28}{300} = 4.27 \times 10^{-3} \text{ statvolts/cm}$$

$$n_s = \frac{3 E_0 d^2}{4e} = \frac{(3)(4.27 \times 10^{-3})(10^{-3})^2}{(4)(4.8 \times 10^{-10})}$$

$$= 6.67 \text{ ions per particle, say 7 ions}$$

Equation 12.18 can be rewritten in terms of the saturation charge n_s. Thus

$$\psi(n) = \pi n_s e \left(1 - \frac{n}{n_s}\right)^2 \quad (12.20)$$

The saturation charge represents the maximum charge a particle can attain with a given field strength. If the field strength is made sufficiently intense, a particle will rid itself of excess charge by the spontaneous emission of either electrons or ions. For electrons a surface field intensity of about 10^7 V/cm is required while for ion emission a field about 20 times greater is needed (Whitby and Liu, 1966). The number of charges which are implied by these fields thus represents the absolute upper limit on particle charging.

Recalling Eq. 12.13, it is seen that the second factor to be evaluated is j, the ion current density in the undistorted field. This is the product of the charge per unit volume and the drift velocity of the ions. The charge per unit volume is $\overline{N}e$, where $\overline{N}$ is the average ion concentration. When the field energy of the ions is small compared with their thermal energy, the drift velocity of the ions in the field direction is proportional to the electric field intensity, i.e.,

$$v_i = ZE_0 \qquad (12.21)$$

where Z, the constant of proportionality, is called the *mobility* of the ions. Because of their difference in size, positive and negative ions have different mobilities. For air a typical value for Z that is often used to represent an average value is 1.4 cm^2 s^{-1} V^{-1} (McDaniel, 1964). In the cgs system of units this is 420 cm^2 s^{-1} statvolt^{-1}. This mobility may not be a representative value, however. Figure 12.6 shows a typical distribution of mobilities in air. It can be seen that although a mobility of 1.4 cm^2 s^{-1} V^{-1} appears to be an average value, it is representative of neither positive nor negative ions. Thus the choice of an average value for mobility can be expected to introduce some error into field charging estimations.

Assuming that an average value for mobility can be used, the current density becomes

$$j = \overline{N}ev_i = \overline{N}eZE_0 \qquad (12.22)$$

Combining the current density with the total electric flux gives

$$\frac{d(ne)}{dt} = \overline{N}eZE_0 \pi n_s e \left(1 - \frac{n}{n_s}\right)^2 / E_0 \qquad (12.23)$$

or

$$\frac{d(n/n_s)}{dt} = \pi \overline{N}eZ \left(1 - \frac{n}{n_s}\right)^2 \qquad (12.24)$$

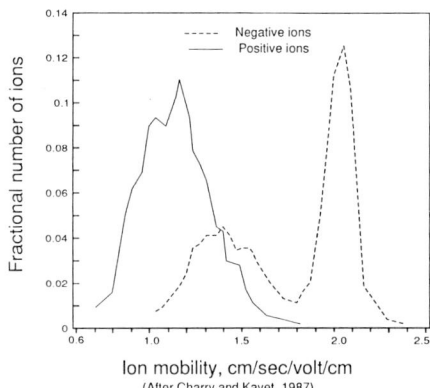

Figure 12.6 Mobility spectra for positive and negative ions in air (T = 20°C, absolute humidity = 4.5 g/cm).

Ion mobility, cm/sec/volt/cm
(After Charry and Kavet, 1987)

which, on integration with the initial condition that $n = 0$ at $t = 0$, gives

$$\frac{n}{n_s} = \frac{\pi \overline{N}eZt}{\pi \overline{N}eZt + 1} \tag{12.25}$$

The factor $\pi \overline{N}eZ$ has the dimensions of the reciprocal of time, so a new time factor t_0 can be denoted as

$$t_0 = \frac{1}{\pi \overline{N}eZ} \tag{12.26}$$

and then

$$\frac{n}{n_s} = \frac{t}{t + t_0} \tag{12.27}$$

The factor t_0 can be considered to be a time constant which determines the rate or rapidity of charging; the smaller the value of t_0, the shorter the time it takes to approach saturation charge. Figure 12.7 shows a plot of Eq. 12.27 indicating that with sufficient time n/n_s reaches the asymptotic value of 1.

One-half the final charge is reached at $t = t_0$ and 91 percent at $t = 10t_0$. Even though larger particles carry much higher saturation charges, the time constant is not size-dependent, and relative charg-

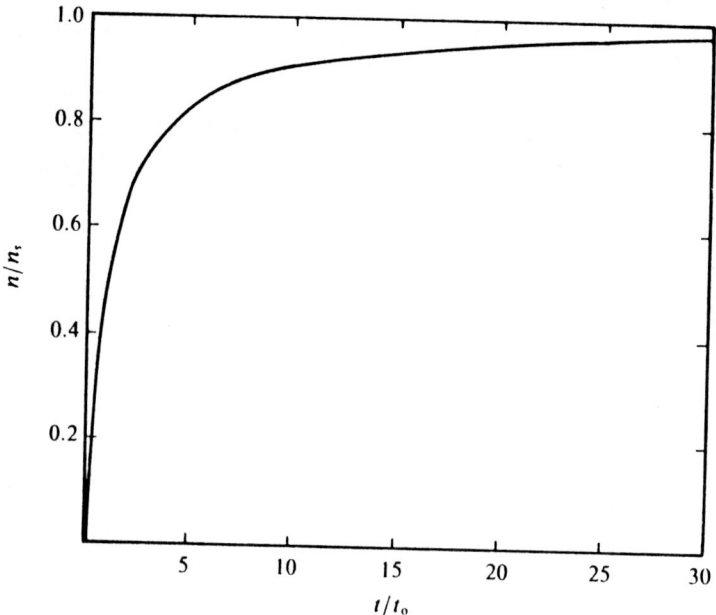

Figure 12.7 Plot of fractional saturation charge as a function of dimensionless time. Note that as t approaches infinity, n/n_s approaches 1.

ing rates of particles of different sizes are the same. Thus in an electrostatic precipitator, particles of various sizes placed in the same electric field will charge to the same degree of charge saturation in the same time.

Example 12.7 The particle residence time in the charging section of an electrostatic precipitator is 0.4 s. If the ion concentration is 10^7 ions per cubic centimeter, what fraction of the maximum charge on the particles will be reached in that time?

$$t_0 = \frac{1}{\pi N e Z} = \frac{1}{(\pi)(10^7)(4.8 \times 10^{-10})(420)}$$

$$= 0.158 \text{ s}$$

$$\frac{n}{n_s} = \frac{t}{t + t_s} = \frac{0.4}{0.4 + 0.158} = \frac{0.4}{0.558} = 0.717$$

i.e., approximately 70 percent of the ultimate particle charge is achieved in 0.4 s. Large particles will carry a much greater charge than their smaller counterparts.

Combined diffusion and field charging

As particle size decreases, charging in an applied electric field results from not only the ordered flow of ions but also the random motion of the ions. Thus a complete charging theory should account for both diffusion and field charging simultaneously. Several difficulties immediately appear. First, diffusion charging places no upper limit on the number of charges a particle may acquire, whereas there is a definite upper limit with field charging. Second, in field charging the particle charge after a given charging period is a function of the square of particle size, while with diffusion charging the charge is approximately a linear function of particle size. With a fairly high applied electric field and small particles, the two mechanisms do give comparable results, although when compared with experimental data, both mechanisms taken separately tend to slightly underestimate particle charge.

Ion production by corona discharge

The most commonly used method for field charging of aerosol particles is by the use of the phenomenon known as *corona discharge*. Corona discharge is discussed in detail by White (1963) and Miller and Loeb (1951b, c) and is considered only briefly here.

Suppose two electrodes are arranged so that the field strength between them is not constant. (This could be done, e.g., with a wire and tube electrode system or a point and plane system.) Then if the potential across the two electrodes is increased, a voltage will be reached where electrical breakdown of the gas occurs nearest either the wire or the point. This breakdown is usually manifested by a blue glow, called a corona discharge. With a corona discharge two distinct electrical zones are produced (Fig. 12.8). In the first zone, immediately around the corona wire and containing the corona glow, local electrical breakdown of the gas takes place, caused by collisions with gas molecules of ions leaving the corona wire. If these ions are sufficiently accelerated, the collisions will free additional ions from the molecules. These new ions are also accelerated and, in turn, produce even more ions by collision. Oppositely charged ions are accelerated toward the corona wire, where they produce additional ions on impact. This process produces a large number of ions of one sign which rapidly move out of the zone of corona glow toward the other electrode (Fig. 12.9).

As the ions leave the zone of high field strength, they tend to attach themselves to gas molecules, producing a cloud of slow-moving ions all having the same sign of charge as the center electrode, either positive or negative. The corona is said to be negative if a cloud of negative ions is formed and positive if positive ions are formed. The ions moving toward the passive electrode thus make up the unipolar charging

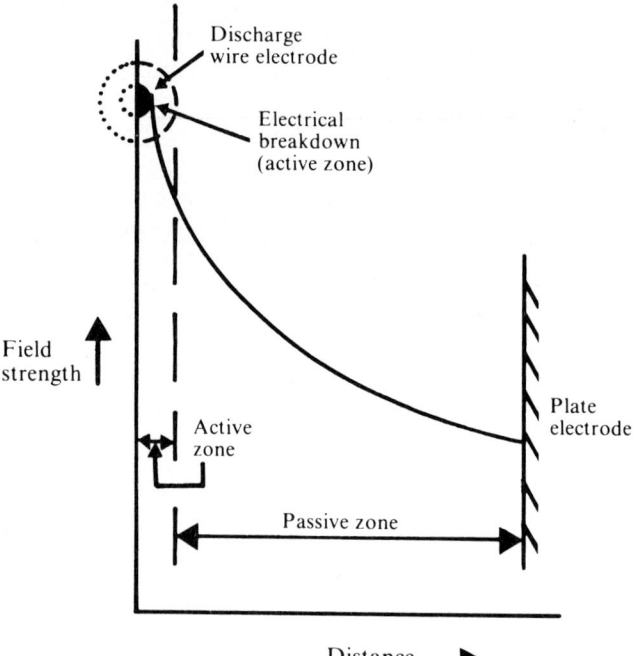

Figure 12.8 Plot of field strength as a function of distance from the discharge wire electrode for a corona discharge.

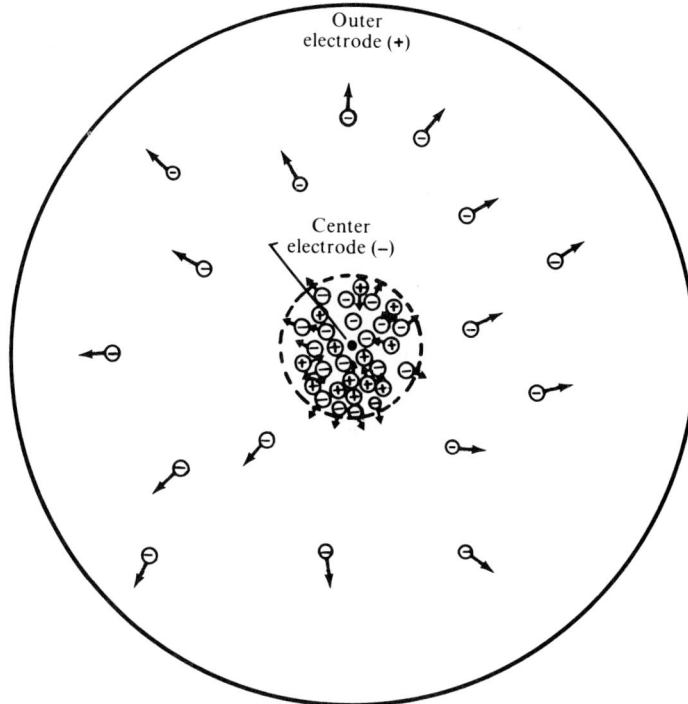

Figure 12.9 Schematic diagram of negative corona discharge showing negative ion motion away from center electrode, positive ion motion toward center electrode.

field for aerosol particles. Ion concentrations are typically on the order of 10^7 to 10^9 ions per cubic centimeter. Since electron attachment coefficients and ion mobilities vary greatly from gas to gas, corona characteristics will differ greatly, depending on the predominant gas and impurities present. For example, with nitrogen alone, negative ion formation is not possible, so that the oxygen component of air is necessary for effective negative particle charging.

With a wire-in-cylinder arrangement, a negative corona produces tufts or beads of glow along the wire length while a positive corona produces a continuous glow along the wire. Generally a negative corona is preferred for particle charging because it is more stable than the positive corona and can be operated at higher potentials and current flow before sparking occurs, both of which are favorable to electrostatic precipitation. On the other hand, with high electrical potentials, ozone is produced. It appears that a positive corona produces less ozone than a negative one. Thus for cleaning air that will subsequently be supplied to a room or building, a positive corona for particle charging is preferred, since less ozone will be produced, even though the air-cleaning efficiency will be somewhat lower.

Charge density in the zone of low field strength depends on ion mobility, which depends on the constituents of the gas being ionized. Nitrogen, hydrogen, and the inert gases absorb few electrons on collision ionization, so charges present in these gases are electrons, having high mobilities and hence a high corona current. Gases such as oxygen, water vapor, sulfur dioxide, and carbon dioxide have a high electron affinity so that negative ions consist almost entirely of gas ions. These gases are called electronegative gases. The corona current for these gases is relatively low.

Maximum attainable particle charge

In deriving the field charging equation, it was shown that for a given field strength and particle size there exists a maximum possible particle charge. It was pointed out that when the field strength reaches the surface field strength for spontaneous emissions of electrons, then the upper limit of particle charging is established. For a solid spherical particle, this limit n_m is given by

$$n_m = \frac{E_s d^2}{4e} \tag{12.28}$$

where E_s is the surface field intensity at which emission of ions or electrons occurs. For electrons $E_s \approx 3.3 \times 10^4$ statvolts/cm while for ion emission $E_s \approx 6.67 \times 10^5$ statvolts/cm.

Example 12.8 Determine the maximum positive charge on a 0.01-μm-diameter sphere.

$$n_m = \frac{E_s d^2}{4e} = \frac{(6.67 \times 10^5)(10^{-6})^2}{4 \times 4.8 \times 10^{-10}}$$

$$= 347.4 \text{ units of charge}$$

With a liquid droplet this maximum charge cannot be reached except in the case of extremely small droplet sizes. This is because of an additional charge limitation placed on liquid aerosols, known as the *Rayleigh limit*. It has been known for many years that as a highly charged droplet evaporates, a point will be reached where the outward force of the electric field at the drop surface exceeds the inward force of the droplet's surface tension. At this point, the drop will be torn apart by the close proximity of like charges and will produce a number of smaller drops in order to create more surface area for the charge. The number of electrons necessary for droplet disintegration was deduced by Rayleigh to be

$$n_r = \frac{1}{e}\sqrt{2\pi\gamma d^3} \tag{12.29}$$

where γ is the surface tension of the liquid. Several experiments have confirmed the validity of this expression (Whitby and Liu, 1966).

Example 12.9 Determine the Rayleigh limit for charge on a 1-μm-diameter water droplet (γ = 72.7 dyn/cm).

$$n_r = \frac{1}{e}\sqrt{2\pi\gamma d^3}$$

$$= \frac{1}{4.8 \times 10^{-10}}\sqrt{2\pi(72.7)(10^{-4})^3}$$

$$= 4.45 \times 10^4 \text{ units of charge}$$

Table 12.3 lists the approximate maximum number of elementary charges on particles of various sizes for the ion, electron, and Rayleigh limits. For comparison a 1-cm-diameter raindrop in a thunderstorm carries about 4×10^8 charges (Sartor and Atkinson, 1967) or about 1 percent of its maximum possible charge. Since for all but the smallest particle sizes the Rayleigh limit gives the lowest charge, highly charged drops which can evaporate will disintegrate until drop diameters on the order of 0.01 μm are reached.

Example 12.10 Considering both the ion and electron limits, find the droplet diameters where the Rayleigh limit just equals these limits. Hence, find the droplet diameter which cannot disintegrate upon evaporation.

$$n_{\text{ion}} = \frac{E_s d^2}{4e}$$

$$n_{\text{Rayleigh}} = \frac{1}{e}\sqrt{2\pi\gamma d^3}$$

TABLE 12.3 Approximate Maximum Number of Elementary Charges on Particles

Limit	Particle diameter, μm		
	0.01	1.0	100
Ion limit	3.47×10^2	3.47×10^6	3.47×10^{10}
Electron limit	1.72×10^1	1.72×10^5	1.72×10^9
Rayleigh limit			
γ = 21 dyn/cm	2.39×10^1	2.39×10^4	2.39×10^7
γ = 72.7 dyn/cm	4.45×10^1	4.45×10^4	4.45×10^7

Equating and solving for d_{eq} give

$$d_{eq} = \frac{32\pi\gamma}{E_s^2}$$

from which the following table can be computed:

		Smallest droplet diameter, μm	
		Alcohol, γ = 21 dyn/cm	Water, γ = 72.7 dyn/cm
(−)	Electron limit	0.019	0.067
(+)	Ion limit	0.00005	0.0002

Positive charge will continue to disintegrate the droplet to molecular size, negative charge will indeed produce a droplet with a finite lower diameter limit.

Charge equilibrium

In previous sections, charging of aerosols by ions of one sign has been discussed. Often, however, ions of both signs are present in essentially equal numbers. In this case extremely high charges of one sign are not likely to be found on any aerosol particles. However, the presence of free ions suggests that some particles will carry charge. This is particularly true for atmospheric aerosols since there are always free ions available for particle charging. Ion concentrations in the atmosphere can vary over a wide range from about 200 up to 3000 ions per cubic centimeter or more, of both polarities. Near ground level the ratio of positive to negative ions is approximately equal, being about 1.2 (Bracken and Johnson, 1987).

Table 12.4 lists typical ambient ion concentrations over land for

TABLE 12.4 Measured Fair-Weather Ion Concentrations

Location	n_+, ions/cm³	n_-, ions/cm³	Reference
Boston, Mass.	210–400	180–345	Yaglou et al. (1931)
Bozeman, Mont.	770	520	Sharp (1972)
England	50–2000	50–2000	Hawkins (1981)
France	220	180	Schreiber and Peyrous (1979)
Georgia	300–400		Perkins and Eisele (1984)
Haifa, Israel	700–1500	575–1100	Robinson and Dirnfeld (1963)
Minnesota	500	360	Hendrickson (1985)
Minnesota	380–800	40–1000	O'Brien (1983)
New Mexico	540	440	Wilkening (1984)
Uppsala, Sweden	700–1925	600–2350	Norinder and Siksna (1949)
Wisconsin	1030	790	Hawkinson and Barber (1981)

SOURCE: Bracken and Johnson (1987).

fair-weather conditions. Maximum ion production in the atmosphere tends to occur more in the warm summer months than in the colder winter. With rain there will usually be more negative ions than positive ions. During thunderstorms the air ion concentration can increase sharply to values on the order of 10^4 ions per cubic centimeter for negative ions and slightly less for positive ions, while during rainfall the air ion concentration can range from around 1×10^3 to 2×10^3 ions per cubic centimeter of either sign. Ion production rates 1 m above the land portion of the earth's surface have been estimated by Wait (1934) to be about 10 ions/(cm^3 · s), with 2 ions/(cm^3 · s) coming from cosmic radiation and the remainder from the decay of natural radioactivity emanating from the ground. Since these emanations are not present over oceans, ion concentrations over oceans are much lower than those over land.

When ions are associated with molecular clusters, they are called *small* ions; when attached to small aerosol particles, they are often called *large* or *Langevin* ions (Fleagle and Businger, 1963). The average life of a small ion is roughly 100 s, that of a large ion about 10-fold longer, or about 1000 s.

The relatively short lifetime of a charge on an aerosol particle implies charge transfer or neutralization, whereas the continued production of ions suggests a replenishment of the particle charge. Thus if there is an equilibrium value of small ions in the atmosphere, there should also be an equilibrium value of charge on aerosol particles present. This equilibrium condition implies that for a given size aerosol particle, there should be a definite fraction having no charge, another fraction having 1 charge, another having 2 charges, etc. Although any given particle may be gaining or losing charge continually, under equilibrium conditions the aerosol as a whole should maintain the same proportion of charged particles.

Steady-state theory of charge equilibrium

A theoretical approach defining bipolar charge equilibrium has been developed by Keefe et al. (1959), and comparison with experimental data suggests that it provides a reasonable model for particle sizes from about 0.05 to at least 2 μm. Keefe et al. applied Boltzmann's law to the distribution of particle charges in dynamic electrical equilibrium. The usual statement of this law is that the number of particles per unit volume having an energy E, denoted $c(E)$, is given by

$$c(E) = A \exp\left(\frac{-E}{kT}\right) \qquad (12.30)$$

where A is a normalization constant. In the case of a charged spherical particle carrying n unit charges with a diameter d,

$$E = E_0 + \frac{n^2 e^2}{d} \tag{12.31}$$

Here E_0 represents the energy of the particle in the absence of any charge whereas the second term represents the additional electrostatic energy. A particle will have the same energy whether it carries a positive or negative charge since the square of the charge is used in Eq. 12.31.

Substituting E given by Eq. 12.31 into Eq. 12.30 gives c_n, the number of particles per unit volume having n elementary units of charge (of one sign):

$$c_n = c_0 \exp\left(-\frac{n^2 e^2}{dkT}\right) \tag{12.32}$$

The term c_0 represents the number of neutral particles per unit volume, given by

$$c_0 = A \exp\left(\frac{-E_0}{kT}\right)$$

The number of particles per unit volume carrying n charges of both signs is twice that given in Eq. 12.32, assuming the numbers of positive and negative particles are equal.

The total number of positively charged particles c_+ or negatively charged particles c_- per unit volume is

$$c_+ = c_- = \sum c_1 + c_2 + c_3 + \cdots \tag{12.33}$$

and the total number of particles per unit volume is

$$c_T = c_0 + c_+ + c_- \tag{12.34}$$

The fraction of particles having n units of charge of one sign, denoted $f(n)$, is

$$f(n) = \frac{c_n}{c_T} = \frac{c_0 \exp[-n^2 e^2/(dkT)]}{c_0 + \sum_1^\alpha 2c_0 \exp[-n^2 e^2/(dkT)]} \tag{12.35}$$

or

$$f(n) = \frac{\exp[-n^2 e^2/(dkT)]}{\sum_{-\alpha}^\alpha \exp[-n^2 e^2/(dkT)]} \tag{12.36}$$

It is interesting to note that according to Eq. 12.36, the equilibrium charge distribution on aerosols is independent of both ion concentration and aerosol concentration. These factors are important, however, in establishing the length of time necessary for equilibrium conditions to develop.

Example 12.11 Determine the fraction of 0.5-μm-diameter aerosol particles (assume spherical shape) at charge equilibrium which carry 2 units of positive charge.

$$\exp\left(\frac{-n^2 e^2}{dkT}\right) = \exp\left[-\frac{n^2(4.8 \times 10^{-10})^2}{(5 \times 10^{-5})(1.38 \times 10^{-16})(293)}\right]$$

$$= \exp[0.114 n^2]$$

n	$\exp[n^2 e^2/(dkT)]$
6	0.017
5	0.058
4	0.161
3	0.359
2	0.634
1	0.892
Σ	2.121

From Eq. 12.36

$$f(n) = \frac{c_n}{c_T} = \frac{\exp[-n^2 e^2/(dkT)]}{\sum_{-\infty}^{\infty} \exp[-n^2 e^2/(dkT)]}$$

The denominator will equal 2.121 for all positively charged aerosol particles $n = +$, 2.121 for all negatively charged aerosol particles $n = -$, and 1.0 for all neutral aerosol particles $n = 0$. Then

$$f(n) = \frac{0.634}{1 + 2(2.121)} \approx 0.121$$

or about 12 percent of the particles carry 2 units of positive charge.

If the term $e^2/(dkT)$ in the exponential of Eq. 12.32 is set equal to y, Eq. 12.33 can be written as

$$\frac{c_+}{c_0} = \frac{c_-}{c_0} = e^{-y} + e^{-4y} + e^{-9y} + \cdots \qquad (12.37)$$

When y is less than 1 (particles with diameter greater than 10^{-2} µm at normal temperature and pressure), the series in Eq. 12.37 can be approximated by

$$\frac{c_+}{c_0} = \frac{c_-}{c_0} = \frac{1}{2}\left(\sqrt{\frac{\pi}{y}} - 1\right) \tag{12.38}$$

and the ratio of uncharged particles to total particles becomes

$$\frac{c_0}{c_T} = \frac{c_0}{c_0 + 2c_+} = \sqrt{\frac{y}{\pi}} \tag{12.39}$$

Equation 12.36 can be rewritten by utilizing Eq. 12.39 as

$$f(n) = \sqrt{\frac{e^2}{dkT\pi}} \exp\left(\frac{-n^2e^2}{dkT}\right) \tag{12.40}$$

a much more convenient form for computing the equilibrium fraction of charge on various aerosol particles. At temperatures roughly equal to room temperature, this equation is applicable to all particles having diameters greater than 5×10^{-2} µm.

Table 12.5 shows the equilibrium charge distribution on various monodisperse aerosols as computed from Eq. 12.40.

As in the case of unipolar charging, when the particle size is roughly equal to or less than the ionic mean free path, the Boltzmann approach given above underestimates the equilibrium charge distribution. Theories have been developed by Fuchs (1964) and Hoppel (1977), among others, that correct for the failure of the Boltzmann approach at small particle sizes. For example, Fig. 12.10 shows the steady-state charge distribution computed from the Boltzmann, Fuchs, and Hoppel approaches for small particles. It can be seen that although all theories tend toward each other as particle size increases or charge number increases, for singly charged particles in the

TABLE 12.5 Equilibrium Charge Distribution Fraction of Charge of Either Sign

	Number of charges on particle								
d	0	1	2	3	4	5	6	7	8
0.05	0.602	0.385	0.013	0.000	0.000	0.000	0.000	0.000	0.000
0.10	0.426	0.482	0.087	0.005	0.000	0.000	0.000	0.000	0.000
0.20	0.301	0.453	0.193	0.046	0.006	0.000	0.000	0.000	0.000
0.50	0.190	0.340	0.241	0.137	0.062	0.022	0.006	0.001	0.000
1.00	0.135	0.254	0.214	0.161	0.108	0.065	0.035	0.017	0.007
2.00	0.095	0.185	0.170	0.147	0.121	0.093	0.068	0.047	0.031
5.00	0.060	0.119	0.115	0.109	0.100	0.091	0.080	0.069	0.058
10.00	0.043	0.085	0.083	0.081	0.078	0.074	0.069	0.064	0.059

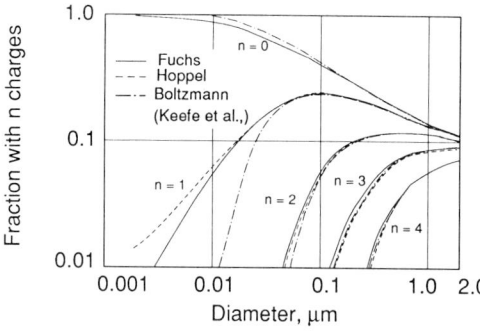

Figure 12.10 Steady-state charge distribution. (*After Hoppel and Frick, 1986.*)

0.001- to 0.1-μm-diameter range, there is a significant difference between the Boltzmann (Keefe et al.) prediction and the more accurate predictions of Fuchs or Hoppel.

Unfortunately the more accurate predictions of Fuchs or Hoppel require quite intricate calculations compared to using the Boltzmann approach. This has discouraged their use.

Wiedensohler (1988) developed an approximation for the Fuchs model, taking advantage of the observation that for an aerosol in charge equilibrium, the fraction of particles of any size with 3 or more elementary units of charge of the same sign can be calculated from Eq. 12.40. Then, for small particles with up to 2 elementary units of charge, he proposed the empirical equation

$$f(n) = 10^{\sum a_i(n)(\log d)^i} \quad (12.41)$$

Values for the approximation coefficients $a_i(n)$ are given in Table 12.6. Equation 12.41 is valid over the range of diameters 0.001 μm < d < 1 μm for $n = -1, 0, 1$ and 0.02 μm < d < 1 μm for $n = -2$ and 2. As can be seen in Fig. 12.10, for d less than 0.02 μm, particles carry at most 1 elementary charge.

TABLE 12.6 Approximation Coefficient $a_i(n)$

$a_i(n)$	$n = -2$	$n = -1$	$n = 0$	$n = 1$	$n = 2$
a_0	−26.3328	−2.3197	−0.0003	−2.3484	−44.4756
a_1	35.9044	0.6175	−0.1014	0.6044	79.3772
a_2	−21.4608	0.6201	−0.3073	0.4800	−62.8900
a_3	7.0867	−0.1105	−0.3372	0.0013	26.4492
a_4	−1.3088	−0.1260	0.1023	−0.1544	−5.7480
a_5	0.1051	0.0297	−0.0105	0.0320	−0.5059

SOURCE: After Wiedensohler (1988).

Example 12.12 Using Wiedensohler's approximation, compute the fraction of 0.1-μm-diameter aerosol particles carrying 1 positive charge under equilibrium conditions.

Recalling Eq. 12.41,

$$f(n) = 10^{\Sigma a_i(n)(\log d)^i}$$

we construct the following table.

i	$a_i(+1)$ (from Table 12.6)	$(\log d)^i$	$a_i(+1)(\log d)$*
0	−2.3484	1	−2.3484
1	0.6044	2	1.2088
2	0.4800	4	1.9200
3	0.0013	8	0.0104
4	−0.1544	16	−2.4704
5	0.0320	32	1.0240
			−0.6556

*d expressed in nm.

Then

$$f(n) = 10^{-0.655} = 0.221 \text{ fraction carrying 1 plus charge}$$

The average number of charges per particle can be determined by adding the charges on all the particles and dividing by the total number of particles; or in terms of the fraction of charged particles $f(n)$, the average number of charges per particle $\bar{n}$ is

$$\bar{n} = \sum_{-\infty}^{\infty} |n| f(n) \qquad (12.42)$$

By replacing the summation with an integral, Eq. 12.42 becomes

$$\bar{n} \approx \int_{-\infty}^{\infty} |n| f(n) \, dn \qquad (12.43)$$

which yields, on integration,

$$\bar{n} = \sqrt{\frac{dkT}{\pi e^2}} \qquad (12.44)$$

a convenient form for determining the average charge for all particles whose diameters are larger than 10^{-1} μm. Keep in mind that $\bar{n}$ represents the average number of charges, regardless of sign. The average number of positive or negative charges is one-half this value.

Example 12.13 Determine the average charge per particle for an aerosol comprised of 0.5-μm-diameter spheres.
For these particles,

$$\bar{n} = \sqrt{\frac{dkT}{\pi e^2}} \approx 2.37\sqrt{0.5} = 1.67$$

$$= \sqrt{\frac{(5 \times 10^{-5})(1.38 \times 10^{-16})(293)}{\pi(4.8 \times 10^{-10})^2}}$$

$$= 1.67 \text{ charges per particle}$$

Transient approach to charge equilibrium

Experimental data indicate that given enough time and otherwise optimum conditions, an equilibrium charge will eventually develop on an aerosol. Often, however, it is of interest to know whether this charge distribution has in fact developed and to gain insight into the factors which could be changed to hasten or retard its development. Exact calculation of the transient approach to charge equilibrium is extremely difficult. It is more appropriate to use an equilibrium half-time, similar to the half-life in radioactive decay, to describe the rate at which charge equilibrium is being reached. This represents the time necessary for one-half the equilibrium charge to be attained and is (Flanagan and O'Connor, 1961)

$$t_{1/2} = \frac{0.693 c_T}{4 q_i} \tag{12.45}$$

where q_i is the ion production rate. This equation indicates that with increased ion production rates or decreased aerosol concentration, charge equilibrium is more quickly reached, a fact borne out by experiment.

With an ion production rate of 10^4 ions/(cm³ · s) and an aerosol concentration of 5×10^4 particles/cm³, equilibrium would be achieved in about 2 s. For atmospheric aerosols where the ion production rate may be only 10 ions/(cm³ · s), even though aerosol concentrations of 5×10^4 particles/cm³ are not uncommon, it takes approximately 1700 s (or about 30 min) for equilibrium to be achieved. O'Connor and Sharkey (1960) report that equilibrium conditions usually prevail in air coming from the ocean. Over an industrial city, however, measurements indicated that the equilibrium charge distribution is not attained (Nolan and Doherty, 1950). This difference is attributed to the shorter time span between the production of the aerosol over the city and its measurement.

Since charge equilibrium can be quickly attained by using high ion production rates or large ion concentrations, it is not surprising to find this method employed for aerosol charge neutralization. Here the idea is to use the large number of free ions to reduce the excess charge on highly charged aerosol particles to as low a value as possible. With a mixture of bipolar ions, charge equilibrium as discussed in the previous sections will be rapidly attained. This method was developed by Whitby (1961) and Whitby and Peterson (1965), and it has been subsequently applied with great success.

Radioactive sources can also be used for charge neutralization, since these produce large numbers of bipolar ions that can then rapidly neutralize highly charged aerosols (Cooper and Reist, 1973).

Problems

1 A 5-μm-diameter unit-density sphere carries a negative charge equal to 200 electrons. If it is placed in an electric field having a strength of 1000 V/cm, determine the force in dynes acting on the particle.

2 Determine the mobility of a 5-μm-diameter unit-density sphere when it carries 200 unit charges.

3 Lead particles flow through a rubber tube. Estimate the sign of the charge produced by static electricity on the particles and the tube.

4 Estimate the charge which will develop in 60 s by diffusion charging if a 0.25-μm-diameter spherical particle is placed in an ion field containing 3×10^8 ions per cubic centimeter. Assume 20°C temperature.

5 The particle residence time in the charging section of an electrostatic precipitator is 0.6 s. What is the ion concentration such that one-half the maximum charge on the particles is reached during this residence time?

6 What fraction of 0.5-μm particles will have an average of 3 charges on them at equilibrium? What fraction will have an average of 4 charges?

7 Plot a curve showing charge as a function of time for diffusion charging, using the terms $2e^2n/(dkT)$ on the y axis and the corresponding dimensionless term on the x axis.

8 Plot a curve for field charging of a 1-μm sphere showing the fraction of total charge as a function of dimensionless time.

9 Using Wiedensohler's approximation, Eq. 12.41, compute the fraction of 0.25-μm-diameter aerosol particles carrying 1 negative charge under equilibrium conditions. Then compare this estimate to one made by using Eq. 12.40.

Chapter 13

Electrostatic Controlled Aerosol Kinetics

Electric Fields

As pointed out in Chap. 12, both the number of electric charges carried by the particle and the strength of the electric field acting on these charges must be known to determine the electric force acting on an aerosol particle. Electric field strength is a vector quantity having both magnitude and direction. The strength of the field is indicated by the number of lines of force passing through each unit area of orthogonal surface. As an example, Fig. 13.1 shows field direction lines (solid lines) and lines of force (dotted for several different geometries). The

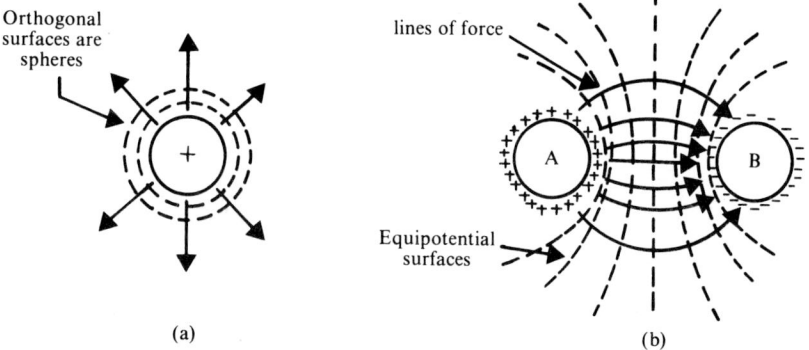

Figure 13.1 (a) Lines of force for a singly charged aerosol particle. Since field strength is force per unit area, field strength decreases as the square of the distance from the particle in this case. (b) Lines of force for two aerosol particles; one with a positive charge, one with a negative charge. Orthogonal surfaces are no longer spheres.

number of lines of force through a unit area is called the *flux* or *induction* through that area.

Field Strength of a Point Charge

The flux through an arbitrarily oriented element of area ds can be shown to be

$$d\phi = \vec{E}\, ds \qquad (13.1)$$

so that the flux through any finite surfaces is

$$\phi = \oint \vec{E}\, ds \qquad (13.2)$$

When this integral is taken over a closed surface, there may be an excess of lines of force leaving the enclosed volume compared to the number entering. This indicates that a field is originating from within the closed volume. If the integral is taken around an aerosol particle having a charge q, then

$$q \propto \oint \vec{E}\, ds \qquad (13.3)$$

indicating that the electric charge on a particle plays a double role. Besides being the object on which an electric field acts, it is also active as the generator of an electric field. This point is important in practical considerations of electrostatic precipitation.

When the particle is represented by a single point charge (Fig. 13.1), the lines of force are radial and equal in all directions. The orthogonal surfaces of equal field strength are spherical surfaces with a common center at the center of the particle, with the flux through any of these spheres of radius r being

$$\phi = \oint \vec{E}\, ds \qquad (13.4)$$

Since the density of the lines of force is the same everywhere, the field strength E must be constant over the surface of the sphere so that

$$\phi = \oint \vec{E}\, ds = 4\pi r^2 |E| = \gamma q \qquad (13.5)$$

Hence, the field strength for a point charge a distance r from the charge is

$$|E| = \frac{\gamma q}{4\pi r^2} \qquad (13.6)$$

where γ is a factor of proportionality.

Coulomb's Law

Suppose a second particle of charge q is situated a distance R from the first particle. Then the force acting on the second particle because of the field generated by the first would be, from Eq. 12.1,

$$F = q'E = \frac{q'\gamma q}{4\pi R^2} \quad (13.7)$$

This is *Coulomb's law*. The units for charge, field strength, and force are made compatible by specifying the units of the factor of proportionality γ. For example, if $\gamma = 4\pi/\epsilon$, where ϵ is the dielectric constant of the medium, the units are in terms of cgs or absolute electrostatic system (esu). Since the dielectric constant for air is essentially 1, for aerosols using the cgs system of units, $\gamma = 4\pi$.

Example 13.1 Two 0.1-μm-diameter unit-density spheres, each carrying 1 positive charge, are situated in air a distance 1 cm apart. Estimate the repelling force between these two particles.
From Eq. 13.7, recalling that $q = ne$,

$$F = \frac{\gamma e e}{4\pi R^2}$$

With cgs units

$$F = \frac{e^2}{\epsilon R^2}$$

for air, $\epsilon = 1$ and so

$$F = \frac{(4.8 \times 10^{-10})^2}{(1)^2} = 2.3 \times 10^{-19} \text{ dyn}$$

Electric forces between particles are negligibly small until the particles are almost touching. For comparison, the gravitational force on these particles is almost 6 orders of magnitude larger than this result. Hence interparticle electric forces can generally be neglected in aerosol computations.

Electrical Units

Very often "absolute" units, rather than electrostatic or cgs units, are used in dealing with electrical quantities. This is done to do away with the very small and large numbers which occur with cgs units. For the absolute system, γ is defined as

$$\gamma = \frac{1}{K_0 \epsilon} \quad (13.8)$$

The constant K_0 has a value of 8.849×10^{-12} A · s/(m · V). Table D.1 (see App. D) lists conversion factors for various electrical parameters to convert from absolute to electrostatic units. For example, the unit of charge of an electron, 4.80294×10^{-10} statcoulomb (esu), becomes, in absolute units, 1.60219×10^{-19} coulomb.

General Equations for Field Strength

The electric field strength at any point is the spatial derivative or gradient of the electrostatic potential at that point. The electrostatic potential for various geometries and boundary conditions for regions with no charge is given by *Laplace's equation*

$$\nabla^2 V = 0 \tag{13.9}$$

or for regions having a charge density

$$\nabla^2 V = -\gamma \rho_s \tag{13.10}$$

known as *Poisson's equation,* where ρ_s is the space charge per unit volume. The symbol ∇^2 represents the Laplacian operator. The problem of calculating the electrostatic field strength is solved by first finding the distribution of potential within the field. Then the derivative of this solution with respect to distance gives the field strength, i.e.,

$$E = -\operatorname{grad} V \tag{13.11}$$

Example 13.2 Write an equation for the electrostatic potential that would exist within a wire and tube type of electrostatic precipitator (see Fig. 13.2) for (*a*) no charge density and (*b*) charge density.

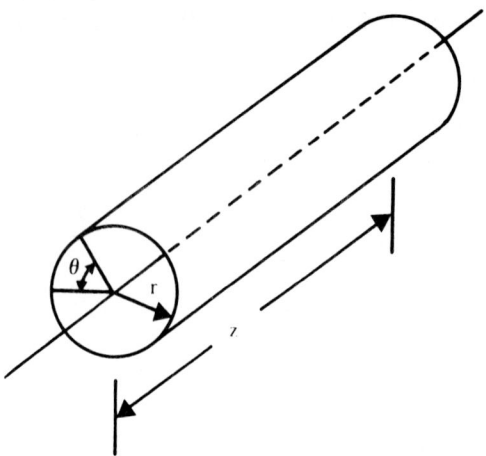

Figure 13.2 Schematic of wire and tube type of electrostatic precipitator.

a. Since a wire and tube type of precipitator is cylindrical, the choice of a cylindrical coordinate system is appropriate. Then

$$\nabla^2 V = 0 = \frac{\partial^2 V}{\partial r^2} + \frac{1}{r}\frac{\partial V}{\partial r} + \frac{1}{r^2}\frac{\partial^2 V}{\partial \theta^2} + \frac{\partial^2 V}{\partial z^2}$$

It is not expected that the potential will vary with the tube length. Hence $\partial^2 V/\partial z^2 = 0$. Also, potential will not vary with θ. Thus $\partial^2 V/\partial \theta^2 = 0$, and since only one independent variable remains, the equation becomes

$$\frac{d^2 V}{dr^2} + \frac{1}{r}\frac{dV}{dr} = 0$$

b. Here the solution is the same except that a space charge is now present:

$$\frac{d^2 V}{dr^2} + \frac{1}{r}\frac{dV}{dr} = -\gamma \rho_s$$

These equations can be expressed in terms of the field strength by applying Eq. 13.11. Then

$$\frac{dE}{dr} + \frac{1}{r}E = 0 \qquad (13.11a)$$

and

$$\frac{dE}{dr} + \frac{1}{r}E = \gamma \rho_s \qquad (13.11b)$$

Constant Field Strength

The field strength between two parallel plates is a constant, being equal to the potential difference across the plates divided by distance between them. Near the surface of the earth, an essentially constant field exists between the negative earth and the positive ionosphere, with field strengths ranging on the order of 0.67 to 3.17 V/cm over land and about 1.3 V/cm over sea (Mason, 1971; Pruppacher and Klett, 1978).

Computation of the Electric Field for Simple Geometries

Often the electric field is not constant (as in the case of parallel plates) but is spatially dependent. Then it is necessary to determine the field strength as a function of some characteristic distance. Consider a cylinder of radius R having a fine wire running down its axis. This could be a tube, e.g., through which an aerosol is flowing. A potential V is established across the wire-tube geometry.

Negligible ionic space charge

When the charge on the center wire is relatively low, the ionic space charge density is assumed to be negligible, and Laplace's equation is applicable. In cylindrical coordinates assuming cylindrical symmetry, with the axis along the axis of the two cylinders, Laplace's equation can be written as

$$\frac{dE}{dr} + \frac{E}{r} = 0 \tag{13.12}$$

(See Example 13.2.) Integration gives

$$E = \frac{C}{r} \tag{13.13}$$

where the constant C has the value

$$C = \frac{V}{\ln(r_o/r_i)} \tag{13.14}$$

and r_i is the radius of the inner electrode, r_o the radius of the outer electrode, and V the potential across the electrodes.

Example 13.3 An electrostatic precipitator sampler consists of a 0.020-in-diameter wire placed along the axis of a 1.5-in-diameter tube. What is the maximum field strength (assuming negligible space charge) at the outer edge of the tube when the precipitator voltage is 20 kV?
From Eq. 13.13,

$$E = \frac{C}{r} = \frac{1}{r_o} \frac{V}{\ln(r_o/r_i)}$$

$$= \frac{(20 \times 10^3)/300}{(1.5 \times 2.54/2)\ln[1.5 \times \tfrac{1}{2}/(0.02 \times \tfrac{1}{2})]}$$

$$= 8.11 \text{ statvolts/cm}$$

In practical units the field strength would be about 2430 V/cm. Under the conditions of the problem, a corona discharge is likely to be found around the center wire, so that the assumption of negligible space charge will not be met. However, the example does illustrate the calculation.

Ionic space charge present

Now suppose the center wire charge is considered, and it is sufficient to produce a corona discharge. This corona—or more exactly the resulting ions produced—gives rise to an ionic space charge within the outer cylinder. Assuming that the wire acts only as an ion source, the current applied to the wire will be used to maintain this space charge, or ionic current, which can be given as

$$i = 2\pi r \rho_s ZE \tag{13.15}$$

where Z is the ionic mobility, and ρ_s the ion density.

Using Poisson's equation with cylindrical coordinates gives (from Example 13.2)

$$\frac{dE}{dr} + \frac{E}{r} - \frac{2i}{ZrE} = 0 \qquad (13.16)$$

since

$$\gamma\rho_s = \frac{2i}{rZE}$$

Integrating gives

$$E = \left(\frac{2i}{Z} + \frac{C^2}{r^2}\right)^{1/2} \qquad (13.17)$$

The constant C depends on corona voltage and current as well as on the inner and outer cylinder diameters. For large values of i and r, Eq. 13.17 reduces to

$$E = \sqrt{\frac{2i}{Z}} \qquad (13.18)$$

implying a constant field strength over most of the cross-section away from the inner electrode.

An approximation for the corona current i has been given by White (1963) as

$$i = V(V - V_0)\frac{2Z}{r_o^2 \ln(r_o/r_i)} \qquad (13.19)$$

Equation 13.19 represents a reasonably good approximation for relatively low corona currents when V, the operating voltage, is slightly above the corona starting point. The corona starting voltage can be estimated from the expression

$$V_0 = 100\delta f r_i \left(1 + \frac{0.3}{\sqrt{r_i}}\right) \ln\frac{r_o}{r_i} \qquad (13.20)$$

where δ is a correction factor for temperature and pressure

$$\delta = \frac{293}{T} \times \frac{P}{760} \qquad (13.21)$$

Temperature is expressed in kelvins and pressure in millimeters of mercury. The factor f is a wire roughness factor, equal to 1 for a perfectly smooth round wire, but usually in practice having a value lying somewhere between 0.5 and 0.7 (White, 1963).

Example 13.4 Determine the field strength for the sampler in Example 13.3 considering ionic space charge. The precipitator voltage is 20 kV. Assume 20°C, standard pressure, $f = 0.6$. Use $Z = 2.2$ cm^2/(Vs ·).
From Example 13.3, $r_i = 0.02 \times 2.54 \times \frac{1}{2} = 0.025$ cm and $r_o = 1.5 \times 2.54 \times \frac{1}{2} = 1.905$ cm.

The corona starting voltage is

$$V_0 = 100\delta f r_i \left(1 + \frac{0.3}{\sqrt{r_i}}\right) \ln \frac{r_o}{r_i}$$

$$\delta = 1$$

$$V_0 = 100(1)(0.6)(0.025)\left(1 + \frac{0.3}{\sqrt{0.025}}\right) \ln \frac{1.905}{0.025}$$

$$= (1.52)(2.88)(4.32)$$

$$= 18.97 \text{ statvolts}$$

$$i = V(V - V_0)\frac{2Z}{r_o^2 \ln(r_o/r_i)}$$

$$= \frac{20{,}000}{300}\left(\frac{20{,}000}{300} - 18.97\right)\frac{2(2.2 \times 300)}{1.905^2 \ln(1.905/0.025)}$$

$$= (66.67)(47.70)(84.25)$$

$$= 2.68 \times 10^5 \text{ statamps/cm}$$

Then

$$E = \sqrt{\frac{2i}{Z}} = \sqrt{\frac{2 \times 2.67 \times 10^5}{660}} = 28.49 \text{ statvolts/cm}$$

Electric field—particles present

Finally, consider the case when there are particles present in the electric field. How is the field modified by the particle space charge? By neglecting the ion space charge as compared to the particle space charge, White (1963) showed by solution of Poisson's equation that the corona starting voltage V_0 would be increased by an amount equal to $\pi \rho_0 r_o^2$. Then, if V'_0 is the corona starting voltage when particles are present,

$$V'_0 = V_0 + \pi \rho_0 r_o^2 \qquad (13.22)$$

For c_T spherical particles of diameter d per cubic centimeter carrying the saturation charge,

$$\rho_0 = N_T n_s e \qquad (13.23)$$

Since $m = c_T(\pi/6)d^3\rho$, where m is the particle mass per cubic centimeter, ρ the particle density, and $n_s e = \chi E_0 d^2/4$, Eq. 13.23 is equivalent to

$$\rho_0 = \frac{3\chi m E_0}{2\pi d\rho} \qquad (13.24)$$

Example 13.5 How much will the corona starting voltage increase in the precipitator of Example 13.4 when fly ash particles having an average diameter of 0.1 μm are present if they have been fully charged in a 5 kV/cm field (assume $\chi = 3$ and $\rho_p = 1$ g/cm^3).

The particle mass concentration is 0.5 g/m^3.

$$\rho_0 = \frac{9}{2} \frac{m E_0}{\pi d\rho}$$

$$= \frac{(9)(0.5 \times 10^{-6})(5/0.3)}{2\pi(10^{-5})(1)}$$

$$= 1.19 \text{ esu/cm}^3$$

$$\pi \rho_0 R_0^2 = (\pi)(1.19)\left(1.5 \times \frac{2.54}{2}\right)^2$$

Increase = 13.6 statvolts = 4080 V

The effect of the particle space charge is to reduce corona current by increasing the corona starting voltage. Increases in aerosol mass concentrations will increase the effective corona starting voltage, as will decreases in aerosol particle size for a given mass concentration. Thus very fine fumes in high concentration can be quite difficult to remove by electrostatic precipitation.

A second space charge effect is the mutual repulsion by particles carrying charges of similar sign. The effect results in an apparent increase in field strength near the collecting surface which can be approximated by the factor

$$\sqrt{1 + \frac{3mr_0}{d\rho}}$$

so that the field strength becomes

$$E = \sqrt{\frac{2i}{Z}} \sqrt{1 + \frac{3mr_0}{d\rho}} \qquad (13.25)$$

In general, this increase in field strength does not offset the reduction in corona current. Hence the net effect of small particles in an electrostatic precipitator is a lowering of collection efficiency.

Example 13.6 Compute the field strength for the precipitator in Example 13.4 when the fly ash of Example 13.5 is included in the calculations.

Corona starting voltage $V_0 = 18.97 + 13.6$, from Example 13.4 and Example 13.5:

$$V_0 = 32.57 \text{ statvolts}$$

$$V = 20 \text{ kV} = \frac{20}{0.3} = 66.67 \text{ statvolts}$$

$$i = V(V - V_0)\frac{2Z}{r_o^2 \ln(r_o/r_i)}$$

$$= 66.67 (66.67 - 32.57)\frac{(2)(660)}{1.905^2 \ln (1.905/0.025)}$$

$$= 1.91 \times 10^5 \text{ statamps/cm}$$

$$E = \sqrt{\frac{2i}{Z}}\sqrt{1 + \frac{3mr_0}{d\rho_p}} = \sqrt{\frac{3.82 \times 10^5}{660}}\sqrt{1 + \frac{(3)(5 \times 10^{-7})(1.91)}{0.1 \times 10^{-4}(1)}}$$

$$= \sqrt{579}\sqrt{1.29} = 27.3 \text{ statvolts/cm}$$

Perturbations in the Electric Field Caused by a Particle or Other Object

Up to now, only coulombic force has been considered. This is the force between a particle and collecting surface due to the net charge on each surface and assuming that the charge on each surface is constant and stationary. Additional electric forces can also be present. Consider two conducting spherical particles, one with a net positive charge and the other with no net charge. As the first particle approaches the second, the positive charge attracts electrons from the back side of the second to its front (Fig. 13.3), forming a dipole with a net negative charge nearest the oncoming particle. This net negative charge sets up an attracting force between the two particles. This force, which can also arise between a charged particle and uncharged collecting surface or vice versa, is known as a *polarization, induction,* or *image* force. It generally tends to enhance the collection of charged particles by *any* surface. It also enhances the collection of uncharged particles by a surface placed in an electric field, although the enhancement is poor for very small uncharged particles since this force is proportional to the volume of each particle.

In many cases it is valid to neglect all electric forces acting on an aerosol with the exception of the coulombic force. This greatly simplifies most problems, but, if not used with care, can produce significant errors or lead one to erroneous conclusions. For example, Fig. 13.4 shows the trajectories of small positively charged particles in an electric field as they flow around an uncharged fiber, in Fig. 13.4a where the electric field tends to move the particles along with the air flow and in Fig. 13.4b where the field imparts a force on the particles in an opposite direction to the airstream. In the former case deposition can take place; in the latter example it does not. Neglect of image forces

Electrostatic Controlled Aerosol Kinetics

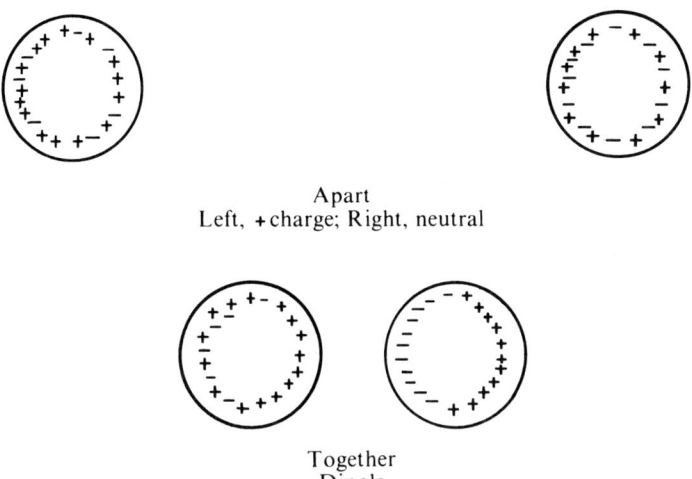

Figure 13.3 Mechanism of charge alignment of conducting aerosol particles as they are brought together by an external force.

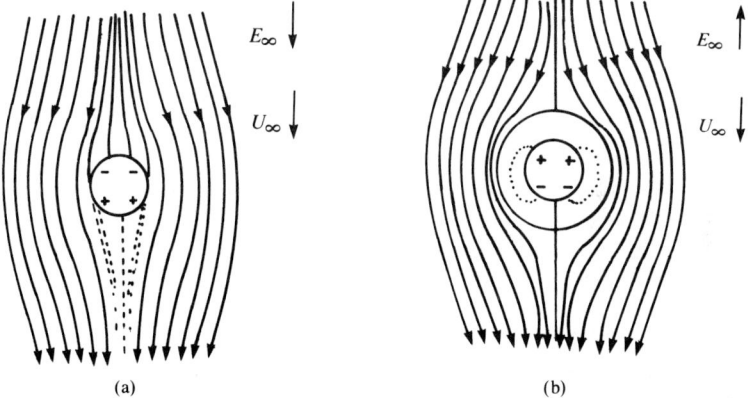

Figure 13.4 (a) Field in direction of particle motion. (b) Field in opposite direction of particle motion.

would have these two cases equivalent, with deposition occurring only through aerodynamic forces (Hochrainer et al., 1969).

Particle Drift in an Electric Field

The main reason for evaluating the charges on aerosol particles and the electric fields that act on these charges is to develop models which describe the effect on particle motion of the electric force.

The equation of motion for an aerosol particle including an electric force present F_E can be written as

$$m \frac{dv}{dt} = \vec{F}_D + \vec{F}_G + \vec{F}_E \tag{13.26}$$

which becomes

$$\tau \frac{d\vec{v}}{dt} = (\vec{u} - \vec{v}) + \tau g \vec{G} - \vec{E} q B \tag{13.27}$$

where B is the particle mobility. When $\vec{u}$ is a constant, $\vec{u}_0$ is the sum of the constant vectors $\vec{u} + \tau g \vec{G}$, and Eq. 13.27 becomes

$$\tau \frac{d\vec{v}}{dt} + \vec{v} = \vec{u}_0 - \vec{E} q B \tag{13.28}$$

In terms of a dimensionless velocity $v' = v/u_0$, we can write, for Eq. 13.28,

$$\tau \frac{dv'}{dt} + v' = 1 - \Gamma \tag{13.29}$$

where the dimensionless parameter Γ, which can be either positive or negative, is equal to EqB/u_0 and indicates the ratio of the particle velocity in an electric field to the constant velocity u_0. If $|\Gamma| \gg 1$, then electric forces predominate, whereas when $|\Gamma| \ll 1$, gravity and inertial effects predominate and electric forces can be neglected. Since in general τ is quite small, when $|\Gamma| \gg 1$, the inertia term $\tau\, dv'/dt$ can be ignored and Eq. 13.29 can be written simply as

$$\vec{v}' \simeq -\Gamma \tag{13.30a}$$

or

$$v \simeq -EqB \tag{13.30b}$$

Consider the case where $|\Gamma| \gg 1$. The electrical drift velocity is given by Eq. 13.30b. Denoting the particle velocity in an electric field as w and assuming a saturation charge, for field charging Eq. 13.30b becomes

$$w = E\left(\frac{\chi E_0 d^2}{4}\right)\left(\frac{C_c}{3\pi \mu d}\right)$$

$$= \frac{\chi E E_0 d C_c}{12\pi \mu} \tag{13.31}$$

The term E_0 is the field generating the particle charge, and E is the collecting field strength.

Example 13.7 Determine the electrical drift velocity of the fly ash particles in previous examples.

From Example 13.6, $E = 27.3$ statvolts/cm. From Example 13.5, $E_0 = 5$ kV/cm $= 16.67$ statvolts/cm. Also $\chi = 3$, $d = 10^{-5}$ cm, and $C_c = 2.97$.

$$w = \frac{\chi E E_0 d C_c}{12\pi\mu}$$

$$= \frac{(3)(27.3)(16.67)(10^{-5})(2.97)}{(12\pi)(1.81 \times 10^{-4})}$$

$$= 5.92 \text{ cm/s}$$

Efficiency of an Electrostatic Precipitator

The utility of the concept of aerosol particle electrical drift velocity can be shown by using it to estimate the theoretical efficiency of an electrostatic precipitator. For simplicity it is assumed that the collector is cylindrical, having a radius R (although this assumption does not affect the results), and that an aerosol is uniformly distributed across the entrance of the collector. In addition, turbulent flow in the collector is assumed such that the uncollected aerosol remains uniformly distributed at any distance from the entrance of the tube. If the electrical drift velocity is constant, the chance of a particle ϕ being collected in a time Δt is

$$\phi = \frac{w(2\pi R)}{\pi R^2} \Delta t = \frac{2w}{R} \Delta t \qquad (13.32)$$

and the chance of its not being collected is $1 - \phi$.

In n intervals of time, the chance of not being collected is $(1 - \phi)^n$. When n is allowed to approach infinity during a residence time period t, $(1 - \phi)^n$ approaches a value of $\exp(-\phi t)$. Denoting ϵ as the collection efficiency,

$$\epsilon = 1 - \exp\left(\frac{-2wt}{R}\right) \qquad (13.33)$$

In terms of the volumetric gas flow through the tube Q, the efficiency is

$$\epsilon = 1 - \exp\left(\frac{-Aw}{Q}\right) \qquad (13.34)$$

where A is the total collecting area of the precipitator. Equation 13.34 is applicable to both tube and plate type of precipitators. It is known as the *Deutsch equation (1922)*, and its general form has been verified many times in practice.

Example 13.8 If the precipitator of Example 13.7 is 6 in long and air flows through it at a rate of 1 ft³/min, determine the efficiency of collection of this unit for 1-μm fly ash particles.

$$\epsilon = 1 - \exp\left(\frac{-Aw}{Q}\right)$$

$$Q = 1 \text{ ft}^3/\text{min} = 1 \times 28.3 \text{ L/min} = 472 \text{ cm}^3/\text{s}$$

$$A = (2)(3.14)\left(\frac{1.5}{2} \times 2.54\right)(6 \times 2.54)$$

$$= 182.4 \text{ cm}^2$$

$$\epsilon = 1 - \exp\left(-\frac{182 \times 5.92}{472}\right)$$

$$= 1 - 0.101 = 0.849 = 89.9\% \text{ efficient}$$

It should be kept in mind that the derivation of Eqs. 13.33 and 13.34 contains many simplifying assumptions which may or may not be valid, depending on aerosol and precipitator characteristics. For example, it is assumed that once a particle is collected, it remains collected. This is not the case except for a liquid aerosol. Also, for dry aerosols, when the particles are good conductors, they rapidly lose their charge to the collecting electrode and pick up a new charge of opposite sign from the electrode, causing them to be repelled. But if the particles are poor conductors, they will lose their charges so slowly that the rain of new charges should be sufficient to maintain the charge on the particles and hold them to the collecting surface.

Theoretical calculations will always overestimate precipitator efficiencies, probably because of reentrainment. This overestimation could be as large as a factor of 2 or more (Rose and Wood, 1966). Even so, drift velocity or "effective migration velocity" is the basis for all precipitator calculations and does provide a good base for the comparison of various designs.

Problems

1 A 0.1-μm-diameter unit-density sphere and a 0.2-μm-diameter unit-density sphere, each carrying 2 positive units of charge, are spaced in air a distance 1 cm apart. Estimate the repelling force between these two particles.

2 An electrostatic precipitator sampler consists of a 0.015-in-diameter wire placed along the axis of a 1-in-diameter tube. What is the maximum field strength (assuming negligible space charge) at the outer edge of the tube when the precipitator voltage is 15 kV?

3 Determine the field strength for the sampler in Prob. 2 considering ionic space charge. The precipitator voltage is 15 kV. Assume 20°C, standard pressure, and $f = 0.6$. Use $Z = 2.2$ cm^2/(Vs ·).

4 Determine the electrical drift velocity of 0.1-µm-diameter spheres having a density of 2.65 g/cm^3 if they are carrying 200 units of charge each and are placed in a collecting field of 70,000 V/m.

5 An electrostatic precipitator is to be used to control emission of 0.5-µm-diameter particles from a paper mill. An efficiency for the collector of 99.6 percent is desired. If the total design flow through the unit is to be 6500 ft^3/min, how many square feet of collector surface are required? Assume $w = 7.5$ cm/s.

Chapter

14

Condensation and Evaporation Phenomena in Aerosols

Condensation and evaporation of aerosols play a great part in human existence. The cycle of water in nature relies on the condensation of water to form cloud droplets, some of which then return to earth in the form of rain or snow. Photographs of the earth's surface taken from outer space reveal that the most distinguishing characteristic of the earth is its cloud cover. Clouds and fogs lower visibility and can have a marked effect on air temperatures at the earth's surface. Fogs in combination with air pollution created by people can result in aerosols which are quite irritating to humans as well as being toxic to some forms of plant life (and, in some cases, to human life as well). Many industrial pollutants appear as aerosols made up of condensed liquids.

Evaporation of liquid drops is equally important. For example, in the application of a pesticide by spraying, it is desired that evaporation be minimized to increase the amount of pesticide reaching the plants. Yet in the production of such foodstuffs as powdered milk or powdered coffee, product quality is improved when evaporation proceeds as quickly as possible. In sampling aerosols, evaporation or condensation may alter aerosol size distribution and affect operation of the sampling instrument. In this case it is desired that static conditions be maintained if at all possible.

Early Observations

Early investigators such as Coulier (1875) and Aitken (1880) found that when they produced clouds by the adiabatic expansion of moist air (no heat transfer between the system and surrounding container), the presence of small dust particles was necessary for cloud formation. If the air were first made dust-free, clouds would not form. In this case clouds appeared only when the expansion was very large. Wilson (1897) extended these studies by defining the conditions under which clouds could be formed without dust particles: spontaneously with very high supersaturations or at lower supersaturations when ions were present. It was these observations that led to the development of cloud chambers for ion track visualization.

Types of Nucleation

Early investigators determined that the formation of an aerosol initially required a surface for condensation. This surface could be made up of a small cluster of vapor molecules, an ion or ionic cluster, or it could be a small particle of some other material, termed a *condensation nucleus*. When condensation of a vapor takes place solely on clusters of similar vapor molecules, it is called *spontaneous* or *homogeneous nucleation*. When condensation occurs on a nucleus or dissimilar material, it is called *heterogeneous nucleation*.

In the case of homogeneous nucleation, supercooling of the liquid making up the drop is common when the drop temperature is lowered below the freezing point, since there are no foreign bodies present in the liquid. For water droplets, supercooling to temperatures as low as $-40°C$ is possible. With a single condensation nucleus in the drop, its purity is such that supercooling is still quite common. This implies that in the formation of any particle by condensation (solid or liquid) it goes through a liquid phase (although the time the particle remains in this phase might be very short), and thus the theory developed for condensation and evaporation of liquid aerosols can also be applied to formation of solid aerosols by gas-phase reactions (Amelin, 1967).

Homogeneous nucleation is thought to take place in three steps. First, the vapor must be supersaturated to an extent that condensation will take place; second, small clusters of molecules or *embryos* must form; third, the vapor must condense on these embryos so that the embryo grows into a full-fledged nucleus which subsequently becomes a droplet. For heterogeneous nucleation only two steps take place, the first and third.

Saturation Ratio

The saturation ratio of a vapor in a gas can be given by the equality

$$S \equiv \frac{p}{p_s(T)} \quad (14.1)$$

where p is the partial pressure of the vapor in the gas and $p_s(T)$ is the saturated vapor pressure of the vapor over a plane of the liquid at a temperature T. When $S > 1$, the gas is said to be supersaturated with vapor; when $S = 1$, the gas is saturated; and when $S < 1$, the gas is unsaturated with vapor. For adiabatic expansion of a gas-vapor system, by using the first law of thermodynamics the saturation ratio of a gas saturated prior to expansion can be given by (Amelin, 1967)

$$S = \left(\frac{V_2}{V_1}\right)^{-K} \exp\left\{\frac{B}{T_1}\left[\left(\frac{V_2}{V_1}\right)^{K-1} - 1\right]\right\} \quad (14.2)$$

where V_1 and V_2 are the volumes before and after expansion, T_1 is the gas temperature in kelvins prior to expansion, K is the ratio of the constant-pressure specific heat to the constant-volume specific heat, and B is a coefficient which comes from the integrated term of the Clausius-Clapeyron equation.

Over a temperature range of -20 to $60°C$ for water vapor, K has a value of 1.4 and B a value of 5367. Table 14.1 lists K and B values for several other vapors in addition to water. The term B is also used in the equation for approximating vapor pressure:

$$\ln P_s(T) = A - \frac{B}{T} \quad (14.3)$$

Additional values for A and B are given in Table 15.2.

TABLE 14.1 Constants for Eq. 14.2 for Selected Vapors–Noncondensing Gas Mixtures

Material	K	B
Air-water vapor	1.40	5367
Air-ethanol vapor	1.40	5200
Argon-water vapor	1.66	5367
CO_2-water vapor	1.31	5367
N_2O-iodine	1.30	7155

Example 14.1 In an experiment, a chamber holding air saturated with water vapor is rapidly expanding adiabatically to 1.25 times its volume. Determine the value of S following the expansion. The initial temperature T_1 is 0°C.

$$S = \left(\frac{V_2}{V_1}\right)^{-K} \exp\left\{\frac{B}{T_1}\left[\left(\frac{V_2}{V_1}\right)^{K-1} - 1\right]\right\}$$

$$= \left(\frac{V_2}{V_1}\right)^{-K} \exp\left\{\frac{B}{T_1}\left[\left(\frac{V_2}{V_1}\right)^{K-1} - 1\right]\right\}$$

$$= \left(\frac{1.25 V_1}{V_1}\right)^{-1.4} \exp\left\{\frac{5367}{273}\left[\left(\frac{1.25 V_1}{V_1}\right)^{0.4} - 1\right]\right\}$$

$$= 0.732 \exp[(19.66)(0.093)]$$

$$= 0.732(6.27)$$

$$= 4.59$$

It has long been recognized that a small droplet will evaporate even when the gas surrounding it is fully saturated. Supersaturation of the gas is necessary to maintain the drop in equilibrium. Supersaturation is required because the probability of a net loss of a molecule from a convex surface is greater than the probability of net loss from a flat surface of infinite extent. A molecule that has left a small spherical droplet has a much more difficult time finding its way back than it would in finding its way back to a flat surface of infinite extent. Thus the high supersaturations necessary for spontaneous condensation are related to the size of the drop produced.

From cloud chamber studies it was found that with dust-free air, expansion ratios of about 1.35 or so were required for cloud formation to take place. Expansion ratios in this range imply saturation ratios or supersaturations on the order of 700 to 800 percent. There have been a number of theories advanced to explain the process of self-nucleation, and although none is completely acceptable in all cases, theory is sufficiently adequate to permit a prediction of aerosol parameters for practical implications.

Homogeneous Nucleation—Kelvin's Equation

Consider the energy balance of a nucleating (or condensing) drop. As the droplet (or embryo) is formed, its surface free energy goes from 0 to $\pi d^2 \gamma$, where d is the diameter of the drop and γ is the liquid surface tension. If the free-energy potential per molecule is ϕ_a in the vapor phase and ϕ_b in the liquid phase, and n is the total number of molecules contained in the drop growing to a diameter d, then the total change in free energy ΔG of the droplet is

Total change in free energy
$$\Delta G = (\phi_b - \phi_a)n + \pi d^2 \gamma \tag{14.4}$$

Now suppose the partial pressure of the vapor near the droplet is changed by a small amount dp (keeping the temperature constant). This produces a corresponding change in the free energy per molecule of vapor $d\phi_a$ and in the free energy per molecule of droplet $d\phi_b$. If V_a is the volume occupied per molecule in the vapor phase and V_b the volume occupied per molecule in the liquid phase,

$$d\phi_a = V_a\, dp$$

and

$$d\phi_b = V_b\, dp$$

Since $V_a \gg V_b$,

$$d\phi_b - d\phi_a = -V_a\, dp = d(\phi_b - \phi_a) \approx -\frac{kT}{p}\, dp \tag{14.5}$$

In this expression k is Boltzmann's constant.

Integrating Eq. 14.5 with the pressure varying from $p_\infty(T)$ to p gives

$$\phi_b - \phi_a = -kT \ln \frac{p}{p_\infty(T)} = -kT \ln S \tag{14.6}$$

The mass of a spherical drop is $(\pi/6)d^3\rho$. Hence the number of molecules n in the drop is

$$n = \frac{N_A}{M} \frac{\pi}{6} d^3 \rho \tag{14.7}$$

where N_A is Avogadro's number and M is the molecular weight of the liquid making up the drop. Substituting Eqs. 14.6 and 14.7 in Eq. 14.4 gives

$$\Delta G = \pi d^2 \gamma - (kT \ln S)\frac{N_A}{M}\left(\frac{\pi}{6}d^3\rho\right) \tag{14.8}$$

an expression for the total free-energy change of the droplet as a function of both drop size and the saturation ratio.

Example 14.2 Compute the free energy change of a 10-Å-diameter water droplet when $S = 4$. Assume $T = 0°C$, $\rho = 1$ g/cm^3, and $\gamma_{water} = (76.1 - 0.155T)$ dyn/cm, where T is in degrees Celsius.

$$\Delta G = \pi d^2 \gamma - (kT \ln S)\frac{N_A}{M} \frac{\pi}{6} d^3 \rho$$

$$= (3.14)(10^{-7})^2(76.1) - (1.38 \times 10^{-16})(273 \ln 4)$$

$$\times \frac{6.02 \times 10^{23}}{18}\left(\frac{3.14}{6}\right)(10^{-7})^3(1)$$

$$= 2.39 \times 10^{-12} - 9.15 \times 10^{-13} = 1.48 \times 10^{-12} \text{ erg}$$

Figure 14.1 shows a plot of ΔG as a function of the particle diameter for various values of S. It can be seen from this plot that Eq. 14.8 implies the existence of an energy barrier that acts to prevent the growth of droplets smaller than some critical size. Drops greater than this critical size will continue to grow, since with each slight increase in size the free energy of the system decreases (i.e., the droplet gives up energy). On the other hand, drops smaller than the critical size evaporate, since with these very small drops evaporation reduces their free energy.

The critical drop size can be determined by differentiating ΔG with respect to d, setting the result equal to zero, and solving for d. This gives

$$d^* = \frac{4\gamma M}{\rho RT \ln S} \tag{14.9}$$

where R is the universal gas constant. When ρ is in grams per cubic centimeter, γ in ergs per square centimeter (or dynes per centimeter)

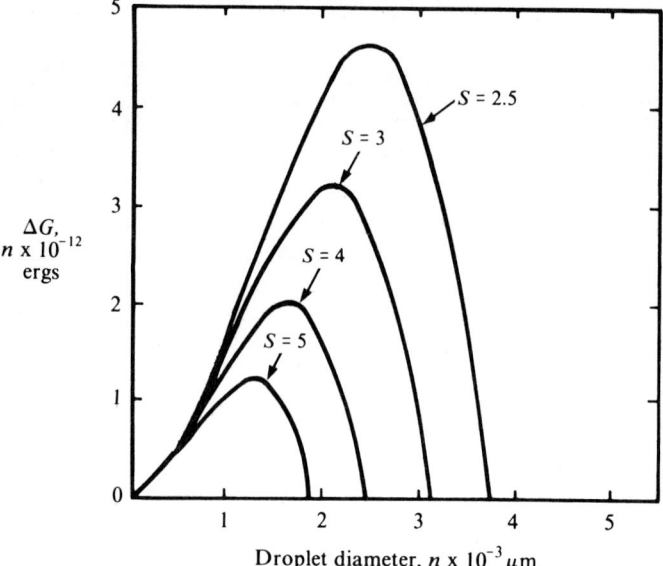

Figure 14.1 Plot of free-energy change as a function of particle diameter for various saturation ratios.

and the temperature is in kelvins, R has a value of 8.3144×10^7 erg · $K^{-1} \cdot mol^{-1}$.

Rearranging terms gives Kelvin's equation

$$\ln S = \frac{4\gamma M}{\rho R T d^*} \tag{14.10}$$

A plot of Kelvin's equation is given in Fig. 14.2.

Example 14.3 Compute the value of S for a droplet diameter d^* of 0.01 μm. Assume water at 0°C.

$$\ln S = \frac{4\gamma M}{\rho R T d^*}$$

$$= \frac{(4)(76.1)(18)}{(1)(8.314 \times 10^7)(273)(10^{-6})} = 0.241$$

$$S = 1.27$$

The curve shown in Fig. 14.2 is an equilibrium line. If for a given drop of diameter d the value of S associated with it produces a point lying to the left of the line, the drop will evaporate. If the point lies to the right of the line, the drop will grow. It is not necessary for a drop of a given size to be associated with a value of S which places a point directly on the curve. The curve indicates the conditions of S and d

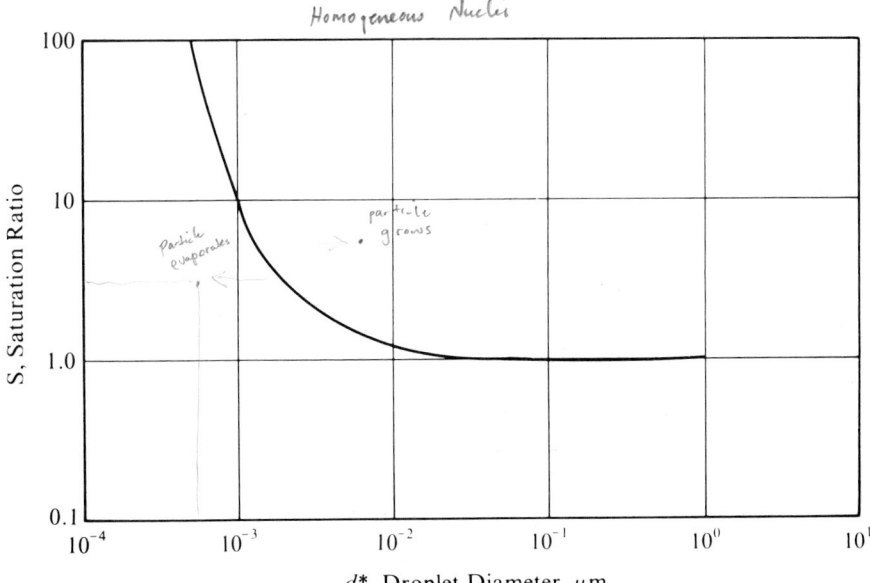

Figure 14.2 Plot of Kelvin's equation, Eq. 14.10.

under which a droplet will evaporate or grow. According to Kelvin's equation, a pure liquid drop will always evaporate when $S < 1$, that is, for water drops in air, if the relative humidity of the air is less than 100 percent. Even with supersaturation, droplets smaller than the critical size will also evaporate. This implies that small droplets of pure liquids have short lifetimes under normal circumstances. With a monodisperse cloud containing many small drops, lifetimes would be longer since the evaporation of some drops results in increased supersaturation, leading to growth of other drops.

In determining droplet free energy, it was assumed that the bulk value for surface tension was applicable to all droplet diameters. When the drop is very small, it is difficult to envision the meaning of surface tension as it is usually defined, and this point is still the subject of much scientific speculation (Sutugin, 1969). Some authors still consider the use of bulk values for very small droplets to be appropriate (Mason, 1971).

Rate of Formation of Critical Nuclei

As mentioned earlier, experiments indicate that spontaneous condensation is not significant until fairly high supersaturations are achieved. For example, supersaturations of slightly less than 5 are necessary with water vapor in particle free air for the formation of a visible fog by adiabatic expansion of moist air at 0°C. This supersaturation implies a critical droplet diameter of about 0.0015 μm and a cluster of several hundred molecules.

Nucleation embryos for homogeneous nucleation are aggregates of vapor molecules which are constantly being formed and disintegrated by random processes. When a cluster is formed which exceeds the critical size, it grows; the likelihood of its formation is a function of the degree of supersaturation. An expression for the number of clusters reaching critical size per unit time is given by Pruppacher and Klett (1978) as

$$J = \frac{\alpha_c}{\rho_w} \left(\frac{2N_A^3 M \gamma}{\pi} \right)^{1/2} \left(\frac{p_x}{RT} \right)^2 S \exp\left(\frac{-\Delta G}{kT} \right) \qquad (14.11)$$

Table 14.2 shows an estimate of the number of water embryos produced per cubic centimeter per second for various saturation ratios according to Eq. 14.11. For these computations the following constants were used: $\alpha_c = 1$, $\rho_w = 1$ g/cm^3, $T = 273$ K, $p_x = 4.58$ mmHg, and $\gamma = 76.1$ dyn/cm. According to Pruppacher and Klett (1978), a value of $J \approx 1$ is necessary for spontaneous condensation to occur.

In aerosol development, formed nuclei and embryos compete for

TABLE 14.2 Nucleation Rates and Molecules per Embryo

S	3	4	5	6
ΔG	3.85×10^{-12}	2.42×10^{-12}	1.79×10^{-12}	1.45×10^{-12}
J, embryos/(cm$^3 \cdot$ s)	1.41×10^{-18}	5.78×10^{-2}	1.10×10^6	1.29×10^{10}
d^*, μm	2.20×10^{-3}	1.74×10^{-3}	1.50×10^{-3}	1.35×10^{-3}
Molecules per embryo	1388	691	442	320

available vapor molecules. The depletion of vapor caused by the growth of small droplets reduces supersaturation, halting nucleation. There have been a number of attempts to model aerosol formation during the expansion of a gas containing a condensable substance (Amelin, 1967), but most require simplifying assumptions and predict droplet number and mean size, saying nothing about the resulting aerosol size distribution. The effect of droplet coagulation is usually neglected, although it is coagulation that leads to the variety of particle sizes formed, and not condensation alone (Fox et al., 1976).

Ions as Nuclei

As mentioned earlier, Wilson (1897) observed that condensation of water droplets in dust-free air took place at lower expansion ratios when ions were present. As charge is placed on a droplet, the free energy of that surface is increased approximately by a factor

$$\frac{q^2}{d}\left(\frac{1}{\epsilon_0} - \frac{1}{\epsilon}\right)$$

where q is the charge on the droplet or ionic cluster, ϵ_0 and ϵ are the dielectric constants of the gaseous medium and liquid, respectively, and d is the droplet diameter. The total change in free energy becomes

$$\Delta G = -\frac{\pi}{6}d^3\rho\frac{RT}{M}\ln S + \pi d^2\gamma + \frac{q^2}{d}\left(\frac{1}{\epsilon_0} - \frac{1}{\epsilon}\right) \quad (14.12)$$

Similar to the case for homogeneous nucleation, Eq. 14.12 can be differentiated, set equal to zero, and used to determine an expression for the saturation ratio at critical drop diameter:

$$\ln S = \frac{M}{RT\rho}\left[\frac{4\gamma}{d} - \frac{2q^2}{\pi d^4}\left(\frac{1}{\epsilon_0} - \frac{1}{\epsilon}\right)\right] \quad (14.13)$$

Equation 14.13 is plotted in Fig. 14.3 for a single charge. Unlike the pure-solution case, droplets carrying charges can exist even at satura-

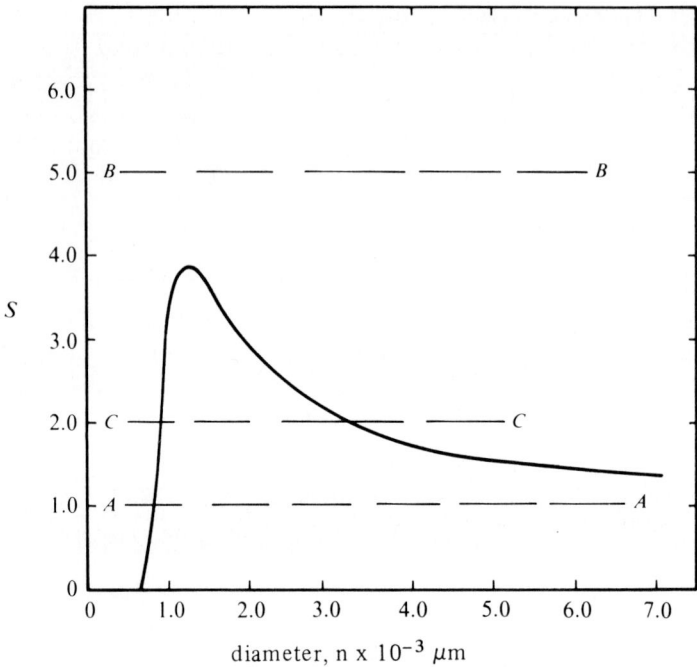

Figure 14.3 Saturation ratio for water as a function of critical particle diameter, single ion, atmospheric pressure, $T = 273°C$.

tion ratios less than 1 (relative humidities less than 100 percent). Under these circumstances the droplet size is quite small.

Example 14.4 Determine the equilibrium droplet diameter for a water droplet containing a single charge at 80 percent relative humidity. Assume $T = 70°F$.

$$\ln S = \frac{M}{RT\rho}\left[\frac{4\gamma}{d} - \frac{2q^2}{\pi d^4}\left(\frac{1}{\epsilon_0} - \frac{1}{\epsilon}\right)\right]$$

If $\epsilon_0(\text{air}) = 1.00$ and $\epsilon(\text{water}) = 80.00$,

$$\ln 0.8 = \frac{18}{(8.31 \times 10^7)(294)(1)}\left[\frac{4(73)}{d} - \frac{2(4.8 \times 10^{-10})^2}{\pi d^4}\left(\frac{1}{1} - \frac{1}{80}\right)\right]$$

$$-3.56 \times 10^8 = \frac{292}{d} - \frac{1.45 \times 10^{-19}}{d^4}$$

$$d = 7.69 \times 10^{-8} \text{ cm} = 7.69 \text{Å}$$

Similar to Fig. 14.2, Fig. 14.3 represents a plot of equilibrium values; droplets can be changing in size either away from or toward the equilibrium line. Three distinct cases are possible for condensation or evaporation on an ion, represented by lines A, B, and C on the curve in Fig. 14.3. To determine whether a droplet will grow or evaporate at a given S, it is helpful to refer to Fig. 14.4, a plot of the equation for the free-energy change on a nucleating ion, Eq. 14.12. Figure 14.4a shows a free-energy plot for case A. Since $\ln S$ is always negative for $S < 1$, the first term in Eq. 14.12 will always be positive. However, when d is very small, the $1/d$ term dominates. As d increases, the importance of the $1/d$ term decreases while the importance of the d^3 term becomes more apparent (the d^2 term also increases but not to as great a degree). Finally a minimum is reached, and subsequently ΔG increases for all increasing values of d. Therefore, droplets whose S value places them along line A will either grow or evaporate toward the equilibrium line, since in this case the equilibrium line represents a stable position.

Case B describes the condition where S is such that the curve of Fig. 14.3 is not intersected at all. The third term in Eq. 14.12 still dominates when d is small, but since the first term is always negative, ΔG

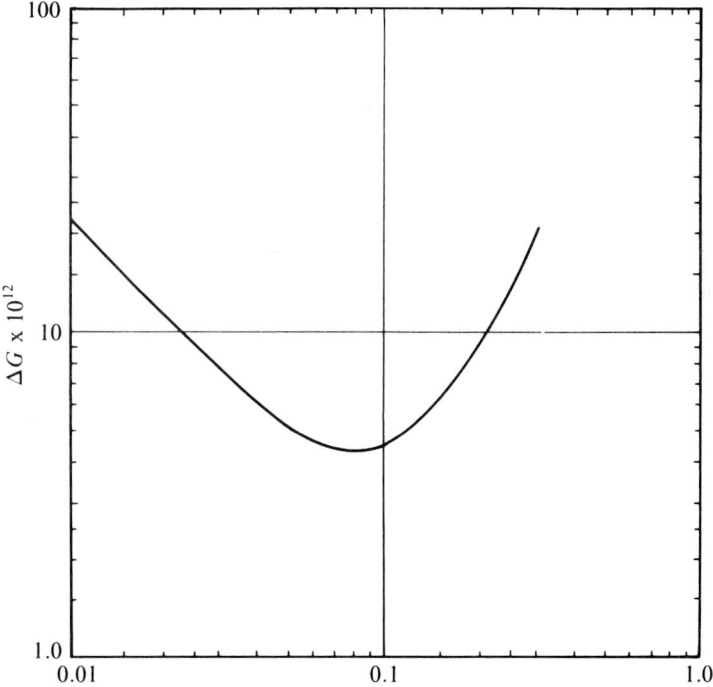

Number represents $d \times 10^{-2}$ μm, i.e., $0.1 = 0.1 \times 10^{-2}$ μm

Figure 14.4a Free-energy change as a function of drop diameter for droplet containing a single ion (line A of Fig. 14.3).

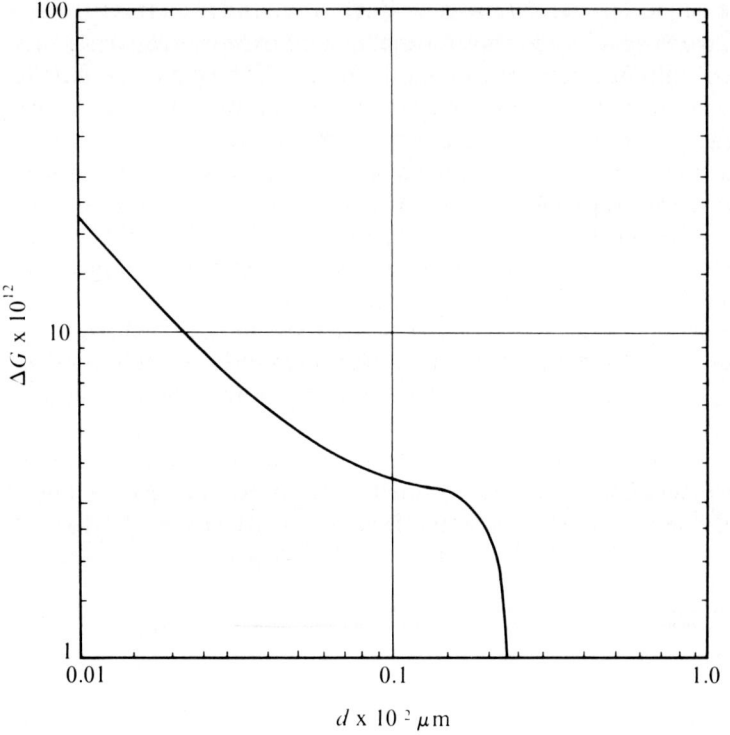

Figure 14.4b Free-energy change as a function of drop diameter for droplet containing single ion (line B of Fig. 14.3).

is always decreasing (Fig. 14.4b). This means that when S exceeds the maximum value as given in Fig. 14.3, any size charged droplet will grow. This explains the formation of clouds in a cloud chamber. The diameter at which this maximum in Fig. 14.3 occurs is given by

$$d = \left[\frac{2q^2(1/\epsilon_0 - 1/\epsilon)}{\gamma\pi}\right]^{1/3} \tag{14.14}$$

Example 14.5 Compute the particle diameter at which S in Fig. 14.3 is maximum. Assume a water droplet at 0°C and 760-mmHg pressure.

From Eq. 14.14

$$d = \left[\frac{2q^2(1/\epsilon_0 - 1/\epsilon)}{\gamma\pi}\right]^{1/3}$$

$$= \left[\frac{2(4.8 \times 10^{-10})^2(1/1 - 1/80)}{76.1(3.14)}\right]^{1/3}$$

$$= 1.24 \times 10^{-7} \text{ cm} = 12.4 \text{ Å}$$

Finally, there is case C, representing a combination of cases A and B. The free energy shows a minimum and then a maximum (Fig. 14.4c) as particle diameter increases. The minimum could be considered a metastable point. Drops tend to grow or evaporate toward this point and away from the maximum free energy. Hence drops at saturation ratios that place them above the curve in Fig. 14.3 will always grow while those lying below the curve will always evaporate.

There has been good gross experimental verification of Eq. 14.13. For example, droplets formed under conditions where the maximum saturation ratio is just exceeded will continue to grow without bound and become easily visible, while those formed when S is less than this maximum will not be seen. Experimentally, the peak saturation ratio in Fig. 14.3 has been found to be about 4.2 for water condensing on negative ions, in good agreement with theory, but much greater, about 6, for water condensing on positive ions. One possible explanation for the difference in the behavior of positive and negative ions has to do with the orientation of the water molecule. Since the water molecule dipole is thought to have its negative end oriented outward in

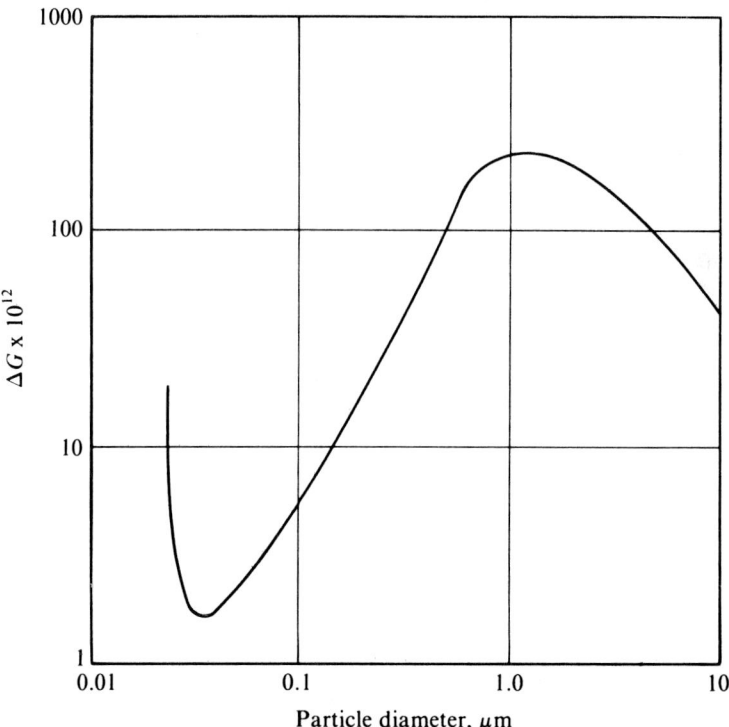

Figure 14.4c Free-energy change as a function of drop diameter for droplet containing a single ion (line C of Fig. 14.3).

the outer several layers of the droplet surface, a negative nucleus would permit the capture of water molecules in the correct orientation, whereas with a positive nucleus the molecules would have to turn themselves around before capture, so that condensation, in this case, would be more difficult.

Example 14.6 Determine the saturation ratio that corresponds to the diameter in Example 14.5. Hence, predict the minimum saturation ratio at which spontaneous condensation on ions will occur. Assume $T = 0°C$. From Eq. 14.13

$$\ln S = \frac{M}{RT\rho}\left[\frac{4\gamma}{d} - \frac{2q^2}{\pi d^4}\left(\frac{1}{\epsilon_0} - \frac{1}{\epsilon}\right)\right]$$

$$= \frac{18}{(8.314 \times 10^7)(273)(1)}\left[\frac{4(76.1)}{1.24 \times 10^{-7}} - \frac{2(4.8 \times 10^{-10})^2}{3.14(1.24 \times 10^{-7})^4}\left(1 - \frac{1}{80}\right)\right]$$

$$= (7.93 \times 10^{-10})(1.84 \times 10^9) = 1.46$$

$S = 4.31$

Heterogeneous Nucleation

Condensation nuclei

In most practical cases, condensation of a vapor takes place in the presence of small dust particles, making unnecessary the extremely high supersaturations required for homogeneous condensation. These small dust particles are given the generic name of condensation nuclei, and they can range in size from near molecular sizes to particles greater than 1 μm. A descriptive classification of these nuclei as often used in atmospheric physics is shown in Table 14.3. Although this classification is arbitrary, it corresponds roughly to particle size ranges for different measurement techniques usually employed.

In the atmosphere, small condensation nuclei greatly exceed the number of large ones, with particle number decreasing roughly as the inverse of the cube of the particle diameter. Number concentration

TABLE 14.3 Condensation Nuclei Size Classification Commonly Used in Atmospheric Physics

Name	Diameter range, μm
Aitken nuclei	0.001–0.4
Large nuclei	0.4–2.0
Giant nuclei	> 2.0

SOURCE: Junge (1953).

and particle size are influenced by such factors as topography, meteorology, elevation, vegetative cover, density of human habitation, and degree of industrialization. In addition, there are diurnal as well as seasonal variations in condensation nuclei levels. Concentrations are also influenced by wind speed and direction. Typical outside air condensation nuclei concentrations can range from as low as 100 to 10^6 particles per cubic centimeters or even higher.

In atmospheric physics the distinction is often made between *condensation nuclei* (CN) and *cloud condensation nuclei* (CCN). Condensation nuclei include the very small particles present in the air whereas cloud condensation nuclei are only those particles on which condensation can take place at relatively low supersaturations (0.1 to 10 percent). There are substantially more CN in the atmosphere at any given time than CCN. For example, Fig. 14.5 shows a comparison between CN (nuclei of 0.05-μm diameter and greater) and CCN (nuclei of 0.1-μm diameter and greater) at various rural and urban locations.

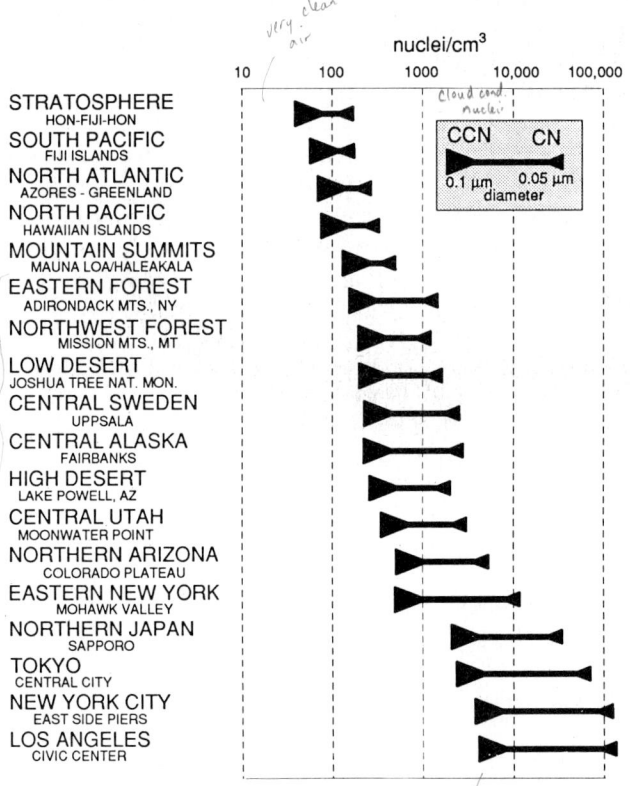

Figure 14.5 Concentration range for CN and CCN. *(Adapted from Schaefer and Day, 1981.)*

The concentration of CN can vary by several orders of magnitude depending on such factors as time of day, arrival of a fresh air mass, precipitation, wind direction, proximity to anthropogenic sources, etc. Thus the values given in Fig. 14.5 are only indicative of the general range of average concentrations possible at a given location.

Sources of condensation nuclei

Condensation nuclei come from a variety of sources. Such processes as photooxidation of natural organic materials over nonurban areas have been suggested as a possible reason for the occurrence of the blue haze usually observed over vegetated areas (Went, 1960). And although a great deal of work remains to be done to explain the mechanisms of photochemical production of aerosols, it is clear that these reactions are also very important in the production of aerosols over urban areas (Goetz and Pueschel, 1967). Photooxidation of organic material may be the most important natural source of condensation nuclei. Other important sources include entrainment of dust particles by the wind, the production of sodium chloride nuclei from sea salt spray, or such spectacular occurrences as forest fires, explosions created by humans, or volcanic eruptions. For example, the eruption of Krakatoa in 1883 released a reported 6.5 km^3 of fine dust (Cadle, 1966), equivalent to approximately 10^{23} particles of 0.1-μm diameter. Meteors and interplanetary dust have also been listed by some authors as sources of condensation nuclei. And finally, organic material, both living and dead (plant spores, microorganisms, feathers, skin tissue, hair, etc.), can act as condensation sites.

Example 14.7 The eruption of Mount St. Helens in May 1980 resulted in the aerosolization of 1 mi^3 of mountaintop. If the average density of the material aerosolized was 2.6 g/cm^3 and 1.0-μm-diameter spheres were produced, determine the number of particles produced.

Volume of material aerosolized:

$$V_A = 1 \text{ mi}^3 = (5280 \text{ ft/mi} \times 30.5 \text{ cm/ft})^3$$

$$= 4.18 \times 10^{15} \text{ cm}^3$$

Volume of one 1.0-μm-diameter sphere:

$$V_r = \frac{\pi}{6}d^3 = \frac{\pi}{6}(10^{-4} \text{ cm})^3 = 5.24 \times 10^{-13} \text{ cm}^3$$

Number of particles produced:

$$\frac{V_A}{V_p} = 7.98 \times 10^{27} \text{ particles}$$

If only 1 percent of the particles were 1.0 μm in diameter, there would have been 7.98×10^{25} particles of this size produced. This is still a lot of particles.

Composition of condensation nuclei

Condensation nuclei can be of organic or inorganic composition, can be soluble or insoluble, or can be insoluble with a thin soluble coating (in which case they are termed *mixed nuclei*). Because of the variety of soluble material existing in the atmosphere, the chemical composition of nuclei is not well defined. Studies of Los Angeles smog collected by electrostatic precipitation indicated that about 60 percent was made up of inorganic substances or minerals, and the remaining 40 percent was a complex mixture of organic compounds, carbon, and pollen (Billings et al., 1980). These percentages would not be the same everywhere. However, a great difficulty in analyzing composition is the relatively small mass of material available for analysis—mass contents in a specific size range of 10 μg or less per cubic meter of air are usual. And there may be different chemical fractions for various size ranges of particles. For example, Junge (1963) found that most of the nuclei with diameters between 0.4 and 2 μm collected in Germany and on the east coast of the United States consisted mainly of ammonium sulfate, whereas the particles whose diameters exceeded 2 μm had a less specific chemical composition, sometimes containing considerable amounts of sodium chloride or sodium nitrate.

Adsorption of atmospheric gases on condensation nuclei can also alter their chemical composition. The pickup of radioactive gases by small particulates is only one form of adsorption, but one which can be easily observed. The exact role of aerosols in the adsorption of gases is an area where little is known at present.

Utilization of nuclei

It should be kept in mind that not all the atmospheric aerosol is available for the condensation process. In fact, it is only a small fraction of the total. As might be expected from reference to Fig. 14.2, the largest (and most soluble) nuclei are activated preferentially. Thus utilization of a given size of nuclei for condensation depends to a large extent on the degree of supersaturation present, and in the atmosphere this, in turn, depends on the rate of cooling of the air.

Utilization also depends on the chemical composition of the nuclei. There are two general classes of condensation nuclei to be considered: soluble nuclei and insoluble nuclei. With soluble nuclei the condensing vapor dissolves the nucleus, changing the properties of the embryo drop from that of a pure liquid. With insoluble nuclei, surface characteristics are important, since once the nucleus is coated with liquid, it behaves in a manner similar to a pure liquid drop. Figure 14.6 shows a schematic illustration of the possible paths for droplet formation by heterogeneous nucleation. These mechanisms are discussed in the following sections.

Insoluble nuclei

The two extremes of insoluble nuclei are nuclei which are easily wetted and those which are not. Nuclei which are easily wetted rapidly take on the appearance of a droplet and subsequently behave as one. To predict droplet growth or evaporation, these particles with easily wettable surfaces can be considered to be pure drop nuclei, and the Kelvin equation can be used directly (but with a lower limit on nucleus size).

In cases where the particle surfaces are not wettable, condensation proceeds with much more difficulty. This is because the condensing liquid tends to pull into small spheres on the particle surface, and only when the entire surface is covered with these spheres is a liquid coating formed. Fletcher (1958a, b) has treated this problem by considering the contact angle between an embryo sphere formed on the particle and the particle surface. His results correspond to what has been observed experimentally—it is very difficult to get condensation to take place on nonwettable particles unless high supersaturations are used. The role of insoluble nuclei in the condensation process is still in question and remains another problem for future investigators to solve.

Soluble nuclei

In many instances condensation takes place on soluble nuclei, producing solution droplets. An example is the condensation of water on a

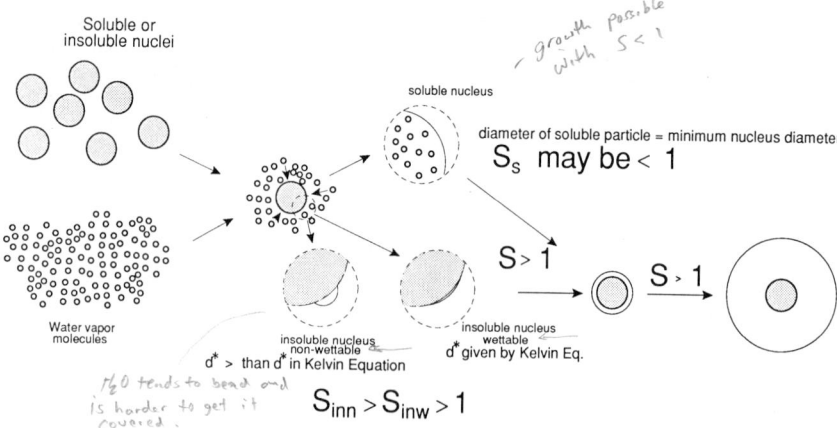

Figure 14.6 Particle formation by heterogeneous nucleation.

sodium chloride nucleus. Initially a saturated NaCl solution is formed. As condensation proceeds, the solution becomes more and more dilute until finally the drop behaves in a manner similar to a droplet of pure liquid. In general, the equilibrium solvent vapor pressure over a solution surface is lower than over a pure solvent surface, the amount of decrease depending on the nature of the solvent and the concentration and nature of the solute. A lower equilibrium vapor pressure means that condensation occurs at lower saturation ratios.

For an electrolyte solution, Robinson and Stokes (1959) give for the reduction in equilibrium vapor pressure over the solution surface the relationship

$$\frac{p'_x(T)}{p_x(T)} = \exp(-\beta b \vartheta M) \tag{14.15}$$

where $p'_x(T)$ is the equilibrium vapor pressure over an infinite plane of solution, $p_x(T)$ is the equilibrium vapor pressure over an infinite plane of pure solvent, M is the molecular weight of the solvent, b is a coefficient known as the molal osmotic coefficient, and β is the number of ions per molecule available for complete ionization. Table 14.4 gives values of β and b for several electrolytes at various concentrations. The factor ϑ is the number of moles of solute per gram of solvent, which for a single spherical drop can be given by

$$\vartheta = \frac{m/W}{(\pi/6)d^{*3}\rho' - m} \tag{14.16}$$

TABLE 14.4 Osmotic Coefficient b of Some Electrolytes at 25°C

Molality	NaCl, $\beta = 2$	MgCl$_2$, $\beta = 3$	(NH$_4$)$_2$SO$_4$, $\beta = 3$	Ca(NO$_3$)$_2$, $\beta = 3$	Al$_2$(SO$_4$)$_3$, $\beta = 5$
0.1	0.932	0.861	0.767	0.827	0.420
0.2	0.925	0.877	0.731	0.819	0.390
0.4	0.920	0.919	0.690	0.821	0.421
0.6	0.923	0.976	0.667	0.831	0.545
0.8	0.929	1.036	0.652	0.843	0.718
1.0	0.936	1.108	0.640	0.859	0.922
1.2	0.943	1.184	0.632	0.879	
1.6	0.96	1.347	0.624	0.917	
2.0	0.983	1.523	0.623	0.917	
2.5	1.013	1.762	0.626	1.001	
3.0	1.045	2.010	0.635	1.051	
3.5	1.080	2.264	0.647	1.103	
4.0	1.116	2.521	0.660	1.157	
5.0	1.192	3.048	0.686	1.263	
5.5	1.231		0.699	1.313	
6.0	1.272			1.361	

SOURCE: Abridged from R. A. Robinson and R. H. Stokes, *Electrolyte Solutions*, Butterworth, London, 1959, p. 483.

where W is the molecular weight of the solute, m the mass of the solute per drop, and d the drop diameter. The primed values refer to the solute-solvent mixture and unprimed values to the pure materials.

Recalling Eq. 14.10 for a pure droplet

$$\ln S = \frac{4\gamma M}{\rho R T d^*}$$

this can be rewritten as

$$\frac{p}{p_x(T)} = \exp \frac{4\gamma M}{\rho R T d^*} \qquad (14.17)$$

For a solution droplet Eq. 14.17 can be written as

$$\frac{p'}{p'_x(T)} = \exp \frac{4\gamma' M}{\rho' R T d^*} \qquad (14.18)$$

Combining Eqs. 14.18 and 14.15 gives

$$\frac{p'}{p_x(T)} = \frac{p'}{p'_x(T)} \frac{p'_x(T)}{p_x(T)} = \exp\left(\frac{4\gamma' M}{\rho' R T d^*} - \beta b \vartheta M\right) \qquad (14.19)$$

an expression for the ratio of the vapor pressure above a solution droplet to the vapor pressure of an infinite plane of pure solvent (see Byers, 1965a and b).

Close examination of various equations which have been proposed for predicting saturation ratios over solution droplets reveals that they differ only in detail and all give essentially the same results. Figure 14.7 is a plot of S versus d^* for NaCl masses of various sizes with water as the solvent, computed from Eq. 14.19. Curves similar to these are very often referred to as *Köhler curves*.

Example 14.8 Determine the value of $p'/p_x(T)$ for a 0.01-μm $(NH_4)_2SO_4$ nucleus when d^* for the solution droplet in which the nucleus is dissolved is 0.1 μm (T = 20°C). Assume a spherical shape, ρ_{solute} = 1.77 g/cm³, W = 132, and M = 18.

$$\frac{p'}{p_x(T)} = \exp\left(\frac{4\gamma' M}{\rho' R T d^*} - \beta b \vartheta M\right)$$

$$\rho' = \frac{\text{mass solute} + \text{mass solvent}}{\text{volume solute} + \text{volume solvent}}$$

$$= \frac{9.268 \times 10^{-19} + 5.231 \times 10^{-16}}{5.236 \times 10^{-16}}$$

$$= \frac{5.240 \times 10^{-16}}{5.236 \times 10^{-16}} = 1.00077$$

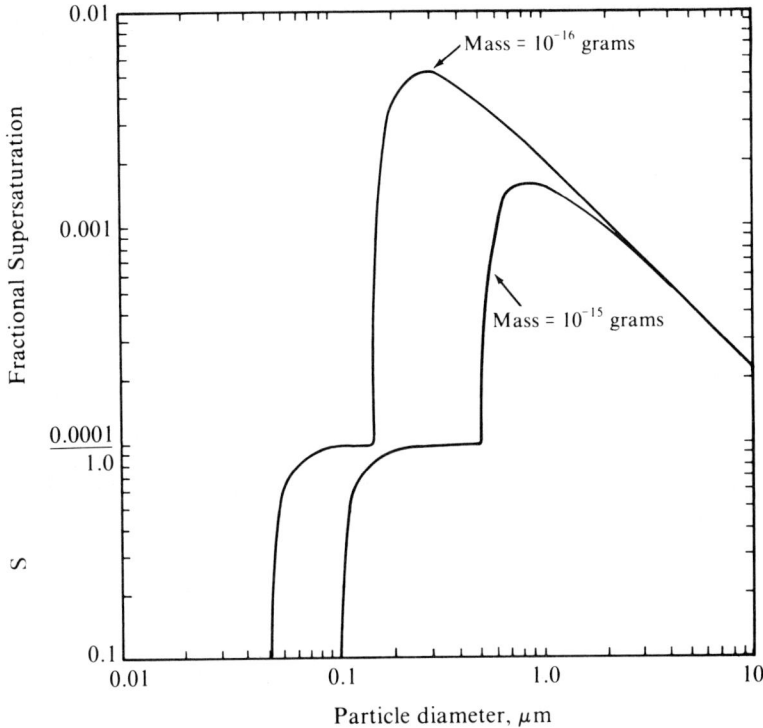

Figure 14.7 Plot of saturation ratio or supersaturation as a function of critical particle diameter for soluble nuclei of 10^{-15} and 10^{-16} g.

$$m' = \left(\frac{\pi}{6}\right)(0.1 \times 10^{-4})^3(1.00077) = 5.240 \times 10^{-16}\,\text{g}$$

$$m = \left(\frac{\pi}{6}\right)(0.01 \times 10^{-4})^3(1.77) = 9.268 \times 10^{-19}\,\text{g}$$

$$\text{Molality} = \frac{10^3 m/m'}{W(1 - m/m')} = \frac{1000(9.268 \times 10^{-19}/5.240 \times 10^{-16})}{132(1 - 9.268 \times 10^{-19}/5.240 \times 10^{-16})}$$

$$= 1.34 \times 10^{-2}\,\text{g} \cdot \text{mol of solute per 1000 g solvent}$$

From Table 14.4, say $b = 0.767$, $\beta = 3$. Then, using 14.16 gives

$$\vartheta = \frac{m/W}{(\pi/6)d^{*3}\rho' - m}\,\text{mol solute/g solvent}$$

$$= \frac{9.268 \times 10^{-19}/132}{5.240 \times 10^{-16} - 9.268 \times 10^{-19}} = 1.345 \times 10^{-5}$$

Assume $\gamma' = 76.1 - 0.155T = 73$ dyn/cm. Finally,

$$\frac{p'}{p_x(T)} = \exp\left[\frac{4(73)(18)}{(1)(8.314 \times 10^7)(293)(10^{-5})} - (0.767)(3)(1.345 \times 10^{-5})(18)\right]$$

$$= \exp(2.156 \times 10^{-2} - 5.569 \times 10^{-4})$$

$$= 1.021$$

Note that this value is the same as for the pure droplet case. There is a difference only when the drop size is close to the nucleus size.

Unlike the curve for condensation on a droplet of pure solvent, when a solute is present, it is possible to have condensation taking place even at relative humidities of less than 100 percent [when $p'/p_x(T) < 1$]. The effect of a solute can be considered to be very similar to the effect of an ion on droplet growth or evaporation except that the basic nucleus size can be much larger.

Analogous to the case for condensation on ions, at a given $p'/p_x(T)$, droplets will grow or evaporate away from the portion of the curve to the right of the maximum and toward that portion lying to the left of the maximum unless $p'/p_x(T)$ is so great that they grow without bound. As a result, it is possible to have stable solution droplets whose sizes are a function of only the mass of solute and the ratio $p'/p_x(T)$. For example, a 1-μm-diameter droplet containing 10^{-15} g of NaCl will rapidly evaporate to a diameter of about 0.6 μm in an atmosphere where $p'/p_x(T)$ is 1.001, whereas if it were initially 3 μm in diameter on formation, the drop would grow without bound until it eventually depleted the water vapor around it or was removed by some process such as sedimentation.

The injection of soluble particles into humid air results in the almost immediate generation of stable droplets of a much larger size. Figure 14.8 shows a plot of stable droplet diameter as a function of NaCl particle diameter (assuming spherical particles) for various relative humidities. At 100 percent relative humidity, particle size is increased about 5 times for NaCl masses of about 10^{-16} g and about 10-fold for masses of about 10^{-13} g. It is this increase in particle size that is responsible for the evolution of haze in the atmosphere when adequate numbers of soluble nuclei are present in conjunction with high humidities. Unlike completely pure droplets, because of hysteresis effects, slight changes in humidity will not significantly alter stable drop size for some solutes.

Hysteresis in evaporation and condensation

Hysteresis describes a process in which a phase change occurs at one humidity when the humidity is rising with the reverse change not oc-

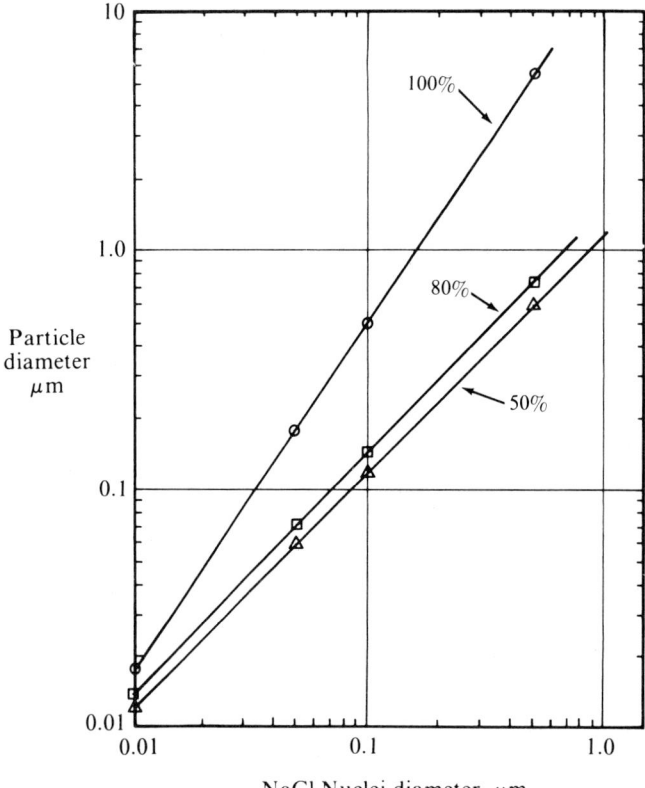

Figure 14.8 Stable droplet diameter as a function of soluble nuclei diameter (NaCl) for various relative humidities.

curring at the same humidity value but at some different humidity when the humidity is falling. A soluble hygroscopic particle in an atmosphere of vapor-laden solvent will initially pick up a solvent envelope by adsorption. At some minimum "relative humidity," the quantity adsorbed becomes such that the soluble particle is dissolved and becomes a liquid droplet. If the humidity is reduced to dry the droplet, it has been observed that the drop remains a liquid even at relative humidities less than that required for initial solution, implying supersaturation of the solution making up the drop. With continued reduction of the "relative humidity" the solute in the drop suddenly crystallizes (Fig. 14.9).

This hysteresis effect was studied by Orr and his colleagues (1958a), who found that solution takes place over a range of 68 to 80 percent relative humidity for various inorganic salts, while recrystallization does not occur until relative humidities are about 30 percent lower.

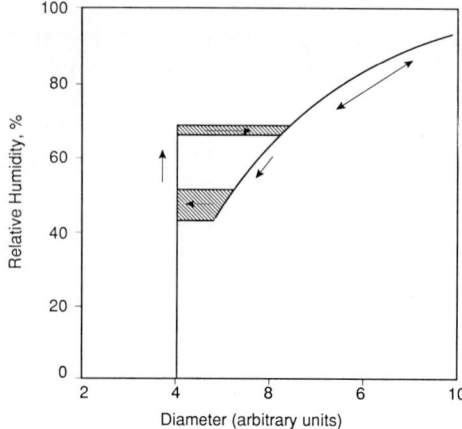

Figure 14.9 Phase transition curve for a droplet of NaCl solution.

For example, for NaCl solutions, crystallization occurs at a relative humidity of 70 percent and recrystallization appears at 40 percent. For nonhygroscopic materials, the effect does not occur. This phenomenon helps explain why smogs and hazes persist at relative humidities well below those at which they originally were formed.

Problems

1 For a 0.01-μm water droplet, compute the value of ΔG when $S = 4$.

2 A volume of air at 80°F at sea level is expanded by rising to an elevation 750 ft above sea level. If the expansion is adiabatic and the air is initially saturated with water vapor, what is the resulting value of S?

3 Determine the value of S at which a 0.03-μm-diameter water droplet will just continue to grow.

4 It was found by Wilson (1897) that when air at 20°C, initially saturated with water vapor and free of any condensation nuclei, was expanded with an expansion ratio in excess of 1.37, homogeneous nucleation occurred. What is the value of S implied by this expansion ratio?

5 Wilson (1897) found that the condensation of water vapor occurred on a negative ion with an expansion ratio of 1.25, whereas for condensation on a positive ion an expansion ratio of 1.31 was necessary. What is the expansion ratio equivalent to the maximum of the S versus d^* plot? (How well does theory agree with experiment?) What is the value of d^* associated with this expansion ratio?

6 A cloud contains 3×10^4 condensation nuclei per cubic centimeter. The condensation nuclei are 0.01-µm-diameter unit-density spheres. How many cubic meters of air must be sampled to get 1 mg of condensation nuclei?

7 What supersaturation is necessary for a 0.05-µm-diameter ammonium sulfate particle (sphere) to grow without limit? What is this value for a 0.5-µm-diameter ammonium sulfate particle?

8 Determine the equilibrium droplet diameter for a water droplet containing three positive charges at 90 percent relative humidity. Assume $T = 70°F$.

9 How many particles would have been produced from the eruption of Mount St. Helens (see Example 14.7) if the average density of the material was 3 g/cm^3 and the particles were 0.01 µm in diameter?

10 A sodium chloride solution droplet has a diameter of 1 µm. If the solubility of sodium chloride in water is 35.7 g/100 cm^3 and sodium chloride content of the drop is 25 percent of the solubility, determine the diameter of the resulting sodium chloride particle if all water is evaporated. Assume spheres in both cases.

Chapter 15

Evaporation and Growth

Besides knowing under what circumstances a droplet will form, it is important to be able to estimate rates of droplet growth or evaporation. For example, soluble particles inhaled into the respiratory system will rapidly become humidified and may grow to such an extent that their aerodynamic properties in the lung change appreciably. Or insecticide droplets sprayed from the air with the hope that they will reach the ground may actually evaporate and never carry out their intended purpose. Or the drying time of droplets containing bacteria might give an estimate of the viability of these microorganisms.

Maxwell's Equation

The original equation for the evaporation of a droplet as a function of time was first derived by James Maxwell in 1877. Although his derivation contains a number of simplifying approximations, Maxwell's equation gives reasonable results for fairly large droplets of pure substances. For this equation, it is assumed that the vapor pressure at the droplet temperature is equal to the partial pressure at the surface of the drop, that is, $p_{\text{surface}} = p_\infty(T_{\text{surface}}) = p_s$. In terms of concentrations of molecules, this means that the vapor concentration at the surface just equals the concentration of saturated vapor, the saturation determined at the droplet temperature. This assumption is valid when the droplet size is not too small compared to the mean free path of the vapor molecules.

Considering that the molecules can leave the droplet surface by diffusion,

$$J = -D \frac{\partial c}{\partial \mathcal{R}}$$

where J is the number of grams of solvent vapor passing through unit area in unit time and $\mathcal{R}$ is a distance measured from the center of the droplet. The term c represents the concentration of solvent vapor (in grams per cubic centimeter). The number of grams of vapor lost or gained per second I through a spherical surface of radius $\mathcal{R}$ enclosing the droplet is

$$I = -4\pi \mathcal{R}^2 D \frac{\partial c}{\partial \mathcal{R}} \qquad (15.1)$$

Integration of Eq. 15.1 with respect to $\mathcal{R}$ gives

$$c = \frac{I}{4\pi \mathcal{R} D} + \mathcal{C} \qquad (15.2)$$

where $\mathcal{C}$ is a constant which depends on the conditions of the problem. If c_l is the concentration of vapor a large distance away from the drop, then

$$c \approx \frac{I}{4\pi \mathcal{R} D} + c_l \qquad (15.3)$$

For practical purposes this approximation can be considered to be an equality. However, by the initial assumption that the concentration of vapor at the droplet surface ($\mathcal{R} = d/2$) is equal to the saturation concentration c_s, then

$$c_s = \frac{I}{2\pi d D} + c_l \qquad (15.4)$$

and

$$I = 2\pi D d(c_s - c_l) \qquad (15.5)$$

Equation 15.5 is known as *Maxwell's equation*. The evaporation rate I is a function of droplet diameter d, diffusion coefficient of the solvent vapor D, and the difference between the solvent vapor pressure at the droplet surface and ambient partial pressure of the vapor. A relationship between vapor concentration and vapor pressure can be found by considering the vapor to be an ideal gas. Then

$$c = \frac{pM}{RT} \qquad (15.6)$$

where, if p is a pressure in millimeters of mercury, M is a molecular weight in gram-moles, and T a temperature in kelvins, then R has a value of 62,360 cm$^3 \cdot$ mmHg/(K $\cdot$ mol). This gives c the units of grams per cubic centimeter. Maxwell's equation can now be written in terms of vapor pressure as

$$I = \frac{2\pi DMd}{RT}(p_s - p_l) \qquad (15.7)$$

The diffusion coefficient for water vapor in air is 0.219 cm^2/s at 0°C

TABLE 15.1 Values of Diffusion Coefficient for Various Gas Mixtures, STP

Gases	D, cm^2/s
H$_2$O-air	0.219
Ethanol-air	0.099
Ethyl-ether air	0.070
Benzene-oxygen	0.080
Mercury-nitrogen	0.119
Iodine-nitrogen	0.070
Iodine-air	0.069

SOURCE: Compiled from W. Jost, *Diffusion in Solids, Liquids and Gases*, Academic Press, New York, 1952, p. 412.

and 760-mmHg pressure (STP). Values for other air-vapor mixtures are given in Table 15.1. For other temperatures the diffusion coefficient can be estimated by the relationship

$$D \approx D_0 \left(\frac{T}{T_0}\right)^{1.94} \qquad (15.8a)$$

whereas for other pressures

$$D = D_0 \left(\frac{p}{p_l}\right) \qquad (15.8b)$$

[see Chap. 3 and Pruppacher and Klett (1978)]. The vapor pressure over an infinite plane of liquid at various temperatures can be estimated by

$$\ln p_x(T) = A - \frac{B}{T} \qquad (15.9)$$

where T is in kelvins and p in millimeters of mercury, and constants A and B are given in Table 15.2.

Example 15.1 A 30-μm-diameter water droplet is evaporating in a chamber. The chamber temperature is 20°C, and the pressure is 760 mmHg. The chamber relative humidity is 50 percent. Find the droplet evaporation rate in grams of water lost per second.

From Eq. 15.7

$$I = \frac{2\pi DMd}{RT}(p_s - p_l)$$

From Table 15.1 and Eq. 15.8a

$$D = 0.219 \left(\frac{293}{273}\right)^{1.94} = 0.251 \text{ cm}^2/\text{s}$$

TABLE 15.2 Values of Constants A and B for Various Liquids, $T \approx 20°C$ (for Eq. 15.9)

Liquid	A	B
Bromine	18.47	3915
Water	21.18	5367
Iodine	22.96	7155
Uranium hexafluoride	22.95	5344
Acetone	18.51	3906
Benzene	20.60	4759
Bromobenzene	19.44	5364
Decane	19.67	5694
Diethylether	17.97	3488
Ethanol	21.24	5200
n-Hexane	18.09	3896
Methyl iodide	17.54	3439
Naphthalene	25.87	8423
n-octane	23.98	6537

SOURCE: Adapted from F. Daniels and R. Alberty, *Physical Chemistry*, 2d ed., Wiley, New York, 1961, p. 126.

From Table 15.2 and Eq. 15.9

$$\ln p_s = 21.18 - \frac{5367}{293} = 2.86$$

$$p_s = 17.50 \text{ mmHg} \qquad p_l = \frac{17.50}{2} = 8.75 \text{ mmHg}$$

Then

$$I = \frac{(2)(3.14)(0.251)(18)(30 \times 10^{-4})}{(62{,}360)(293)}(17.50 - 8.75)$$

$$= 4.08 \times 10^{-8} \text{ g/s}$$

Maxwell's equation is derived specifically for evaporation or condensation at small Knudsen numbers. If the droplet were evaporating into a vacuum or near vacuum (large Knudsen numbers), the evaporation rate would be a function of the square of the droplet diameter. Hence, with high Knudsen numbers Eq. 15.5 is not valid without a correction for the free molecular nature of the medium.

In addition, Maxwell's equation does not allow for temperature changes in the droplet due to condensation or evaporation. The temperature within a droplet at steady-state conditions (constant evaporation rate and temperature) T_s can be found from the expression

$$T_l - T_s = \frac{DML}{RK_t}\left(\frac{p_s}{T_s} - \frac{p_l}{T_l}\right) \qquad (15.10)$$

where T_l and p_l are the temperature and partial pressure of vapor, respectively, away from the drop. The value of p_l is given by $p_l = Sp_x(T_l)$. Equation 15.10 indicates that the temperature of the droplet is independent of droplet size. Here L is the latent heat of evaporation of water (in calories per gram), and K_t is the thermal conductivity of air in calories per centimeter per second per kelvin. These two factors can be approximated for water over the temperature range -10 to $40°C$ by

$$K_t = (5.74 + 0.018T) \times 10^{-5} \qquad (15.10a)$$

and

$$L = 597 - 0.55T \qquad (15.10b)$$

where for both Eqs. 15.10a and 15.10b, T is expressed in degrees Celsius (Ludlum, 1980).

Equation 15.10 is the fundamental psychrometric equation which permits wet-bulb temperatures to be calculated, as pointed out by Davies (1978) and others. Thus a psychrometric chart can be used to estimate steady-state droplet temperature by finding the wet-bulb temperature corresponding to a given ambient temperature and relative humidity. This wet-bulb temperature is *the evaporating droplet temperature!*

Example 15.2 Using Example 15.1, estimate the droplet evaporation rate, taking the temperature of the droplet into account.

As the droplet evaporates, it cools itself. Therefore the saturation vapor pressure p_s should be determined at the equilibrium droplet temperature, *not* at the ambient temperature.

For a chamber temperature of 20°C (68°F) and relative humidity of 50 percent, Eq. 15.10 gives

$$T_l - T_s = \frac{DML}{RK_t}\left(\frac{p_s}{T_s} - \frac{p_l}{T_l}\right)$$

$$= \frac{(0.251)(18)(590)}{(62{,}323)(610 \times 10^{-5})}\left[\frac{p_s}{273 + T_s} - \frac{0.50(17.50)}{273 + 20}\right]$$

By trial and error, guessing T_s, and computing L and p_s, the two sides of the equation are compared until they are equal. This gives

$$T_s = 13.26°C$$

Alternatively, the psychrometric chart in App. F indicates a wet-bulb (droplet) temperature of approximately 13.9°C (57°F). The slight difference between the calculated value and that given in the psychrometric chart is thought to be due to small differences in the values for various parameters in Eq. 15.10, as well as differences in the manner of computing the wet-bulb temperature.

Using the calculated value for T_s, from Eq. 15.9 the vapor pressure associated with this temperature is

$$\ln p_s = A - \frac{B}{T}$$

$$= 21.18 - \frac{5367}{286.3} = 2.43$$

$$p_s = 11.37 \text{ mmHg}$$

For ambient conditions,

$$p_\infty(T_l) = 17.50 \text{ mmHg} \qquad p_l = \frac{17.50}{2} = 8.75 \text{ mmHg}$$

$$I = \frac{(2)(3.14)(0.240)(18)(30 \times 10^{-4})}{(62{,}360)(293)}(11.37 - 8.75)$$

$$= 1.17 \times 10^{-8} \text{ g/s}$$

This is a slower evaporation rate than would be estimated if the lower temperature within the droplet had not been taken into account.

Growth or Lifetime of Drops— Langmuir's Equation

In the previous section the evaporation rate of a droplet of a specific diameter was considered. But as the droplet evaporates, its diameter decreases. Since a large number of molecules are required to affect the droplet size significantly, a quasi-stationary condition can be assumed to crudely estimate the drying time, or *lifetime*, of the drop. Then

$$I = \frac{dm}{dt} \tag{15.11}$$

Since

$$dm = \frac{\pi}{2} d^2 \rho \, dd \tag{15.12}$$

and, from Eq. 15.7

$$\frac{dm}{dt} = \frac{2\pi DMd}{RT}(p_s - p_l) \tag{15.13}$$

$$\frac{dd}{dt} = \frac{4DM}{d\rho RT}(p_s - p_l) \tag{15.14}$$

Integration and rearrangement of terms give the time for a droplet of diameter d to evaporate to diameter d_0 when the partial pressure in the surrounding medium is p_l. This equation, sometimes known as *Langmuir's equation*, is

$$t = \frac{\rho RT(d^2 - d_0^2)}{8DM(p_s - p_l)} = \rho RT(d^2 - d_0^2)\left\{8DMp_x(T_l)\left[\frac{p_s}{p_x(T_l)} - S\right]\right\}^{-1} \quad (15.15)$$

where S is the saturation ratio. For condensation $p_s \approx p_x(T_l)$ is used in place of computing a value from Eq. 15.10.

Data showing experimental verification of Langmuir's equation are sparse, but some data were recently compiled by Davies (1978) which show reasonably good agreement with Eq. 15.15 when temperature corrections are taken into account. Equation 15.15 can apply equally to either droplet evaporation or droplet growth by condensation. For evaporating droplets, when a polydisperse cloud of droplets dries, the degree of polydispersion is increased both through the differences in drying rates for the various sizes of drops and through enhanced coagulation caused by the polydispersity of the drops.

On the other hand, for a polydisperse collection of condensation nuclei (particles > 0.1 μm in diameter), it can be seen from Eq. 15.15 that the rate of growth of small droplets will be faster than the rate of growth of larger droplets. Thus, as long as there is an excess of vapor for condensation, there will be a tendency for condensing droplet size distributions to become more homogeneous.

Example 15.3 Estimate the time required for a 20-μm-diameter water droplet to completely evaporate when the relative humidity is 30 percent. Assume the ambient temperature is 20°C.

A temperature of 20°C is equivalent to 68°F. From App. F the wet-bulb temperature is estimated to be 51.5°F, equivalent to 10.8°C. The vapor pressure of water at 10.8°C is

$$\ln p_s = 21.18 - \frac{5367}{273 + 10.8} = 2.269$$

$$p_s = 9.668 \text{ mmHg}$$

$$p_l = 17.50 \text{ mmHg}$$

Adjusting D for the droplet temperature gives

$$D \approx 0.219\left(\frac{283.8}{273}\right)^{1.94} = 0.236 \text{ cm}^2/\text{s}$$

Using Eq. 15.15 gives

$$t = \frac{\rho RT(d^2 - d_0^2)}{8DM(p_s - p_l S)}$$

$$= \frac{(1)(62{,}360)(293)(20 \times 10^{-4})^2}{(8)(0.236)(18)[9.668 - 17.50(0.3)]}$$

$$= 0.487 \text{ s}$$

Modifications to Langmuir's Equation

Calculations using Langmuir's equation show that the lifetime of very small volatile droplets in air is surprisingly short. But for large drops the lifetimes predicted by Langmuir's equation appear to be approximately correct, as indicated by experimental measurements.

It is possible to extend the useful range of this equation to somewhat smaller drops. Fuchs (1959) initially pointed out that Langmuir's equation could not be correct for small particles having diameters approaching the mean free path of the gas since the equation predicted rates of molecular escape which exceeded the evaporation rate into a vacuum. To correct this difficulty, Fuchs considered the diffusion process to start a distance of approximately one mean free path from the droplet surface.

Applying Fuchs' modification to Langmuir's equation in the case of evaporation into a space where $p_l \approx 0$ gives

$$t = \frac{\rho RT}{DMp_x(T)} \left(\frac{d^2}{8} + \frac{dD}{2\alpha v_x} - \frac{\Delta d}{2} + \Delta^2 \ln \frac{d + 2\Delta}{2\Delta} \right) \quad (15.16)$$

In Eq. 15.16, Δ represents the distance an evaporating molecule must travel before it strikes a "gas" molecule and can be estimated from

$$\Delta = \lambda \left(\frac{m_1 + m_2}{m_1} \right)^{1/2} \quad (15.17)$$

The term α is an accommodation coefficient which has a value for pure water, according to Fuchs (1959), of 0.034 and for water with impurities can be even smaller; m_1 and m_2 are the masses of single gas and vapor molecules, respectively; λ is the mean free path of the gas molecules; and v_x is the most probable one-dimensional molecular velocity of the vapor molecule, defined as

$$v_x = \left(\frac{kT}{2\pi m_2} \right)^{1/2} \quad (15.18)$$

The effect of Fuchs' modification on Eq. 15.15 is to increase the lifetime of very small drops fairly significantly and to even have some effect on drops having diameters of 10 µm or more. Figure 15.1 shows a plot of the ratio of Eq. 15.16 to Eq. 15.15 for similar conditions as a function of d, the droplet diameter. As an example, for 1-µm-diameter droplets, evaporation times as estimated by using the Fuchs modification are about 20 times greater than those estimated with Langmuir's equation alone.

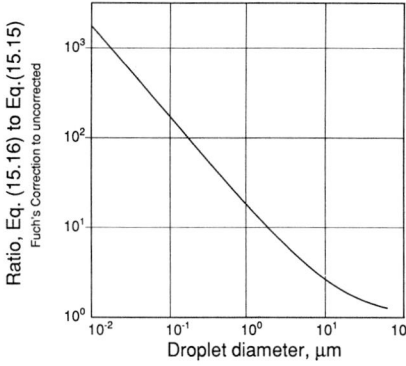

Figure 15.1 Evaporation times with and without Fuchs' correction.

Example 15.4 Compute the value of t, using Fuchs' correction for a 2-µm water droplet at 20°C evaporating into a space where $p_l = 0$. (Use $\alpha = 0.034$.) From Eq. 15.10, $T_s = 4.469°C$ and $p_s = 6.38$ mmHg. Then from Eq. 15.16

$$t = \frac{\rho RT}{DMp_s}\left(\frac{d^2}{8} + \frac{dD}{2\alpha v_x} - \frac{\Delta}{2}d + \Delta^2 \ln\frac{d+2\Delta}{2\Delta}\right)$$

$$v_x = \left[\frac{(1.38 \times 10^{-16})(293)(6.05 \times 10^{23})}{(2)(3.14)(18)}\right]^{1/2}$$

$$= 1.47 \times 10^4 \text{ cm/s}$$

$$\Delta = 6.87 \times 10^{-6}\left(\frac{18+29}{29}\right)^{1/2} = 8.75 \times 10^{-6} \text{ cm}$$

$$t = (7.03 \times 10^5)(5.00 \times 10^{-9} + 4.54 \times 10^{-8} - 8.75 \times 10^{-10} + 1.93 \times 10^{-10})$$

$$= (7.03 \times 10^5)(4.97 \times 10^{-8}) = 0.035 \text{ s}$$

Evaporation Time in a Saturated Medium

Langmuir's equation indicates that a droplet will not evaporate when $S = p/p_s \geq 1.0$. But according to the Kelvin equation, curvature effects will cause small droplets to evaporate, even when S exceeds 1.0. How can this apparent contradiction be resolved? One way is to replace the $p_s - p_l$ term in Eq. 15.14 with an equivalent term from Kelvin's equation which takes curvature into account. Recalling Kelvin's equation

$$\frac{p}{p_z(T)} = \exp\frac{4\gamma M}{\rho RTd} \tag{15.19}$$

This can be written as an expansion

$$\frac{p}{p_z(T)} = 1 + \frac{4\gamma M}{\rho RTd} + \frac{1}{2}\left(\frac{4\gamma M}{\rho RTd}\right)^2 + \frac{1}{6}\left(\frac{4\gamma M}{\rho RTd}\right)^3 + \cdots \tag{15.20}$$

If $p/p_x(T) \approx 1$, the first two terms of Eq. 15.20 are adequate to determine $p/p_x(T)$ so that

$$\frac{p}{p_x(T)} = 1 + \frac{4\gamma M}{\rho R T d} \qquad (15.21)$$

$$p_x(T) - p = -\frac{4\gamma M p_x(T)}{\rho R T d} \qquad (15.22)$$

For the case of complete evaporation, $p = 0$ and then Eqs. 15.14 and 15.22 can be combined,

$$\frac{dd}{dt} = \frac{4^2 D M^2 \gamma p_x(T)}{d^2 \rho^2 R^2 T^2} \qquad (15.23)$$

which gives, on integration and rearrangement of terms,

$$t = \frac{R^2 T^2 \rho^2}{4^2 D M^2 \gamma p_x(T)} \left(\frac{d^3}{3} - \frac{d_0^3}{3} \right) \qquad (15.24)$$

where d is the diameter of a droplet of initial diameter d_0 after a time t. For complete evaporation, $d = 0$ and

$$t = \frac{d_0^3}{3 D \gamma p_x(T)} \left(\frac{\rho R T}{4M} \right)^2 \qquad (15.25)$$

This represents an expression for the total lifetime of droplets in a medium which is essentially saturated with vapor.

Example 15.5 Estimate the lifetime of a 0.1-μm-diameter water droplet in saturated air at 20°C.
From Eq. 15.25

$$t = \frac{d_0^3}{3 D \gamma p_x(T)} \left(\frac{\rho R T}{4M} \right)^2$$

$$= \frac{(10^{-5})^3}{(3)(0.24)(72)(17.49)(1333)} \left[\frac{(1)(8.314 \times 10^7)(293)}{72} \right]^2$$

Note that p had to be converted to units of dynes per square centimeter by the factor 1333 because the surface tension units are also dynes per square centimeter and R is expressed in appropriate units.

$$t = 8.92 \times 10^{-5} \text{ s}$$

Growth and Evaporation of Moving Droplets

As a droplet moves relative to its surrounding medium, its evaporation rate may be speeded up by having the medium sweep away vapor molecules near the drop's surface. Fuchs (1959) examined in detail both theory and experimental data on this question and concluded that from a theoretical standpoint for droplets evaporating in the Stokes region (Re < 1), increased evaporation on the front face of the droplet is balanced by decreased evaporation on the rear. Thus he concluded that the overall rate would be unchanged.

For larger values of Re, Pruppacher and Klett (1978) give the following empirical equations to estimate the increase in evaporation rate caused by the moving medium.

For Re < 2.5,

$$\overline{f}_v = 1.00 + 0.108\,(Sc^{1/3}\,Re^{1/2})^2 \tag{15.26}$$

and for Re > 2.5,

$$\overline{f}_v = 0.78 + 0.308\,Sc^{1/3}\,Re^{1/2} \tag{15.27}$$

Here Re is the Reynolds number and Sc the Schmitt number, defined as the ratio of the kinematic viscosity to the diffusion coefficient

$$Sc = \frac{\nu}{D} \tag{15.28}$$

The term $\overline{f}_v$ is the ratio of the evaporation rate for the particle in moving air to the evaporation rate of the particle in still air. Because in a moving airstream a small aerosol particle rapidly attains the velocity of the medium around it, in most cases droplet motion can be neglected in considering evaporation and condensation estimates, unless the droplet diameters exceed ≈ 40 μm.

Problems

1 Determine the lifetime of a 50-μm-diameter benzene droplet when it is released into an atmosphere of oxygen at 25°C.

2 What would be the value of t for this same benzene droplet, using Fuchs' correction? Make whatever assumptions are necessary.

3 What would be the drying time for a 10-μm water droplet? Assume an air temperature of 20°C and a relative humidity of 20 percent. Also take into account the droplet temperature.

4 A 200-μm droplet is evaporating in air at 25°C and 40 percent relative humidity. At equilibrium, what is the internal temperature of the droplet?

5 Mason (1971) states that droplets with average radii of 30 μm evaporate completely after falling a few meters in unsaturated air. Show that this is true.

6 Zak (1936) found that the time for complete evaporation of a 1-mm diameter water droplet suspended from a thin glass fiber was 605 s at 20°C. Compare this value to one computed from Langmuir's equation, (*a*) assuming no temperature depression in the droplet and (*b*) considering a temperature depression (assume $S \approx 0$).

7 In an experiment water droplets were dried over phosphorous pentoxide to give an atmosphere of 0 percent relative humidity. Using Eq. 15.10, determine the droplet temperature if the ambient temperature is 19°C. Compare this value to one found from use of the psychrometric chart in App. F.

Chapter

16

Optical Properties

Extinction

The optical effects of aerosols are spectacular. Clouds, haze, and smoke all appear as they do because of the optical properties of the individual particles and the effects of these particles on each other. Few people are not affected by the drama of light scattering and absorption; brilliant sunsets inspire, myriad cloud colors excite, and dense, thick fogs disorient.

Most objects that can be seen are visible because they scatter light. A tree is seen with all its shadings of light and dark because of variations in scattering intensities over the different parts of the tree. Otherwise only a silhouette or an outline of the tree would be seen. Selective scattering and absorption are responsible for colors of things. Thus a leaf on a tree illuminated with natural light looks green because it absorbs red light more efficiently than green light. Black smoke appears black because the smoke particles efficiently absorb all visible wavelengths of light. White smoke appears white because the smoke particles efficiently scatter all visible wavelengths.

All aerosol particles scatter radiation. Some can also absorb it. Water, along with many other materials, is transparent to visible radiation but absorbs infrared radiation at some wavelengths.

The combination of scattering plus absorption is called *extinction*. Scattering functions that describe the light scattered or absorbed by a particle can be computed for spherical or cylindrical shapes by using a general mathematical theory known as the *Mie theory* after Mie (1908), although other investigators at the time had proposed essentially identical theories. See, e.g., the discussion by Kerker (1969). Although Mie's theory was formulated for spherical particles, experiments by Napper and Ottewill (1964) and Berry (1962, 1966) indicate that angular scattering patterns and extinction predictions for isometric particles such as cubes or octrahedra differ very little from those for spherical particles of the same equivalent size.

In this chapter overall scattering properties of particles are examined, and the relationship of these properties to visual phenomena is reviewed. In Chap. 17 angular scattering of light from aerosol particles is investigated.

Definition of Terms

For something to be "seen," there must be a *source* to provide the radiant energy and a *receptor* to receive and translate this energy to an image. The source can be some type of radiant energy emitter such as the sun, a light bulb, or an aerosol particle (the carbon particles glowing in a flame, e.g.), or it can be an object from which light from some other source is scattered, such as a building or a tree or an aerosol particle. The receptor is usually the eye, but it could be an instrument such as a photometer or a photomultiplier tube.

It is not enough, however, to have a source and receptor, although these are necessary conditions. For something to be seen, there must also be contrast between the object and its background. Without contrast a receptor cannot distinguish between the two.

Source represents the emission of radiant energy, which may be per unit time or area or both. In the cgs system of units, radiant energy is measured in ergs; and in the mks system it is measured in joules. Other radiometric units evolve as listed in Table 16.1.

For visible light other definitions have evolved with the concept of a *luminator* as the source of luminous energy and *lumination* as the process. The definitions are shown in Table 16.2, arranged in a manner

TABLE 16.1 Radiometric (Physical) Concepts

Name	Symbol	CGS unit	MKS unit
Radiant energy	U	erg	J
Radiant density	u	erg/cm^3	J/m^3
Radiant flux	P	erg/s	W
Radiant emittance	W	erg/(s · cm^2)	W/m^2
Radiant intensity	J	erg/(s · ω(*))	W/ω
Radiance	N	erg/(s · ω · cm^2)	W/(ω · m^2)
Irradiance	H	erg/(s · cm^2)	W/m^2
Spectral reflectance	r		
Spectral transmittance	t		

(*)ω = unit solid angle. The unit is normally the steradian.

SOURCE: W. E. K. Middleton, *Vision through the Atmosphere,* University of Toronto Press, Toronto, 1963, p. 6.

TABLE 16.2 Psychophysical (Photometric) Concepts

Name	Symbol	CGS unit	MKS unit*
Luminous energy	Q	lumerg	$lm \cdot s$
Luminous density	q	lumerg/cm^2	$lm \cdot s/m^2$
Luminous flux	F	lumerg/s	lm
Luminous emittance	L	lumerg/(s $\cdot$ cm^2)	lm/m^2
Luminous intensity	I	lumerg/(s $\cdot$ ω)	lm/ω
Luminance	B	lumerg/(s $\cdot$ ω $\cdot$ cm^2)	cd/m^2
Illuminance	E	lumerg/(s $\cdot$ cm^2)	lm/m^2 = lx

*lm $\cdot$ s = lumen $\cdot$ sec.
SOURCE: W. E. K. Middleton, *Vision through the Atmosphere,* University of Toronto Press, Toronto, 1963, p. 7.

similar to Table 16.1. Units such as the lumerg are hardly ever used because the cgs system has fallen into disfavor among optical physicists, but they are listed here because of the attempt in this book to stay with a consistent set of units throughout.

Fortunately most problems involving aerosol optics do not require the use of absolute quantities but instead make use of such unitless radiometric indicators as absorbance, reflectance, and the like. Table 16.3 lists several of the more common of these indicators. Figure 16.1 illustrates some of the more common definitions of light intensity.

In some works involving aerosol scattering the term *albedo* is used to describe the extinction of light by a particle or a system of particles. *Albedo* is defined as the fraction of incident light or radiant energy that is reflected or scattered by the particle or system of particles. It is a dimensionless fraction commonly used to describe the light reflected from the earth back into space.

Example 16.1 A satellite is used to relay television signals to the earth. If scattering and absorption by the earth's atmospheric aerosols cause the earth's albedo to be 0.60, how much of the signal from the satellite reaches the earth?

Fraction of signal reaching earth = 1.00 − 0.6 = 0.4

TABLE 16.3 Unitless Radiometric Indicators

Quantity	Symbol	Defining equation	Unit
Absorbance	a, A	a = (*)absorbed/(*)incident	Numeric
Reflectance	ρ, R	ρ = (*)reflected/(*)incident	Numeric
Transmittance	τ, T	τ = (*)transmitted/(*)incident	Numeric

*Represents the appropriate quantity such as U, P, B, etc.
SOURCE: D. Sliney and M. Wolbarsht, *Safety with Lasers and Other Optical Sources,* Plenum, New York, 1980, p. 937.

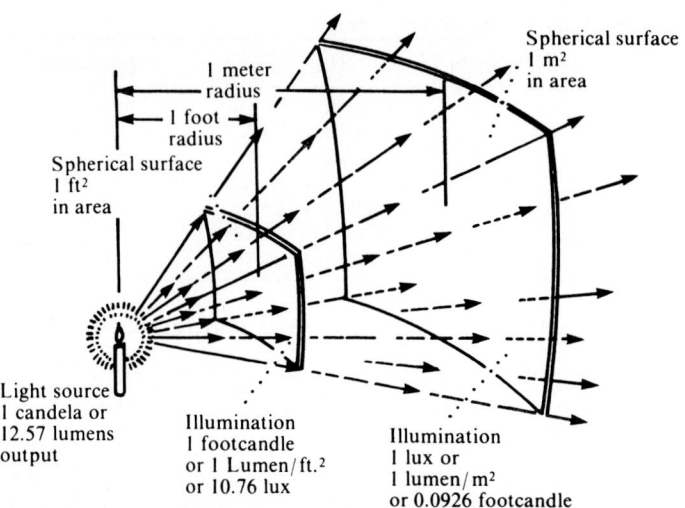

Figure 16.1 Definitions of light intensity.

A fundamental factor required for consideration of the optical properties of aerosols is the particle refractive index m. The *refractive index* of a material is defined as the ratio of the speed of light in a vacuum to the speed of light in the material. When there is appreciable absorption of radiation in the aerosol particle as well as scattering, it is necessary to express the refractive index of a material as a complex number of the form

$$m = \nu - i\kappa \tag{16.1}$$

Here the parameter ν represents the real part of the index and κ the imaginary part. The real part represents scattering and the imaginary part, absorption. Carbon particles, e.g., have a refractive index of about $(2 - i)$. In general, for airborne particles, the real part of the refractive index for most materials lies between 1.0 and 1.6, and the imaginary part between 0 and 1. Refractive indices of various materials are given in Table 16.4.

Most aerosol materials will vary in their refractive index depending on the wavelength of light used, their chemical composition, and, in some cases, their orientation with respect to the light source and receptor. Since complex indices of refraction are not well established for most materials (Deirmendjian, 1969), optical models of aerosols may contain errors because of the uncertainty of these values.

Extinction of Light—Bouguer's Law

It has been established theoretically and experimentally that a thin section dl of a particle-containing medium such as air will both scatter

TABLE 16.4 Refractive Indices for Selected Materials

Material	Density, g/cm^3	m	Wavelength
Vacuum	—	1.000	Visible light
Air	0.0012	1.0002918	Visible light
Alumina	3.9	1.67	Visible light
Ice	1.00	1.31	Visible light
Water	1.00	1.333	Visible light
Water	1.00	1.153 − 0.0968i	11 μm
H_2SO_4	1.841	1.430	550 nm
38% H_2SO_4, 62% H_2O	1.32	1.394	550 nm
38% H_2SO_4, 62% H_2O	1.32	1.46 − 0.38i	9.5 μm
Diamond	3.51	2.417	Visible light
Glass	2.45	1.51–2.00	Visible light
NaCl	1.33	1.5443	Visible light
Polystyrene latex	1.05	1.5	Visible light
$(NH_4)_2SO_4$	1.769	1.528	550 nm
NH_4HSO_4	1.780	1.482	589 nm
NH_4NO_3	1.725	1.559	550 nm
$CaCO_3$	2.930	1.586	550 nm
SiO_2	2.17–2.66	1.478	550 nm
Atmospheric aerosol		1.5 − 0.02i	Visible light
Urban aerosol	1.60	1.5 − 0.1i	Visible light
Soot aggregates	1.00	1.56 − 0.47i	Visible light

and absorb light in an amount proportional to the flux of light entering the section. If F is the luminous flux as defined in Table 16.2 and b a factor of proportionality for scattering, this can be written as

$$dF = -bF\,dl \tag{16.2}$$

Equation 16.2 integrates to

$$F = F_0 e^{-bl} \tag{16.3}$$

where l is the path length from source to receptor. For absorption, with k as the factor of proportionality, an equation similar to Eq. 16.2 gives

$$F = F_0 e^{-kl} \tag{16.4}$$

The coefficients b and k are the scattering coefficient and absorption coefficient, respectively, and can be combined into a single extinction coefficient $\gamma = b + k$, so that

$$F = F_0 e^{-(b+k)l} = F_0 e^{-\gamma l} \tag{16.5}$$

The factor γ is also known as the *turbidity* or *attenuation coefficient*. Equations 16.3 through 16.5 are all forms of what is sometimes known as *Beer's law* but should more properly be called *Bouguer's law*, in honor of the person who empirically established it in 1760.

Example 16.2 In a certain experiment the luminous flux F is reduced to 36.8 percent of its original value when a beam of light is passed through an aerosol over a path length of 10 m. Determine the numerical value of γ in 1/meter.

From Eq. 16.5

$$\frac{F}{F_0} = e^{-(b+k)l} = e^{-\gamma l} = 0.368$$

$$-\gamma l = \ln 0.368$$

$$-\gamma l = -1.00$$

$$\gamma = 0.100 \text{ m}^{-1}$$

For an aerosol containing many particles of the same size,

$$\gamma = nQ_{\text{ext}}A \tag{16.6}$$

where n is the number of particles per unit volume of medium, A is the particle cross-sectional area, and Q_{ext} is an extinction efficiency factor defined as

$$Q_{\text{ext}} = \frac{\text{total energy flux extinguished by single particle}}{\text{total energy flux geometrically incident on particle}}$$

The term Q_{ext} rises in value as d increases from near 0 for very small particles to 2 for larger particles. A scattering efficiency factor Q_{scat} and an absorption efficiency factor Q_{abs} can each be defined in a manner similar to Q_{ext}. Then from Eq. 16.5

$$Q_{\text{ext}} = Q_{\text{scat}} + Q_{\text{abs}} \tag{16.7}$$

For spherical particles Eq. 16.6 becomes

$$\gamma = nQ_{\text{ext}}\frac{\pi}{4}d^2 \tag{16.8}$$

With a polydisperse aerosol having n_i particles of cross-sectional area A_i, it is necessary to sum γ for each particle size over all particle sizes, or

$$\gamma = \sum_{i=1}^{\infty} n_i Q_{\text{ext},i} A_i \tag{16.9}$$

Equation 16.9 can be treated as an integral by considering $n(d)\,dd$ particles per unit volume having diameters in the interval d to $d + dd$. The total number of particles per unit volume is

$$n = \int_0^{\infty} n(d)\,dd \tag{16.10}$$

For spheres, remembering that Q_{ext} is also a function of d,

$$\gamma = \int_0^\infty \frac{\pi}{4} d^2 n(d) Q_{ext}\, dd \qquad (16.11)$$

For the case of a thin cloud consisting of spherical particles whose concentration m_p g/cm^3, γ can be expressed in terms of the mass concentration by recalling that $m = (\pi/6)d^3\rho n$ so that γ becomes

$$\gamma = nAQ_{ext} = \frac{3 m_p Q_{ext}}{2 \rho d} \qquad (16.12)$$

When d is expressed in centimeters, γ has the units 1/centimeters. For a constant Q_{ext} and mass of material, Eq. 16.12 indicates that decreasing the particle size increases the extinction of light for an aerosol. That is, for the same amount of mass, small particles produce more haze in the atmosphere than large particles do. This conclusion is valid only over the range where Q_{ext} is essentially constant. For very small particles Q_{ext} is strongly dependent on particle size.

Example 16.3 A typical urban aerosol has an average particle concentration of 75 μg/m^3. Assuming that the average particle can be represented by a 0.6-μm sphere with a density of 2 g/cm^3, determine the value of γ, in 1/meters, if $Q_{ext} = 2$.
From Eq. 16.12

$$75\ \mu g/m^3 = 75 \times 10^{-12}\ g/cm^3$$

$$\gamma = \frac{(3)(75 \times 10^{-12})(2)}{(2)(2)(6 \times 10^{-5})} = 1.88 \times 10^{-6}\ cm^{-1}$$

$$= 1.88 \times 10^{-4}\ m^{-1}$$

For visible light, aerosols are most optically active in the 0.1- to 1.0-μm-diameter range. This can be seen in Fig. 16.2, which is a plot of energy flux extinguished by the particle per unit volume versus par-

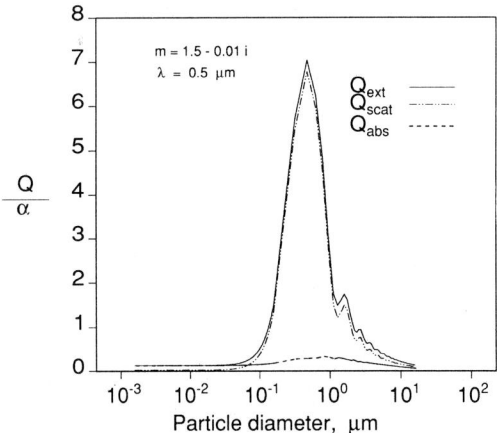

Figure 16.2 Energy flux extinguished by a particle per unit volume.

ticle diameter for light of wavelength 0.5 μm and a typical atmospheric aerosol. It can be seen that particles smaller than about 0.05 μm have the least effect in light attenuation as do those larger than about 3 μm.

Assumptions Implicit in Bouguer's Law

Utilization of Bouguer's law for extinction estimates assumes that certain simplifying assumptions are met. For example, it is assumed that the scattered light will have the same wavelength as the incident light. Although this is not precisely the case, wavelength changes are so small that they can usually be neglected.

A second assumption is that the particles act as independent scatterers, i.e., the scattering of light by one particle does not influence the scattering by another. If particles are separated by more than about 2 diameters, this assumption is met (Van de Hulst, 1957). For 1-μm-diameter particles, this means concentration on the order of 1.25 × 10^{11} particles per cubic centimeter before this assumption breaks down. At a concentration of this magnitude, other factors such as coagulation (see Chap. 18) come into play to reduce the concentration, so the assumption is always valid.

A third assumption is that of single scattering, which is another way of saying that a maximum of one scatter per photon is allowed. This assumption implies that Bouguer's law is valid only for thin clouds or low concentrations; i.e., the product of γl should not exceed 0.1 (Hodkinson, 1966). A simple way to test for the absence of multiple scattering is to observe if the scattering intensity is doubled with a doubling of concentration. If so, single scattering prevails.

Computation of Extinction Coefficient

The term Q_{ext} represents an efficiency factor. In Example 16.3 it was assumed Q_{ext} had a constant value of 2. In actuality for small particles Q_{ext} is a function of d, the particle diameter, whereas for larger particles Q_{ext} is equal to 2. For very small particles with even a little bit of absorption, $Q_{ext} \approx Q_{abs}$. For pure scatterers, $Q_{ext} \approx Q_{scat}$. Then for very small particles

$$Q_{ext} = Q_{scat} = \frac{8}{3}\alpha^4 \left(\frac{m^2 - 1}{m^2 + 2}\right)^2 \qquad (16.13)$$

where m is the refractive index of the particle. The particle diameter and wavelength of the incident light λ are related through the dimensionless factor α, defined as

$$\alpha = \frac{\pi d}{\lambda} \qquad (16.14)$$

Specifically, Eq. 16.13 applies when $\alpha \ll 1$. This special case of light scattering is called *Rayleigh scattering*. One important example of Rayleigh scattering is the scattering of light by molecules making up the earth's atmosphere.

Example 16.4 Compute the value of Q_{scat} for an "air" molecule when illuminated with blue light ($\lambda = 0.4$ μm). Then compare this value to Q_{scat} for the same molecule illuminated with red light ($\lambda = 0.7$ μm). Assume a molecular diameter of 4×10^{-8} cm and $m = 1.000292$ (pure scattering).

$$\alpha_{blue} = \frac{\pi(4 \times 10^{-8})}{0.4 \times 10^{-4}} = 3.14 \times 10^{-3}$$

$$\alpha_{red} = \frac{\pi(4 \times 10^{-8})}{0.7 \times 10^{-4}} = 1.80 \times 10^{-3}$$

$$Q_{scat\ blue} = \frac{8}{3}(3.14 \times 10^{-3})^4 \left(\frac{1.000292^2 - 1}{1.000292^2 + 2}\right)^2$$

$$= 9.84 \times 10^{-18}$$

$$Q_{scat\ red} = \frac{8}{3}(1.80 \times 10^{-3})^4 (3.79 \times 10^{-8})$$

$$= 1.05 \times 10^{-18}$$

Ratio, blue/red = 9.38

Notice that in this case blue light is scattered about 9 times as efficiently as red light. Could this have something to do with the sky being blue and sunsets red?

For larger particles the relationship of Q_{ext} to d is more complex, with Q_{ext} rising as d increases according to the fourth-power relationship of Rayleigh scattering until Q_{ext} finally oscillates around and then becomes asymptotic to a value of $Q_{ext} = 2$, the degree of oscillation depending on the amount of absorption taking place. For absorbing particles there are essentially no oscillations; for pure scatterers the oscillations are quite pronounced.

For water droplets ($m = 1.33$), a plot of Q_{ext} versus α is shown in Fig. 16.3. Oscillations in the value of Q_{ext} are due to internally reflected light being in or out of phase during scattering. Also shown is a plot for a material having the same real refractive index as water but a small and intermediate absorption component ($m = 1.33 - 0.01i$ and $m = 1.33 - 0.1i$). The effect of absorption on oscillations in Q_{ext} as the absorption component increases can be clearly seen.

The asymptotic value of $Q_{ext} = 2$ implies that a particle can remove light from an area equal to twice its cross-section. This "extinction

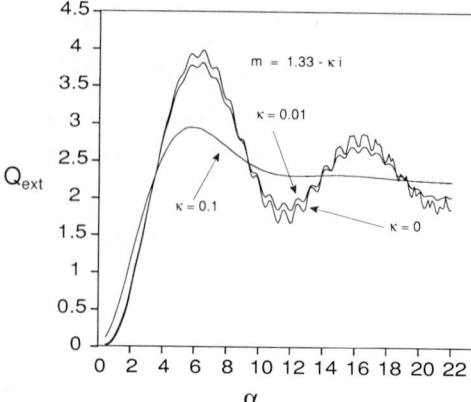

Figure 16.3 Plot of Q_{ext} versus α for $m = 1.33 - \kappa i$.

paradox" arises from light being diffracted by the particle into the space behind it from which other photons had previously been scattered or absorbed. This amount will just be equal to that diffracted through a hole having the same cross-sectional area as the particle. As a result, the total amount of energy removed from the proceeding wave will be twice that shadowed out by the particle itself. Hence $Q_{ext} = 2$.

The asymptotic value of $Q_{ext} = 2$ is important not only for particles which are primarily scatterers but also for irregularly shaped particles or those which are highly absorbing. These types of particles reach the value $Q_{ext} = 2$ at fairly small particle sizes. For example, for an irregular, transparent particle $Q_{ext} = 2$ is reached when $\alpha(m - 1) > 10$, or for a salt crystal, when its edge is approximately 3 μm.

With spheres having some absorption Q_{ext} rises quite sharply from 0 to a value slightly in excess of 2, and then with continued increase in α it slowly falls back to the asymptotic value of 2. This behavior can be seen both in Fig. 16.3 for the case of $\kappa = 0.1$ and in the calculations shown in Fig. 16.4 (Deirmendjian, 1969). In these calculations the real part of m was kept constant with only the imaginary part of the refractive index being varied. The maximum value of Q_{ext} which could be reached is reduced by absorption, and the efficiency of smaller particles is increased. It can also be seen that a highly reflecting sphere ($m = 10$) is not as efficient over some size ranges in removing light as particles which have some absorption.

For ensembles of particles the light extinction is additive, as discussed previously. However, in certain cases simplifying assumptions can be made. For example, suppose there is a polydisperse aerosol having $n(d)$ particles of diameter d per unit volume. From Eq. 16.11

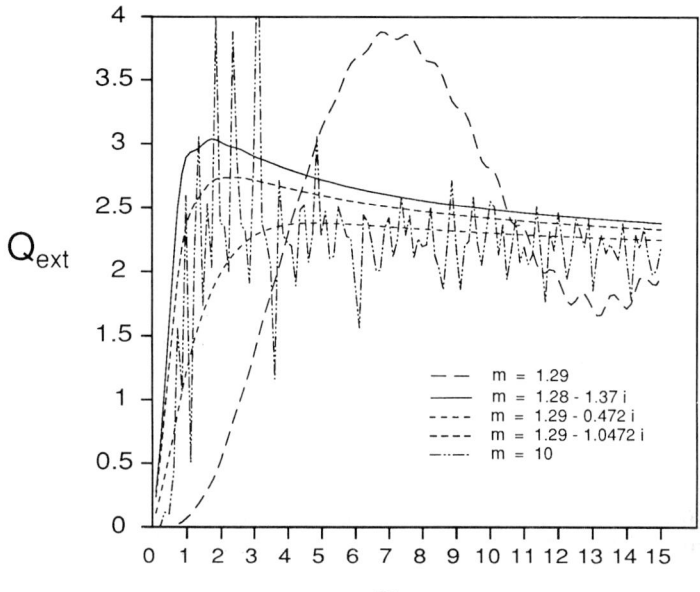

Figure 16.4 Q_{ext} as a function of α for various refractive indices. (*Adapted from D. Deirmendjian, 1969.*)

the extinction coefficient is

$$\gamma = \int_0^x \frac{\pi}{4} d^2 n(d) Q_{ext}\, dd \qquad (16.15)$$

and the mass of material per unit volume is

$$m_p = \frac{\pi}{6} \rho_p \int_0^x d^3 n(d)\, dd \qquad (16.16)$$

Chýlek (1978) has suggested that Eqs. 16.15 and 16.16 can be combined to provide a relationship between aerosol mass and extinction coefficient. Then

$$m_p = \frac{2\rho_p \int d^3 n(d)\, dd}{3 \int d^2 n(d) Q_{ext}\, dd} \gamma \qquad (16.17)$$

Although Q_{ext} is a complicated function of both α and m, in its extremes (e.g., very small d or large d) it can be roughly approximated by the form $Q_{ext} = u + z\alpha$ or $Q_{ext} = 2$, where z is a constant slope and u is a constant. These approximations are illustrated in Fig. 16.5 for a nonabsorbing and an absorbing particle.

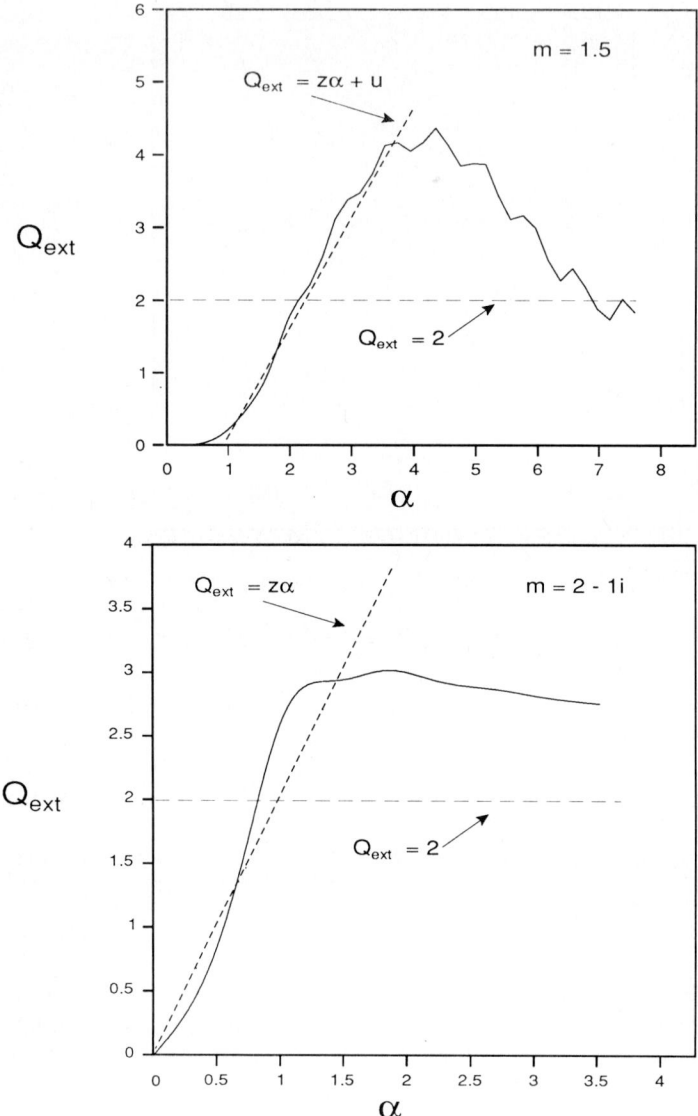

Figure 16.5 Q_{ext} as a function of α.

For the case of $Q_{ext} = 2$, Eq. 16.17 becomes

$$m_p = \frac{\rho_p}{3} \frac{\int d^3 n(d)\, dd}{3 \int d^2 n(d)\, dd} \gamma \qquad (16.18)$$

The ratio

$$\frac{\int d^3 n(d)\, dd}{\int d^2 n(d)\, dd} = d_{sm} \qquad (16.19)$$

is equal to the surface mean diameter (Sauter diameter). Hence

$$m_p = \frac{\rho_p}{3} \gamma d_{sm} \qquad (16.20)$$

For the case of a lognormal aerosol having a geometric mean diameter of d_g and geometric standard deviation of σ_g,

$$m_p = \frac{\rho_p}{3} \gamma d_g \exp(2.5 \ln^2 \sigma_g) \qquad (16.21)$$

Example 16.5 Estimate the value of γ for a lognormally distributed aerosol having $d_g = 2.0$ and $\sigma_g = 1.5$ if the particle density is 2 g/cm³ and $m = 100$ μg/m³. From Eq. 16.21

$$\gamma = \frac{3 m_p}{\rho_p d_g \exp(2.5 \ln^2 \sigma_g)}$$

$$= \frac{3(100 \times 10^{-12})}{(2)(2 \times 10^{-4}) \exp(2.5 \times 0.164)} = 4.97 \times 10^{-7} \text{ cm}^{-1} = 4.97 \times 10^{-5} \text{ m}^{-1}$$

For very small aerosol particles (or large values of λ) when there is some absorption, Eq. 16.17 becomes

$$m_p = \frac{2 \rho_p \int d^3 n(d)\, dd}{3 \int d^2 n(d)(\pi d/\lambda)\, dd} \gamma = \frac{2 \rho_p \lambda \gamma}{3 \pi z} \qquad (16.22)$$

an expression independent of particle size. The constant z is a function of m, the refractive index. Error is minimized when m is complex (some absorption) and α is small. These conditions are often met with infrared illumination.

Receptor—Contrast

To see an object, whether it is emitting light or not, requires a certain level of contrast between the object and its background. Without contrast the object blends into its background and becomes invisible. For an isolated object surrounded by a uniform and extensive background, *contrast* can be defined as

$$C = \frac{B}{B'} - 1 \tag{16.23}$$

where B is the luminance of an object and B' the luminance of the background. When these are equal, the contrast C is equal to zero, indicating that the object cannot be seen. When an object is less luminous than its background, C has a negative value, reaching -1 for a black object against a white background. For the opposite case C can go to infinity. A light at night represents a large positive C value.

The question of how much contrast is necessary for an object to be seen is one of great importance and much speculation. From extensive studies carried out during World War II, it appears that this contrast level depends on such tangibles as the medium through which the object is being viewed (is the object being viewed through air or water?) and on intangibles such as the psychological state of the viewer, eye adaption, etc. For daylight conditions with a black object being viewed against a white background, a value of $C = -0.02$ is often used. Although rough, this value is used as a guide for visibility approximations.

Example 16.6 Determine the value of B/B' which gives rise to $C = -0.02$.
From Eq. 16.6

$$C = \frac{B}{B'} - 1 = -0.02$$

$$\frac{B}{B'} = 1 - 0.02 = 0.98$$

The luminance of the object is 98 percent of the luminance of the background.

Alteration of Contrast

Aerosols in the atmosphere change the contrast of the atmosphere. This is evident to anyone who has viewed objects at a distance. Visibility decreases as the atmospheric load of aerosol material increases.

The first attempt to explain the alteration of contrast was made by Koschmieder (1924a and 1924b). In developing his model for the alteration of contrast, he determined the amount of light from around a black object scattered into the vision of an observer such that the observer would think the scattered light came from the object. To do this, it was necessary for Koschmieder to make a number of simplifying assumptions. He assumed that the atmosphere contains a large number of small particles and each volume element contains a large number of particles, much smaller than the element. He further assumed that the earth's curvature could be neglected, each volume element could be treated as a point source with independent scattering, a cloudless sky (equal illumination to all parts of the atmosphere in a horizontal plane), a constant scattering coefficient b, the observed object small compared to the distance of the object to the observer, and negligible refraction of light by the earth's atmosphere.

A volume element is then defined as shown in Fig. 16.6 where the element has volume

$$d\varphi = x^2 \, d\omega \, dx \tag{16.24}$$

Since this volume element will be illuminated exactly as the object, regardless of x, the luminous intensity scattered from the volume element toward the observer is

$$dI = Ab \, d\varphi \tag{16.25}$$

where A is a constant. The illuminance at the eye of the observer due to the light from $d\varphi$ is

$$dE = dI \, x^{-2} e^{-bx} \tag{16.26}$$

The x^{-2} term arises from the inverse-square law, indicating the amount of light coming from $d\varphi$ at a particular distance x to the ob-

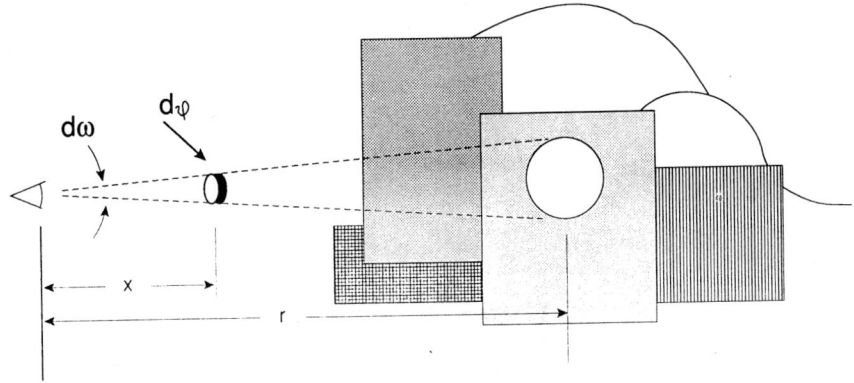

Figure 16.6 Model conditions for Koschmieder's derivation.

server relative to all light coming from that distance x. The e^{-bx} term represents the scattering of light along the path from $d\varphi$ to the eye. Then

$$dB = \frac{dE}{d\omega} = Abe^{-bx} dx \tag{16.27}$$

$$B = \int_0^r Abe^{-bx} dx = B_h(1 - e^{-br}) \tag{16.28}$$

where B_h is the illuminance of the horizon sky and r is the distance from the object to the observer.

From Eq. 16.23

$$C = \frac{B}{B_h} - 1 = -e^{-br} \tag{16.29}$$

Next ϵ is defined as the threshold of brightness contrast. This represents the contrast level where a black object can just be seen against a white background. Using $\epsilon = -0.02$ in Eq. 16.29 and recognizing that this value can vary by as much as an order of magnitude, C can be replaced in Eq. 16.29. Then,

$$-0.02 = -e^{-br} \tag{16.30}$$

Rewriting in terms of r

$$r = -\frac{1}{b} \ln 0.02 = \frac{3.91}{b} \tag{16.31}$$

The term r is known as the *visual range* and has units of length. Since in actuality concentration can vary over the distance between the object and the observer, r should more properly be called the *meteorological range* (Charlson et al., 1967).

Example 16.7 What is the estimated visual range in an aerosol having a concentration of 0.1 mg/m³ if the geometric mean particle diameter is 0.6 µm ($\sigma_g = 2.2$) and $Q_{scat} = 2$? Assume $\rho = 2$ g/cm³.
From Eq. 16.21

$$\gamma = \frac{3m}{\rho_p d_g \exp(2.5 \ln^2 \sigma_g)}$$

$$= \frac{3(0.1 \times 10^{-9})}{(2)(0.6 \times 10^{-4}) \exp(2.5 \times 0.622)} = 5.28 \times 10^{-7} \text{ cm}^{-1}$$

$$r = \frac{3.91}{5.28 \times 10^{-7}} = 7.41 \times 10^6 \text{ cm} = 74.1 \text{ km}$$

It might be thought that the visual range equation could be used to measure aerosol mass concentration in the atmosphere since visual range is a fairly simple measurement. Indeed, some studies comparing predicted mass to measured mass concentrations tend to bear this assumption out. The difficulty lies in choosing a proper average particle diameter d. As pointed out in Chap. 14, the average size of atmospheric aerosol particles can vary markedly, depending on their moisture content. For soluble nuclei this can be further confounded by the hysteresis effect, by which the value of d will be determined by whether the nuclei are in an atmosphere of rising or falling humidity. Since often this fact is difficult to ascertain, especially with a moving air mass measured at a stationary point, mass concentration measurements derived from extinction measurements should be considered valid only for cases where the atmospheric humidity is less than 40 percent.

Problems

1 If the luminance of the sky near the horizon on a moonless clear night is 10^{-3} cd/m^2, determine whether a star of magnitude 5 (0.1 cd/m^2) would be seen.

2 Charlson et al. (1967) give the empirical expression

$$\text{Mass } (\mu\text{g/m}^3) = (3 \times 10^5)b$$

where b is in meters^{-1}. Assuming $Q_{\text{scat}} = 2$, what particle size does this relationship imply?

3 Using the expression in Prob. 2, show that

$$l \times \text{concentration} = 1.2 \text{ g/m}^2$$

What is the physical significance of this product?

4 Determine Q_{scat} for a 0.01- and 0.05-μm-diameter water droplet using blue light.

5 The particulate standard relating visual range to particulate concentration is

$$R = \frac{120}{G}$$

where R is the range in kilometers and G is the particle concentration in micrograms per cubic meter. Assuming that Q_{ext} equals 2, what average particle diameter is implied by this standard?

6 In Prob. 5, what would be the implied geometric mean diameter if the geometric standard deviation were assumed to be 2.0?

Chapter 17

Optical Properties

Angular Scattering

Definitions

The terms Q_{scat}, Q_{abs}, and Q_{ext} represent loss of radiation along the path from source to observer, i.e., the extinction of light by an aerosol. But often interest centers more on the scattering of light in a single direction. The diameter of a particle can be estimated by the quantity of light scattered from the particle into a detector. Or, if a thundercloud is to be tracked by radar, the intensity of radiation backscattering is important.

To establish a frame of reference, θ is defined as the forward angle of scattering, as shown in Fig. 17.1, representing the angle between

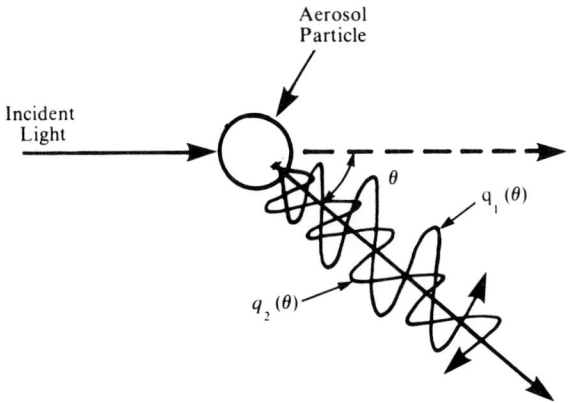

Figure 17.1 Sketch of definitions $q_1(\theta)$ and $q_2(\theta)$ for directional scattering from an aerosol particle.

the direction of propagation of the light and the direction of scattering. An angle of $\theta = 0°$ represents *forward scattering,* i.e., light scattered along the direction of propagation.

A scattering intensity coefficient $q(\theta)$ can be defined as

$$q(\theta) = \frac{\text{flux scattered into } \theta}{\text{flux geometrically incident on particle}} \qquad (17.1)$$

The incident light is considered to be parallel and may or may not be polarized. If it is polarized, then the scattered light must also be polarized in the same plane. If the incident light is not polarized, the scattered light may or may not be polarized. With polarized light $q_1(\theta)$ is defined as the scattering intensity coefficient for light propagated in a plane normal to the plane formed by the incident and scattered light vectors; $q_2(\theta)$ is the scattering intensity coefficient for light propagated parallel to that plane.

For unpolarized incident light

$$q(\theta) = \tfrac{1}{2}[q_1(\theta) + q_2(\theta)] \qquad (17.2)$$

The $q_1(\theta)$ or $q_2(\theta)$ components can be observed by polarizing either the incident or scattered radiation. Then, if I_0 is the intensity of an incident beam of perpendicularly polarized light, the intensity of the beam scattered per unit solid angle by a particle of diameter d will be

$$I = q_1(\theta) I_0 \frac{\pi}{4} d^2 \qquad (17.3)$$

A similar expression can be written for plane polarized light. For n monodisperse particles per unit volume, assuming single scattering,

$$I = n q_1(\theta) I_0 \frac{\pi}{4} d^2 \qquad (17.4)$$

In terms of $q_1(\theta)$ and $q_2(\theta)$, the relationship for the total scattering efficiency of a particle is

$$Q_{\text{scat}} = \pi \int_0^\pi [q_1(\theta) + q_2(\theta)] \sin \theta \, d\theta \qquad (17.5)$$

Example 17.1 For Rayleigh scattering the perpendicularly polarized light is scattered equally in all directions, $q_1(\theta) = c$; the plane polarized light is scattered according to $q_2(\theta) = c \cos^2 \theta$, where the constant c depends on the values of α and m. If c is equal to $\alpha^4/\{\pi[(m^2 - 1)/(m^2 + 2)]^2\}$, use Eq. 17.5 to determine Q_{scat} for Rayleigh scattering.

$$Q_{\text{scat}} = \pi \int_0^\pi (c + c \cos^2 \theta) \sin \theta \, d\theta$$

$$= 2\pi c \left(-\cos\theta - \frac{\cos^3\theta}{3} \right) \Big|_0^{\pi/2}$$

$$= 2\pi c \left(\frac{4}{3} \right) = \frac{8}{3} \alpha^4 \left(\frac{m^2 - 1}{m^2 + 2} \right)^2$$

Mie Scattering—The Mie Theory

Determination of the value of $q(\theta)$ as a function of m and θ for all values of α and the two polarization states can be accomplished through the use of the Mie theory of radiation scattering. Unfortunately solutions of Mie's equations do not lend themselves readily to numerical computation, but they can be solved by using computers. Manageable solutions other than Mie's theory are available for the cases where $\lambda > d$ (Rayleigh scattering) or when $\alpha \gg 1$ (geometric optics). For the intermediate region, representing particle diameters from about 0.1 to 10 μm, Mie solutions must be used.

Although Mie's theory was first published in 1908, computations of scattering coefficients were not tabulated to any extent until the 1940s (Lowan, 1948), and then the available tables were quite limited. Nevertheless, considering that each data point represented many hours of error-free calculation with a desk calculator, the accuracy of these early tables is indeed remarkable.

With the advent of high-speed electronic computers, calculation of Mie scattering functions has become routine, if not commonplace, and library programs are available for this task. Kerker (1969) gives a listing of various sources of computed Mie functions and discusses problems associated with the mechanics of computation. Programs are available for doing Mie computations on a personal computer (e.g., RI, Inc., 1989).

Mie scattering functions are generally presented in terms of the intensity parameters for Mie scattering, also known as the *angular intensity functions* $i_1(\theta)$ and $i_2(\theta)$. The subscripts of these functions indicate perpendicular and plane polarization, respectively. Besides being functions of the scattering angle θ, $i_1(\theta)$ and $i_2(\theta)$ are functions of the particle properties m and α [e.g., Lowan (1948) or Denman et al. (1966)].

For a given α and m, the angular intensity functions are related to the scattered intensity coefficients $q_1(\theta)$ and $q_2(\theta)$ by the expressions

$$q_1(\theta) = \frac{i_1(\theta)}{\pi \alpha^2} \tag{17.6}$$

$$q_2(\theta) = \frac{i_2(\theta)}{\pi \alpha^2} \tag{17.7}$$

Plots of $i_1(\theta)$ and $i_2(\theta)$ as a function of θ for $m = 1.33$ and three values of α are given in Fig. 17.2a to c. Figure 17.3a to c shows plots of $i_1(\theta)/\alpha^2$ as a function of α for 0°, 90°, and 180° scattering, respectively. Although representing only one refractive index, that of water droplets, these figures show the variations in $i_1(\theta)$ and $i_2(\theta)$ expected for many transparent aerosol materials. As α increases, the forms of both $i_1(\theta)$ and $i_2(\theta)$ become more complicated.

The degree of polarization of light scattered from a particle that is illuminated with unpolarized light can be estimated from the ratio $[i_1(\theta) - i_2(\theta)]/[i_1(\theta) + i_2(\theta)]$.

Example 17.2 Perpendicularly polarized white light is used to illuminate a 0.5-μm-diameter sphere. If the sphere has a refractive index of 1.33, what will be the predominant color of light scattered at an angle of 90°?

For visible light (λ = 0.4 to 0.7 μm) and a 0.5-μm-diameter sphere, α can range in value from 2.24 to 3.93. From Fig. 17.3b, $i_1(\theta)$ is a maximum within this range when α is equal to 2.8. This represents a light wavelength of 0.56 μm, a color of violet blue.

Approximations to Mie theory

There have been many attempts to find approximations for the Mie extinction and intensity functions that would be more wieldy, but for the most part the search has been unsuccessful. Penndorf (1962) gives

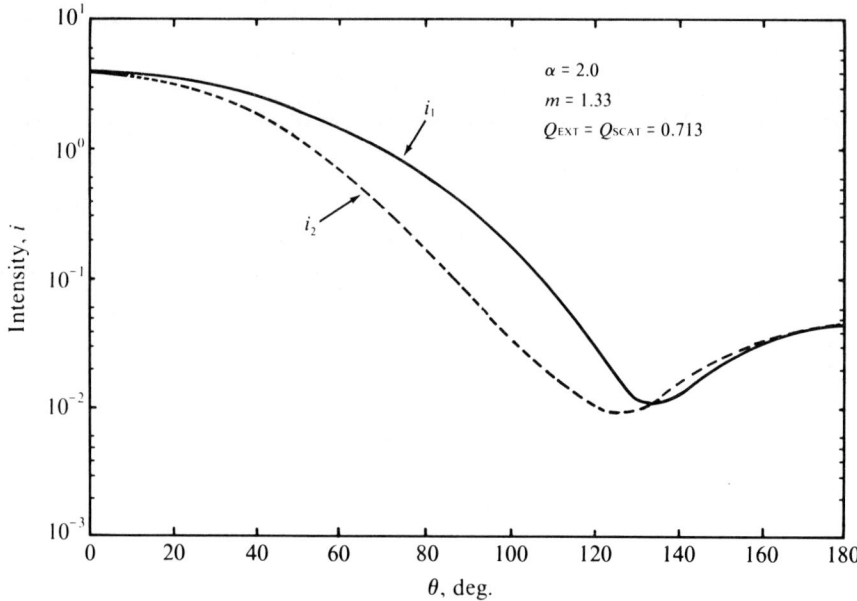

Figure 17.2a Angular intensity functions $i_1(\theta)$ and $i_2(\theta)$ for a sphere; $\alpha = 2.0$, $m = 1.33$.

Optical Properties   285

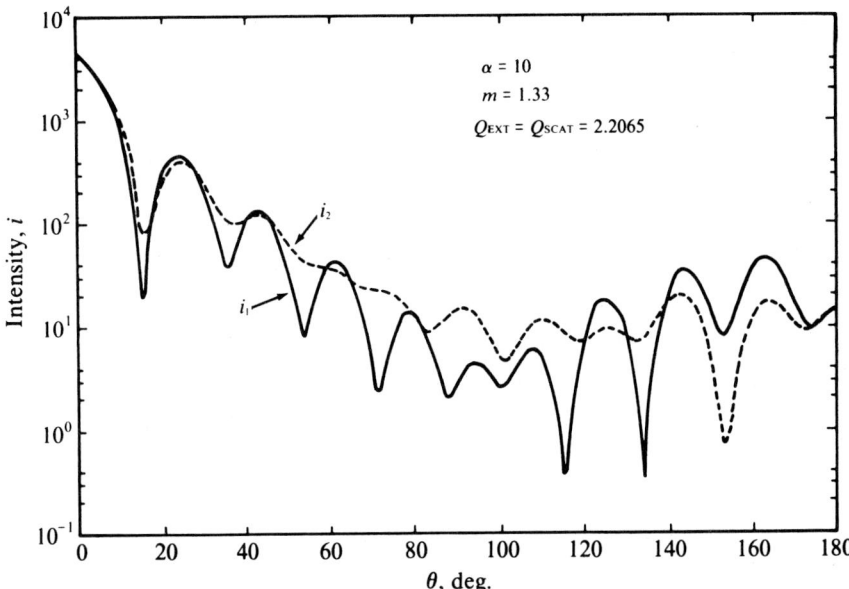

Figure 17.2b Angular intensity functions $i_1(\theta)$ and $i_2(\theta)$ for a sphere; $\alpha = 10$, $m = 1.33$.

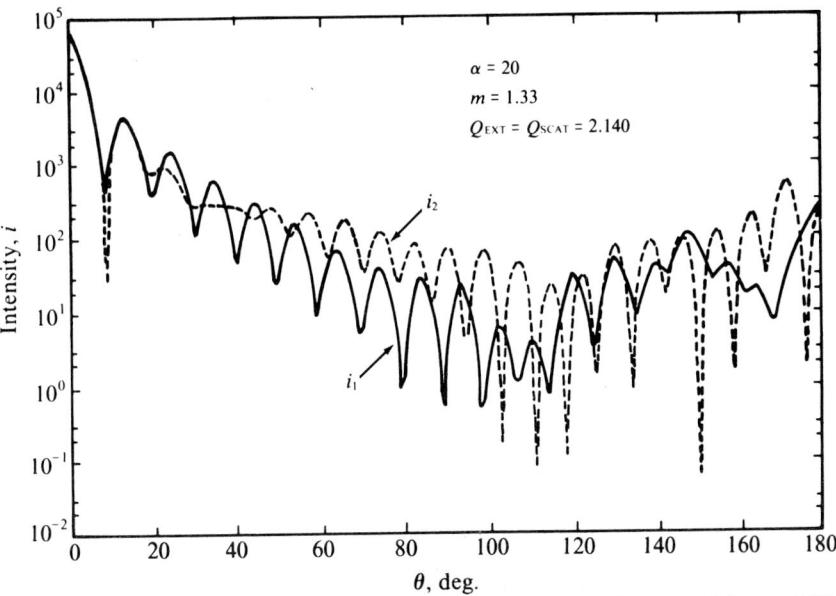

Figure 17.2c Angular intensity functions $i_1(\theta)$ and $i_2(\theta)$ for a sphere; $\alpha = 20$, $m = 1.33$.

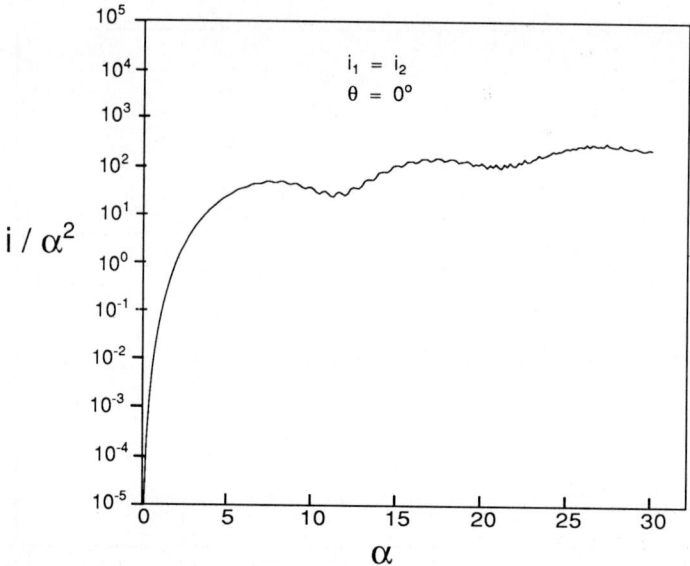

Figure 17.3a Plot of i/α^2 for 0° scattering angle, $m = 1.33$.

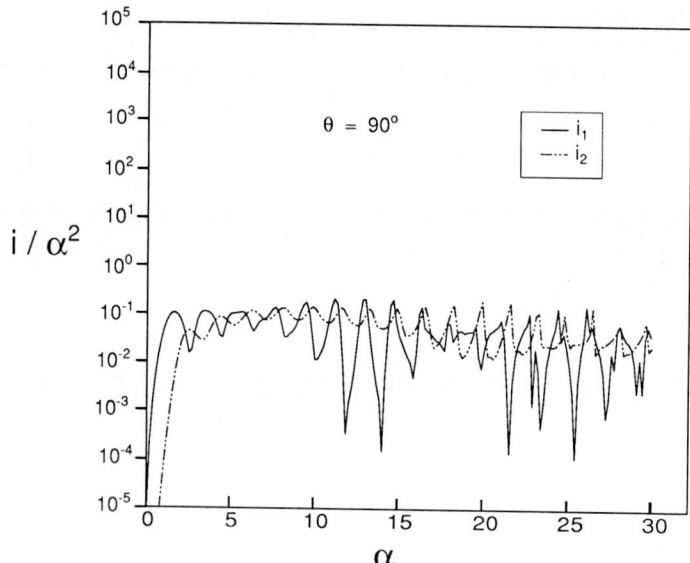

Figure 17.3b Plot of i/α^2 for 90° scattering angle, $m = 1.33$.

Optical Properties 287

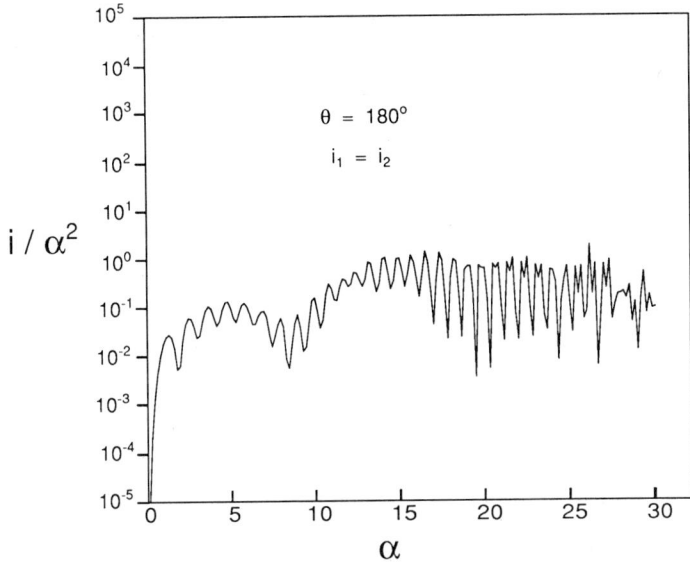

Figure 17.3c Plot of i/α^2 for 180° scattering angle, $m = 1.33$.

an approximation for $i_1(\theta)$ for forward scattering ($\theta = 0°$) which he claims to be valid for $\alpha > 5$ and any m. This equation is

$$i_1(0°) \approx \left(\frac{\alpha^2}{4} Q_{scat}\right)^2 \tag{17.8}$$

Table 17.1 compares some values of $i_1(0°)$ calculated from Eq. 17.8 with values from more exact Mie theory computations.

TABLE 17.1 Comparison of Values of $I_1(0°)$ Calculated from Eq. 17.8 and Mie's Theory

α	Q_{scat}	m	$i_1(0°)$ from Mie's theory	$i_1(0°)$ from Eq. 17.8
5	3.928	1.5	610.1	602.7
10	2.882	1.5	5,209	5,191
15	1.985	1.5	12,490	12,500
20	2.036	1.5	42,570	41,500
25	2.250	1.5	127,500	124,000
30	2.353	1.5	282,400	280,000

SOURCE: 1.5 refractive index data from R. Giese et al., *Tabellen der Streufunktionen $i_1\theta$, $i_2\theta$*, Akademie-Verlag, Berlin, 1962, p. 10.

Example 17.3 Compute the value of $i_1(0°)$ for a 3-μm-diameter sphere ($m = 1.33$) illuminated with blue light ($\lambda = 0.4$ μm). From Mie theory $Q_{scat} = 2.1549$.

$$\alpha = 23.56$$

$$i_1(0°) = \left[\frac{(23.56)^2(2.1549)}{4}\right]^2$$

$$= 8.95 \times 10^4$$

For $\alpha = 23.6$ and $m = 1.33$, Mie theory gives for $i_1(0°)$ a value of 8.99×10^4.

According to Kerker (1969), it is possible to estimate the $i_1(\theta)$ values for scattering in the near forward direction (scattering angles no greater than several degrees) over the range $\alpha = 5$ to 30 by using the approximation

$$i_1(\theta) \approx i_1(0°)\left[\frac{J_1(\alpha \sin \theta)}{\alpha \sin \theta}\right]^2 \tag{17.9}$$

In Eq. 17.9, J_1 represents a Bessel function of the first order and of argument $\alpha \sin \theta$. Appendix G lists tabulated values of positive Bessel functions of order 1.

The value of Q_{scat} can be estimated by using Van de Hulst's (1957) approximation for transparent spheres

$$Q_{scat} = 2 - \frac{4}{\psi}\sin \psi + \frac{4}{\psi^2}(1 - \cos \psi) \tag{17.10}$$

The term ψ is defined as $\psi = 2\alpha(m - 1)$, and in Eq. 17.10, ψ is expressed in radians.

Example 17.4 Denman et al. (1966) give a computer value for $i_1(5°)$ of 1.91×10^4 for a 4-μm-diameter particle ($m = 1.33$) illuminated with light having a wavelength of 0.5 μm. Compare this value to one calculated by using Eqs. 17.8, 17.9, and 17.10.

$$\alpha = 25.1$$

$$\psi = (2)(25.1)(1.33 - 1) = 16.59$$

$$Q_{scat} = 2 - \frac{4}{16.59}\sin 16.59 + \left(\frac{4}{16.59}\right)^2(1 - \cos 16.59) = 2.21$$

$$i_1(0°) = \left[\frac{(25.1)^2(2.21)}{4}\right]^2 = 1.22 \times 10^5$$

$$\alpha \sin \theta = 2.188$$

From App. G, $J_1(2.188) \approx 0.556$ so

$$i_1(5°) = (1.22 \times 10^5)\left(\frac{0.556}{2.19}\right)^2 = 7.84 \times 10^3$$

This result is somewhat lower than that given by the reference above.

Polydisperse aerosol

As an aerosol becomes more polydisperse, there is a tendency for the scattering patterns to become smoother. This can be seen in Fig. 17.4. As the aerosol becomes more polydisperse, the scattering curves tend to lose the large fluctuations seen with monodisperse scattering, although some of the irregularity still remains.

Rayleigh scattering

In the case where the particle size is less than the wavelength of light $\alpha < 0.3$, the electromagnetic field can be assumed to be uniform over

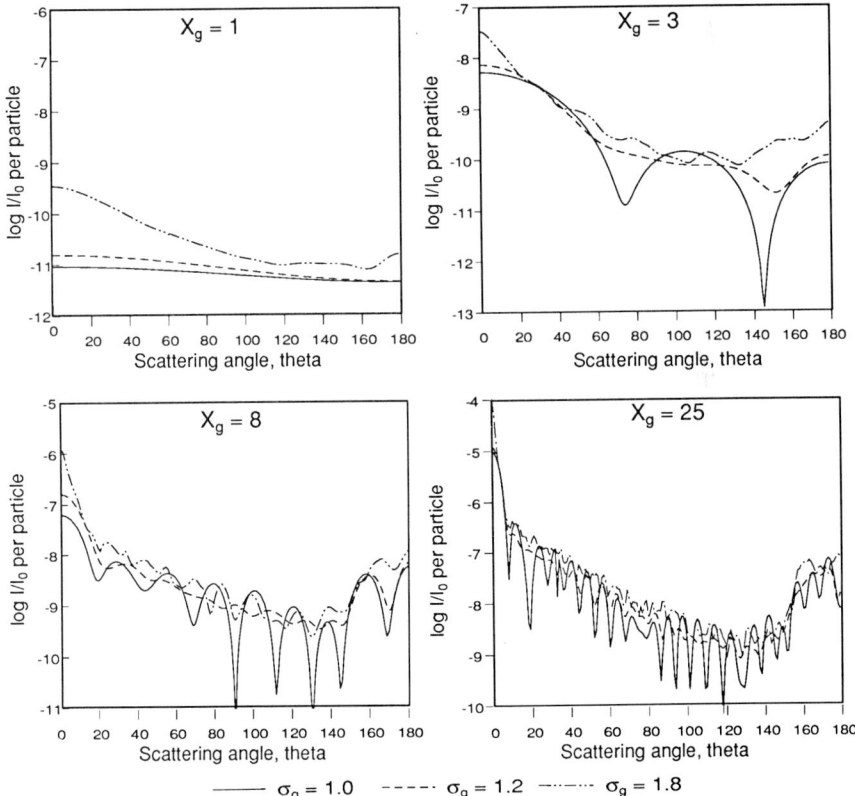

Figure 17.4 Angular intensity ratio, I/I_0, for lognormal aerosols with different degrees of polydispersity, $\lambda = 550$ nm.

the entire particle, giving equations for scattering that are relatively simple. This approach was successfully used by Lord Rayleigh (1899) to explain the scattering of light by "air" molecules. To honor this finding, it is common practice today to refer to the scattering of light when α is very small as *Rayleigh scattering* or scattering in the *Rayleigh region*.

Rayleigh was able to explain that natural light when scattered at right angles to its source is completely polarized. He reasoned that a photon of natural light approaching a particle along the x axis (centered at the origin of cartesian coordinates) can be resolved into waves vibrating in the y and z directions. At the particle this photon produces a dipole oscillating in the same directions. Since there is no forward or x component of wave motion for photons scattered into an angle of 90° to the incident beam, there will be only one component of the wave motion, that being the component perpendicular to the plane formed by the incident and scattered beams. That is, the $i_1(\theta)$ component will remain, and the $i_2(\theta)$ component will disappear. By the definitions given above, $q_2(90°) = 0$ for scattering in the Rayleigh region, and the scattered beam is perpendicularly polarized (see, e.g., Kerker, 1969, or Van de Hulst, 1957).

Rayleigh made the important discovery that the intensity of light scattered by very small particles is proportional to the fourth power of the wavelength of the incident light. This finding explains the blue color of the sky on a clear day as well as the apparent blue color of smoke or fumes made up of very small particles.

For very small nonabsorbing spheres, Rayleigh's theory gives

$$i_1(\theta) = \alpha^6 \left(\frac{m^2 - 1}{m^2 + 2}\right)^2 \tag{17.11}$$

$$i_2(\theta) = \alpha^6 \left(\frac{m^2 - 1}{m^2 + 2}\right)^2 \cos^2 \theta \tag{17.12}$$

Thus for a particle illuminated with natural incident light of intensity I_0, the intensity per unit solid angle is

$$I = \frac{1 + \cos^2 \theta}{2} I_0 \left(\frac{m^2 - 1}{m^2 + 2}\right)^2 \frac{\alpha^6}{k^2} \tag{17.13}$$

where k is the wave number, defined as $k = 2\pi/\lambda$. From the ratio of α^6/k^2, the fourth-power dependence of scattering on wavelength is shown.

Example 17.5 Compute the intensity of natural light scattered at a 45° angle by a 0.01-μm water droplet ($m = 1.33$) when illuminated by red light ($\lambda = 0.7$ μm) and blue light ($\lambda = 0.4$ μm).

$$\left(\frac{\alpha^6}{k^2}\right)_{red} = 1.01 \times 10^{-18}$$

$$I = \frac{1 + 0.500}{2}I_0\left(\frac{1.33^2 - 1}{1.33^2 + 2}\right)^2(1.01 \times 10^{-18})$$

$$\frac{I}{I_0}_{red} = (0.750)(0.042)(1.01 \times 10^{-18}) = 3.15 \times 10^{-20}$$

$$\left(\frac{\alpha^6}{k^2}\right)_{blue} = 9.51 \times 10^{-18}$$

$$I = \frac{1 + 0.500}{2}I_0\left(\frac{1.33^2 - 1}{1.33^2 + 2}\right)^2(9.51 \times 10^{-18})$$

$$\left(\frac{I}{I_0}\right)_{blue} = (0.750)(0.042)(9.51 \times 10^{-18}) = 2.97 \times 10^{-19}$$

The relationships between the values of $i_1(\theta)$ and $i_2(\theta)$ for Rayleigh scattering are sketched in polar coordinates in Fig. 17.5. The factor $i_1(\theta)$ forms a circle centered at the origin while $i_2(\theta)$ produces two circles tangent at the origin and lying on the $\theta = 0°$ line. The disappearance of the $i_2(\theta)$ scattering at $\theta = 90°$ can be seen.

Scattering patterns with increasing α

As the particle size parameter becomes larger, the backscattered component slowly begins to weaken and the forward scattering increases

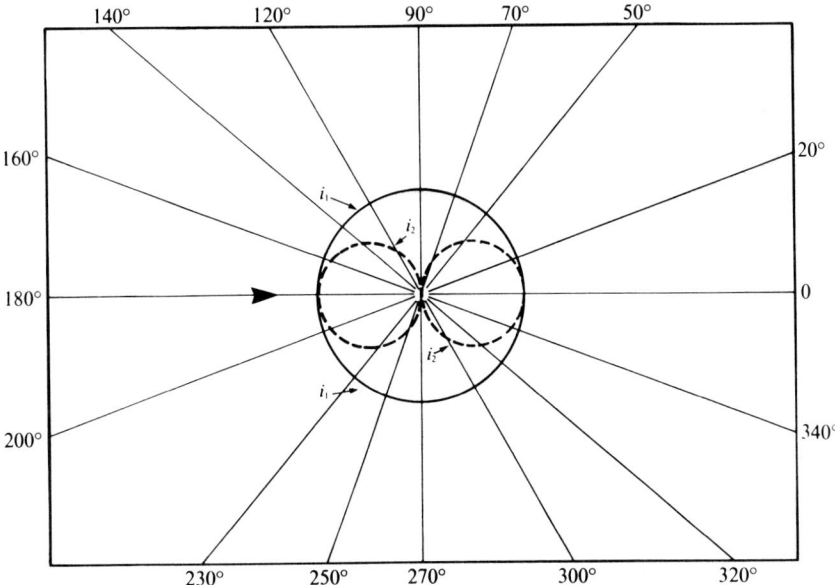

Figure 17.5 Polar diagram of angular scattering functions $i_1(\theta)$ and $i_2(\theta)$ for Rayleigh scattering; $\alpha \ll 1$, $m = 1.33$.

in intensity. With the increasing asymmetry comes the development of maxima and minima which, with increasing α, develop toward the rear of the scattering pattern and then move toward the front, constantly growing. Eventually, for monodisperse aerosols, these maxima and minima form complex scattering patterns having as their most outstanding characteristics irregularity and a strong forward scattering component. The development of a complex scattering pattern with increasing α is illustrated in Fig. 17.6.

Thus for aerosols in the sizes of interest, particle scattering is characterized by the strong forward scattering of both plane polarized and perpendicularly polarized components. For nonspherical or irregularly shaped particles of the same size range, this strong forward scattering tendency is also evident, although there are fewer maxima and minima in the angular scattering pattern. For aerosol particles in the range of interest, there is also a pronounced backscatter (180°) component. As mentioned previously, the degree of irregularity in the scattering curves decreases as the aerosol becomes more polydisperse and the forward scattering component increases.

Smaller sizes of particles (and for a polydisperse cloud, more particles) will be seen when the particles are viewed between the light source and the observer than when viewed with side illumination. This observation can be confirmed easily by comparing the number of

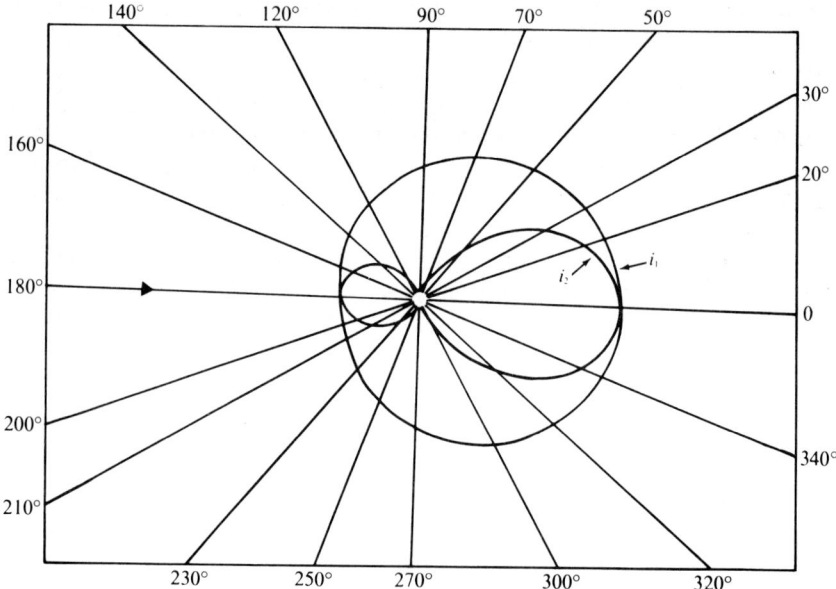

Figure 17.6 Polar diagram of angular scattering functions $i_1(\theta)$ and $i_2(\theta)$ for α = 1.0, m = 1.33.

dust particles seen by looking up a shaft of sunlight with particles viewed at right angles to the sun.

Forward scattering of light by dust particles in the atmosphere is responsible for the red color of the sun as it sets and for the glorious sunsets that herald evening. Just as the sun is setting, when cloud conditions are right, it is possible to observe a red glow in the east, as well as the sunset in the west. In this case the 180° backscattered light is responsible.

The scattering diagrams for particles in the Mie region show many maxima and minima, some quite close together, but slightly altered in angular position with small changes in α. These anomalies are responsible for the strange colors that can appear in the sky or in an experimental chamber from time to time when conditions of particle size and monodispersity are right. For example, Aitken (1892) reported that on rare occasions the sun appeared greenish, at other times it appeared bluish. Following a very large forest fire in Canada in the early 1950s, both the sun and the moon appeared bluish for several days. This led Cadle (1966) to remark that the rare occurrence of such a cloud of particles in the atmosphere could be the source of the expression "once in a blue moon."

Radiative Transfer

If multiple interactions between photons and aerosol particles do take place, the resulting model of this behavior is extremely complicated. This class of problem is often described as the *problem of radiative transfer,* similar to the neutron scattering problem. In concept, the approach is quite simple. For example, in Fig. 17.7 various aerosol particles are shown as dots. These particles may be similar or dissimilar. Light radiating from plane A can reach plane B either by single scattering, as shown by scattering from particle R, or by multiple scattering, shown by scattering from particle P to particle Q and then to plane B. Any number of scattering events are possible before the light

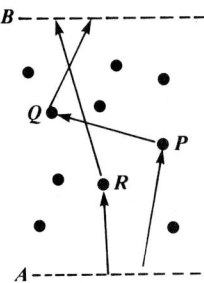

Figure 17.7 Model of multiple scattering.

photon reaches plane B, more scatterings being likely as the aerosol concentration increases.

When all possible photon paths from A are considered and all photons reaching B are summed, the problem is solved. Various approximations have been proposed to reduce this almost infinite task, but a complete solution has not yet been developed. For more information on this mathematically interesting problem, see Kerker (1963).

Applications

Particle size represents an important variable in both the extinction and scattering of light. It is not surprising, therefore, that a wide variety of optical techniques are available for particle sizing. In this section several of the more common methods are discussed.

Diffraction rings

If a light source is viewed through a thin, fairly monodisperse aerosol, a series of bright rings (coronas) will be seen. If d is the diameter of the cloud particles and γ_n the angular radius of the nth minimum between the rings, then the following empirical relationships are valid:

$$\sin \gamma_1 = \frac{1.22\lambda}{d} \quad (17.14a)$$

$$\sin \gamma_2 = \frac{2.24\lambda}{d} \quad (17.14b)$$

$$\sin \gamma_3 = \frac{3.24\lambda}{d} \quad (17.14c)$$

When white light is used (assuming $\lambda = 0.571$ μm), the outer edge of the ring is considered to be the minimum. Often only the light scattered from the first ring is intense enough to be seen. This phenomenon can be seen around the moon on some nights prior to the arrival of a cold front. If the moon is used as the light source, its angular radius should be subtracted from the total angular radius before Eqs. 17.14 are used. This technique has been used to derive particle size information in very high clouds such as altostratus clouds.

Example 17.6 On a cloudy night, the first diffraction ring around a blue light source ($\lambda = 0.4$ μm) having an apparent angular radius of 7° appeared at an angle of 15°. What is the particle diameter implied by these measurements?

$$\sin(15 - 7) = \frac{1.22(0.4)}{d}$$

$$d = 3.51 \, \mu m$$

Higher-order Tyndall spectra

If one illuminates a monodisperse aerosol with white light, a series of well-defined colors are produced at various scattering angles. These colors appear because of the extreme dependence of the scattered intensity at a given angle on the value of α and hence, for a monodisperse aerosol, on the wavelength of light. As a result, at a given angle one particular wavelength will scatter much more light than any of the other visible wavelengths, and one color will predominate. Johnson and LaMer (1947) found that over a fairly wide range of refractive indices, the number and angular positions of a given color (red was chosen because it was considered easiest to see) were a function of particle size. For a traverse of 180° the relationship can be approximated by

$$d \approx \frac{\text{number of reds from } 0° \text{ to } 180°}{5} \tag{17.15}$$

More exact computations of this relationship, based on Mie's theory, are shown in Figs. 17.8 and 17.9. A device known as the *aerosol owl* can be used to make these measurements. It consists of a viewing chamber on which a telescope is mounted that can be rotated through roughly 160°. A protractor is attached to the telescope so that the angle at which a red is observed can be noted. The number of reds found

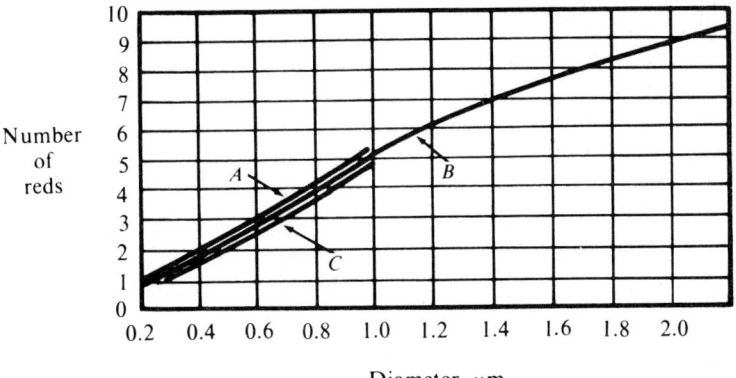

Figure 17.8 Observed and theoretical number of reds. A = sulfur ($m = 2.0$), B = stearic acid ($m = 1.43$), c = calculated (all indices).

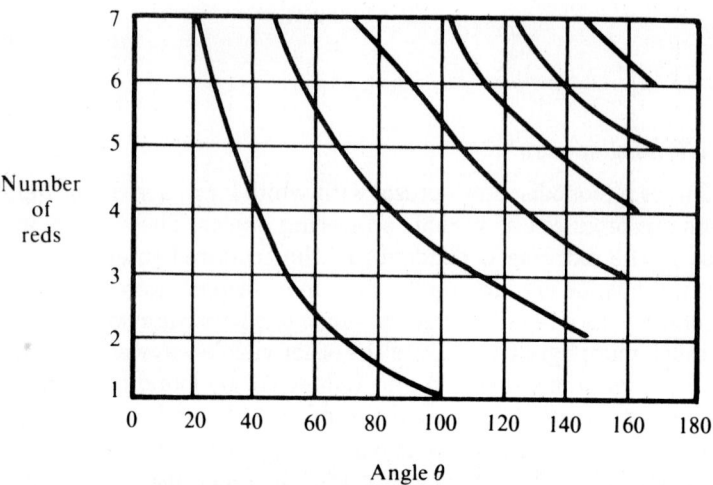

Figure 17.9 Angular position of reds for stearic acid aerosol.

can be used to determine particle size, which can also be confirmed by comparing the angles at which the reds appear with Fig. 17.9. A more complete description of the owl is given by Dennis (1976) or Green and Lane (1964), and discussions of the use of higher order Tyndall spectra are presented by Kitani (1960) and Maron and Elder (1963).

Example 17.7 A monodisperse aerosol is viewed in an aerosol owl, and four red bands are observed as the sample is scanned over approximately 160°. Determine the aerosol particle diameter indicated by these data.
From Eq. 17.15

$$d \approx 4/5 \approx 0.8 \; \mu m$$

Use of the forward scattering lobe

A technique for sizing monodisperse particles of unknown refractive index uses the observation that the intensity of the forward scattering lobe is mainly due to Fraunhofer diffraction and is thus independent of the refractive index of the particle. Hodkinson (1966) computed a family of curves for the ratio of intensity viewed at two different angles. This relationship is shown in Fig. 17.10. If forward angles as small as 5° can be measured, this procedure should be accurate in measuring particle sizes of $\alpha = 1$ to $\alpha = 18$.

Example 17.8 The ratio of the intensity of light scattered at an angle of 15° to light scattered at an angle of 10° when a monodisperse aerosol is illuminated

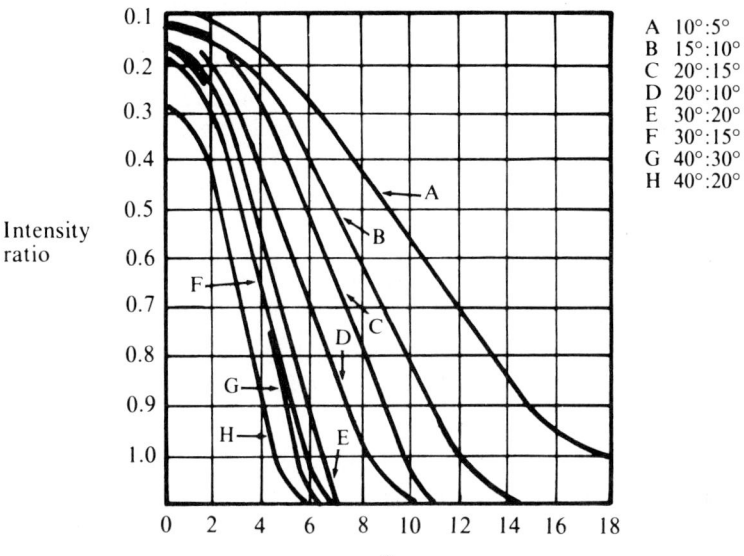

Figure 17.10 Scattering intensity ratio for scattering at indicated angles, as a function of α.

with unpolarized red light (λ = 0.5 μm) is 0.35. What particle diameter is implied by this measurement?

From Fig. 17.10, α = 5.5, so

$$d = \frac{\alpha \lambda}{\pi} = \frac{5.5\,(0.5)}{3.14} = 0.875 \mu m$$

Single-particle scattering measurements

The scattering technique for particle sizing that has gained the most popular acceptance was first developed by Gucker and O'Konski (1949) and since then has been refined to such an extent that now there are many commercially available instruments using this technique. Particle size is determined by passing the particles, one at a time, through an illuminated volume, and the amount of light scattered by each particle into some solid angle is measured. Knowledge of the flow rate and the number of particles counted per unit time gives the aerosol concentration, whereas the amount of light scattered per particle gives an indication of that particle's size. This type of instrument is, in theory, capable of measuring both the concentration and the size distribution of an aerosol. A good discussion of these types of aerosol detectors has been given by Whitby and Willeke (1979).

Single-particle optical counters cannot be easily used for measuring high aerosol concentrations. Whitby and Liu (1967) have shown that if many small particles are in the illuminated sensitive volume of this type of instrument at the same time, significant errors are introduced because the particles are considered by the detector to be fewer in number and greater in size. Thus there is a practical upper limit on the aerosol concentration that can be measured by such an instrument, being on the order of 1000 to 10,000 particles per cubic centimeter.

Example 17.9 An optical particle counter samples an aerosol at a flow rate of 175 cm^3/min into a sensitive volume of 0.01 cm^3. What is the implied maximum particle concentration (in particles per liter) that can be sampled?

To have a maximum of one particle in the sensitive volume at any time, on average there should be no more than 100 particles per cubic centimeter, or 100,000 particles per liter.

Because of the statistical distribution of particles entering the counter, even though an average of only one particle will be in the chamber at any time, actually sometimes there will be more than one particle at other times no particles at all. Hence, the theoretical limit given in Example 17.9 cannot be achieved without some coincidence losses. The sensitive volume of the optical counter could be considered to require a certain time t_R to recover from counting one particle prior to counting another. This is the average time taken for a particle to pass through the sensitive volume

$$t_R = \frac{\text{volume of sensitive volume}}{\text{volumetric flow rate}}$$

The observed counting rate can be increased to account for coincidence losses by adding to the counting rate a factor Θ, given by

$$\Theta = \frac{C^2 t_R}{1 - C t_R} \tag{17.16}$$

Example 17.10 Using the data in Example 17.9, determine the corrected aerosol concentration when the indicated concentration is 23,100 particles per liter.

In 1 min, 23,100 (particles per liter) × 0.175 (L/min) particles are counted, or 4040 counts/min.

$$t_R = \frac{0.01 \text{ cm}^3}{175 \text{ cm}^3/\text{min}} = 5.71 \times 10^{-5} \text{ min}$$

$$\Theta = \frac{(4.04 \times 10^3)^2 (5.71 \times 10^{-5})}{1 - (4.04 \times 10^3)(5.71 \times 10^{-5})} = 1210 \text{ additional counts/min}$$

Thus the corrected counting rate is 4040 + 1210 = 5250 counts per minute, or the corrected concentration is

$$\frac{5250}{0.175} = 30{,}000 \text{ particles/L}$$

Rearrangement of Eq. 17.16 shows that a saturation count will be reached by the counter where increasing particle challenge does not increase particle count. Since this saturation level is fairly low for most optical particle counters compared to typical aerosol concentrations, some sort of dilution system is usually required for field use.

Problems

1 The terms $i_1(\theta)$ and $i_2(\theta)$ are related to $q_1(\theta)$ and $q_2(\theta)$, the scattered intensity coefficients, by the equations

$$i_1(\theta) = q_1(\theta)\frac{\pi}{4}d^2k^2$$

$$i_2(\theta) = q_2(\theta)\frac{\pi}{4}d^2k^2$$

where k is the wave number ($k = 2\pi/\lambda$). If for Rayleigh scattering

$$q_2(\theta) = \frac{\alpha^4}{\pi}\left(\frac{m^2 - 1}{m^2 + 2}\right)^2 \cos^2\theta$$

plot on polar coordinate paper $q_1(\theta) + q_2(\theta)$.

2 The value of $q_1(\theta)$ for perpendicularly polarized light scattered in the Rayleigh region is given by

$$q_1(\theta) = \frac{\alpha^4}{\pi}\left(\frac{m^2 - 1}{m^2 + 2}\right)^2$$

Determine the scattering efficiency for light scattered into the interval $\theta = 0°$ to $20°$.

3 Perpendicularly polarized white light is used to illuminate a 0.25-μm-diameter sphere. If the sphere has a refractive index of 1.33, what will be the predominant color of light scattered at an angle of 90°?

4 Using Eq. 17.8, compute the value of $i_2(0°)$ for a 1-μm-diameter sphere illuminated with light having a wavelength of 0.5 μm.

5 Infrared rays have wavelengths in the range of 0.7 to 5000 μm. For 10-μm-diameter particles, what values of α correspond to the wavelengths of the infrared rays?

6 A 1.0-μm-diameter particle is illuminated with visible light having a wavelength of 0.5 μm and infrared radiation having a wavelength of 10 μm. Assuming equal intensity of the incident radiation, find the ratio of the scattered visible to infrared radiation for forward scattering.

Chapter 18

Coagulation of Particles

Introduction

The aerosol properties discussed in previous chapters relate primarily to individual particles. For the most part, discussions have avoided consideration of interference effects between particles. But one area where interparticle effects cannot be neglected is aerosol coagulation, also known as aggregation or agglomeration.

It is easily observed that at high aerosol concentrations, individual particles coalesce to form larger chains or flocs made up of many particles. The process of coagulation may be brought about solely by the random motion and subsequent collision of particles (often called thermal coagulation) or the collisions could be caused by such external forces as turbulence or electricity. In general, these external forces will act to increase the rate of coagulation.

The process of coagulation has received a great deal of attention, especially with regard to the coagulation of colloidal solutions. Unfortunately, because both particle number and particle size change with time, numerical models that predict these variables as a function of time are highly complex and therefore messy to use or, if simplified, are representative but inexact. In many cases, results derived from the simplified models can be used to predict aerosol coagulation rates and number concentration with reasonable accuracy, although little can be said of the resulting size distributions.

Coagulation of Monodisperse Spherical Particles

The simplest coagulation problem is thermal coagulation of monodisperse spherical particles. Since only the first several particle collisions are considered, the size of the resulting agglomerated particles will not be appreciably different from the size of the initial particles. The model has been used for many years for solid particles and forms the basis for the definition of the coagulation coefficient. It is especially applicable to the coagulation of liquid drops, since the size of the agglomerated drop increases only as the cube root of the number of drops comprising it. The approach was first presented by Smoluchowski (1911) for coagulation in dilute electrolytes, but it was shown to be applicable also to aerosols, within the limits discussed above (Whytlaw-Gray and Patterson, 1932).

In Smoluchowski's approach, a number of spherical particles of diameter d are considered to be randomly separated from each other at $t = 0$. If the particles are also moving about randomly by thermal diffusion, it is desired to know the likelihood that they will collide (and then stick together) within some time t. Smoluchowski first considered the case where one particle was fixed, acting as a sink for the other particles. He then determined the diffusion rate of other particles to this central particle.

For diffusion

$$\frac{\partial c}{\partial t} = D\nabla^2 c \qquad (18.1)$$

where c is the concentration of particles and D is their diffusion coefficient. If r is a distance measured from the center of the fixed particle, and assuming spherical symmetry, Eq. 18.1 can be written as

$$\frac{\partial c}{\partial t} = D\left(\frac{\partial^2 c}{\partial r^2} + \frac{2}{r}\frac{\partial c}{\partial r}\right) \qquad (18.2)$$

or in even more convenient form as

$$\frac{\partial(cr)}{\partial t} = D\frac{\partial^2(cr)}{\partial r^2} \qquad (18.3)$$

Example 18.1 Show that Eq. 18.3 is equivalent to Eq. 18.2.
If c is a function of t and r,

$$\frac{\partial(cr)}{\partial t} = r\frac{\partial c}{\partial t}$$

and

$$\frac{\partial(cr)}{\partial r} = r\frac{\partial c}{\partial r} + c$$

Differentiating with respect to r gives

$$\frac{\partial^2(cr)}{\partial r^2} = r\frac{\partial^2 c}{\partial r^2} + \frac{\partial c}{\partial r} + \frac{\partial c}{\partial r}$$

Then substituting in Eq. 18.3 gives

or

$$r\frac{\partial c}{\partial t} = D\left(2\frac{\partial c}{\partial r} + r\frac{\partial^2 c}{\partial r^2}\right)$$

$$\frac{\partial c}{\partial t} = D\left(\frac{\partial^2 c}{\partial r^2} + \frac{2}{r}\frac{\partial c}{\partial r}\right)$$

Since all particles are the same diameter, it can be assumed that they will intersect the central particle when they come to within a distance d of it. Thus at this point the concentration will be zero, i.e.,

$$c' = 0 \quad \text{at } r = d \text{ (for } t > 0\text{)}$$

In addition, initially particles are assumed to be uniformly distributed throughout the volume of interest with concentration c. Thus

$$c' = c \quad \text{at } t = 0$$

With these conditions, Eq. 18.3 can be solved to give

$$c' = c\left[1 - \frac{d}{r} + \frac{d}{r}\text{erf}\left(\frac{r-d}{2\sqrt{Dt}}\right)\right] \tag{18.4}$$

The number of particles Φ which diffuse to within a distance d of the central fixed particle in unit time is equal to the product of the diffusion current and surface area of the sphere of radius d. The diffusion current is given by

$$J = -D\frac{\partial c'}{\partial r} \tag{18.5}$$

where the slope $\partial c'/\partial r$ is to be evaluated at $r = d$. Thus

$$\Phi = 4\pi d^2 D\frac{\partial c'}{\partial r} \tag{18.6}$$

Since at $r = d$, from Eq. 18.4,

$$\frac{\partial c'}{\partial r} = \frac{c}{d}\left(1 + \frac{d}{\sqrt{\pi Dt}}\right) \tag{18.7}$$

then in the time interval dt the number of particles reaching the surface surrounding the central particle is

$$\Phi dt = 4\pi dDc\left(1 + \frac{d}{\sqrt{\pi Dt}}\right)dt \tag{18.8}$$

Now suppose that the fixed particle is not fixed, but is able to diffuse along with the other particles. The diffusion of this particle must also be taken into account. The combined diffusion coefficient of two particles relative to each other is equal to the sum of the diffusion coefficients of the single particles, so that the moving particle collides with

$$8\pi dDc\left(1 + \frac{d}{\sqrt{\pi Dt}}\right)dt$$

particles in dt, remembering that Eq. 18.8 applies to equal-sized spheres. Also, with c particles per unit volume there will be $c/2$ collisions if every particle collides once, there being two particles involved in each collision. The number of collisions per unit volume which will take place in the time interval dt is

$$\frac{dc}{dt} = -\frac{8}{2}\pi dDc^2\left(1 + \frac{d}{\sqrt{\pi Dt}}\right) \tag{18.9}$$

The second term inside the parentheses can be ignored, since it will generally be much less than 1, especially after some time has elapsed. This can be seen by referring to Table 18.1, which gives values of the term $d/\sqrt{\pi D} = \xi$ for various size particles. Fuchs (1964) points out that ξ represents the possibility of particles initially being quite close to the fixed particle. Thus ξ disappears as a stationary rate develops. Since for all practical cases ξ is quite small, it is ignored. One should keep in mind, however, that there could be conditions where ξ would be important. The effect of this term, if any, is to increase the initial coagulation rate.

Defining the coagulation constant K_0 as

$$K_0 = 8\pi dD = \frac{8kT}{3\mu}C_c \tag{18.10}$$

yields a result which should be, at least for large particles with $C_c \approx 1$, independent of particle size.

TABLE 18.1 Calculated Values of ξ

Particle diameter, cm	$d/\sqrt{\pi D} = \xi$
1×10^{-7}	2.42×10^{-7}
5×10^{-7}	6.01×10^{-6}
1×10^{-6}	2.39×10^{-5}
5×10^{-6}	5.67×10^{-4}
1×10^{-5}	2.12×10^{-3}
5×10^{-5}	3.54×10^{-2}
1×10^{-4}	0.107
5×10^{-4}	1.28
1×10^{-3}	3.65

Example 18.2 Determine the value for the coagulation constant K_0 for air at 20°C and standard pressure for particles having $C_c \approx 1$.

$$K_0 = \frac{(8)(1.38 \times 10^{-16})(293)(1)}{(3)(1.83 \times 10^{-4})}$$

$$= 5.89 \times 10^{-10} \text{ cm}^3/\text{s}$$

In the cgs system the units of K_0 are cubic centimeters per second.

Using K_0 as the coagulation constant and neglecting the second term in Eq. 18.9 yield the usual form of the coagulation equation:

$$\frac{dc}{dt} = -\frac{K_0}{2}c^2 \tag{18.11}$$

Integration of Eq. 18.11 with the initial condition that $c = c_0$ when $t = 0$ gives

$$\frac{1}{c} - \frac{1}{c_0} = \frac{K_0}{2}t \tag{18.12}$$

Equation 18.12 shows that the inverse of the concentration at any time is a linear function of time, the slope of the line being determined by the coagulation constant. Experimental data from both monodisperse and polydisperse aerosols follow this general form, at least initially. As will be discussed later, the coagulation constant may be appreciably larger than the theoretical value. If t_h is defined as the half-value time, i.e., the time in which the concentration decreases by a factor of 2, then

$$c = \frac{c_0}{1 + t/t_h} \tag{18.13}$$

Example 18.3 An aerosol made up of 5-μm-diameter spheres has an initial concentration of 100 million particles per cubic foot (100 mppcf). After 1 day what will be the aerosol concentration (also in mppcf)? How long will it take to reach one-half the initial concentration?

$$K_0 = 6.13 \times 10^{-10} \text{ cm}^3/\text{s}$$

$$100 \text{ mppcf} = \frac{100(10^6)}{28.3(10^3)} = 3.53 \times 10^3 \text{ p/cm}^3$$

$$\frac{1}{c} - \frac{1}{3.53 \times 10^3} = \frac{6.13 \times 10^{-10}}{2}(60 \times 60 \times 24)$$

$$c = 3.23 \times 10^3 \text{ p/cm} = 91.4 \text{ mppcf}$$

To find the time to reach one-half the initial concentration,

$$\frac{2}{c_0} - \frac{1}{c_0} = \frac{K_0}{2} t_h$$

$$t_h = \frac{2}{c_0 K_0} = \frac{2}{(3.53 \times 10^3)(6.13 \times 10^{-10})}$$

$$= 9.23 \times 10^5 \text{ s} = 10.69 \text{ days}$$

Coagulation of Particles of Two Different Sizes

For the case of coagulation of particles of two different sizes, it is possible to follow the same approach as given for the monodisperse case except that d is replaced with $(d_1 + d_2)/2$ and $2D$ is replaced with $D_1 + D_2$. Then the coagulation coefficient K_{12} becomes

$$K_{12} = 2\pi(d_1 + d_2)(D_1 + D_2) \tag{18.14}$$

In terms of the particle mobility, the coagulation coefficient is

$$K_{12} = 2\pi(d_1 + d_2)(B_1 + B_2)kT \tag{18.15}$$

As previously defined, the mobility B is

$$B = \frac{D}{kT} = \frac{C_c}{3\pi\mu d}$$

The minimum coagulation constant occurs for coagulation of equal-size particles. This can be seen in Fig. 18.1 which shows a three-dimensional graph of the coagulation constant matrix. The valley indicated on the plot represents the constants for coagulation of equal-size particles.

Example 18.4 Compute the combined coagulation coefficient for coagulation of 1.0- and 0.1-μm-diameter spheres.
From Table 9.1, D for a 1-μm particle is 2.76×10^{-7} cm^2/s and D for a 0.1-μm particle is 7.01×10^{-6} cm^2/s.

$$K_{12} = 2\pi(10^{-4} + 10^{-5})(2.76 \times 10^{-7} + 7.01 \times 10^{-6})$$

$$= 5.04 \times 10^{-9} \text{ cm}^3/\text{s}$$

Coagulation of many sizes of particles

Tikhomirov et al. (1942) and Gillespie (1963) have shown that if one is interested only in the change in the number of all the particles dc/dt, then it is possible to combine all coagulation coefficients into a single

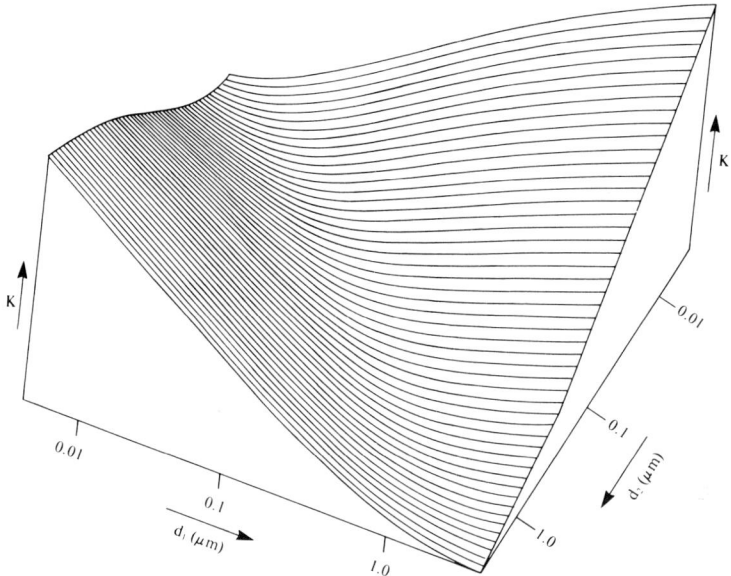

Figure 18.1 Coagulation constant K, matrix symmetric in d_1, d_2. Minimum value is 4.4×10^{-10} cm³/s. Maximum value is 1.28×10^{-5} cm³/s. (*From Kuhlman, 1982.*)

one expressed in terms of the mean values $\overline{d}$, $\overline{(1/d)}$, and $\overline{(1/d^2)}$. This combined coefficient K' is

$$K' = \frac{4kT}{3\mu}\left[1 + \overline{d}\,\overline{\left(\frac{1}{d}\right)} + A\lambda\overline{\left(\frac{1}{d}\right)} + A\lambda\overline{d}\,\overline{\left(\frac{1}{d^2}\right)}\right] \quad (18.16)$$

where A is the Cunningham correction factor constant (see Table 5.1) and λ is the mean free path of the gas. Thus K' can be used as before in Eq. 18.11.

If particle sizes are initially distributed lognormally, an expression can be written for K' in terms of the geometric mean diameter d_g and geometric standard deviation σ_g:

$$K' = \frac{4kT}{3\mu}\left[1 + \exp(\ln^2 \sigma_g) + \frac{2A\lambda}{d_g}\exp\left(\frac{1}{2}\ln^2 \sigma_g\right) + \frac{2A\lambda}{d_g}\exp\left(\frac{5}{2}\ln^2 \sigma_g\right)\right] \quad (18.17)$$

Table 18.2 lists values of K' for various d_g and σ_g. With increasing polydispersity this coefficient can become quite large, implying that extremes in polydispersity are quickly reduced by agglomeration, particularly of the smaller particles. Also, it is clear that the coagulation rate for a polydisperse aerosol is greater than for a monodisperse one. However, coagulation of a monodisperse aerosol initially increases polydispersity, so that for any coagulating aerosol the coagulation coefficient is not constant, but is itself a variable bound to the coagulation rate. There is little wonder that the interpretation of coagulation data is difficult, and it is amazing that the simple coagulation theory, as presented here, is as adequate as it is for many applications. It appears, however, that as time progresses, coagulation tends toward a single size distribution.

Example 18.5 Figure 18.2 shows a plot of $1/c$ as a function of t for a polydisperse aerosol. Using this figure, determine the value of K'.

From the figure, the slope of the line can be found by

$$\text{Slope} = \frac{1/c_2 - 1/c_1}{t_2 - t_1}$$

$$= \frac{3.92 \times 10^{-8} - 0.68 \times 10^{-8}}{85 - 10}$$

$$= 4.32 \times 10^{-10}$$

$$K' = 2 \times 4.32 \times 10^{-10} = 8.64 \times 10^{-10} \text{ cm}^3/\text{s}$$

TABLE 18.2 Values of Coagulation Coefficient K' for Aerosols of Varying Degrees of Polydispersity (Value $\times 10^{-10}$ cm^3/s)

Geometric mean diameter d_g, μm	Geometric standard deviation σ_g			
	1.0	1.5	2.0	2.5
0.02	53.2	67.8	116.4	238.8
0.04	29.6	37.1	62.1	124.3
0.10	15.4	18.7	29.5	55.6
0.16	11.8	14.1	21.3	38.4
0.20	10.6	12.6	18.6	32.7
0.40	8.23	9.49	13.1	21.2
1.00	6.84	7.65	9.88	14.3
1.60	6.48	7.19	9.07	12.6
2.0	6.37	7.03	8.79	12.1
4.0	6.13	6.74	8.25	10.9
10.0	5.99	6.54	7.93	10.2
16.0	5.95	6.50	7.85	10.1

Computed with the following constants: $\rho = 1$ g/cm^3, $T = 293°C$, $\mu = 1.83 \times 10^{-4}$ P, $\lambda = 0.0653$ μm.

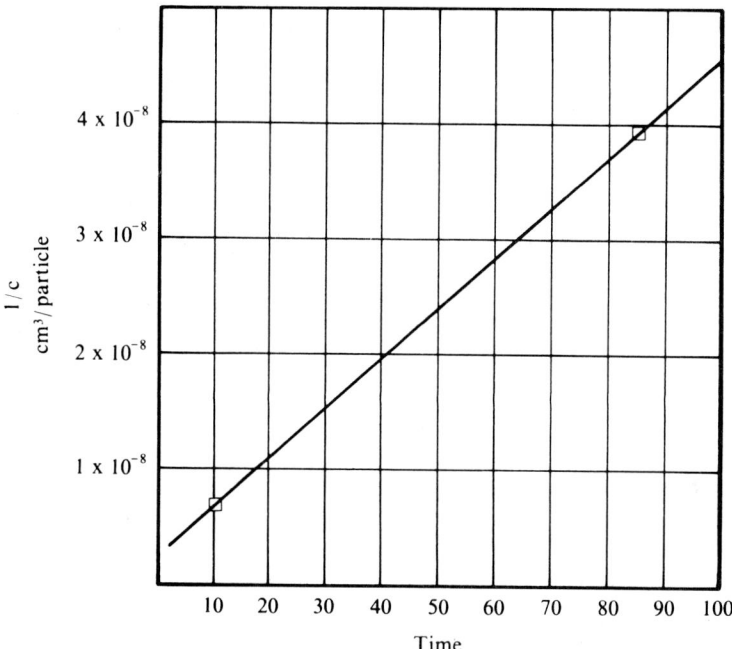

Figure 18.2 Example plot of $1/c$ versus t for a polydisperse aerosol.

Differential Equation Form

Suppose there exists an aerosol which initially consists of particles of sizes 1, 2, 3,... up to size m. The change in concentration of the kth-size particle results not only from a loss of particles from this size group by combination with other particles, but also from a gain in particles into this size group by the right combination of smaller particles. This can be written as

$$\Omega = \Psi - \Phi$$

where Ω = change in number in the kth size, Ψ = increase due to combination making a kth-size particle, and Φ = loss of kth-size particle by combination, or more formally

$$\frac{dc_k}{dt} = \frac{1}{2}\sum_{j=1}^{k-1} K_{j(k-j)} c_j c_{k-j} - \sum_{j=1}^{x} K_{kj} c_k c_j \qquad (18.19)$$

Example 18.6 Given an aerosol made up of particles of discrete sizes d_1, d_2, d_3,..., d_n, write an equation expressing the change in the concentration of size d_4 particles with time, if no particles are lost from this size interval.

With no particle loss, the second term of Eq. 18.19 is zero. Then

$$\frac{dc_4}{dt} = \frac{1}{2}(K_{13} c_1 c_3 + K_{22} c_2 c_2)$$

The total change in the number of particles of all sizes is equal to the sum of the change of the numbers of particles in the individual sizes, which can be written as

$$\frac{dC}{dt} = -\frac{1}{2}\sum_{k=1}^{\infty}\sum_{j=1}^{\infty} K_{kj} c_k c_j \qquad (18.20)$$

which is just one-half the sum of the second part of Eq. 18.19 since there is no overall increase in the number of particles during coagulation. For Eq. 18.20, no sources or sinks have been assumed for the coagulating particles.

Limitations of the Differential Equation Form

In Eq. 18.19 the coagulation rates of particles of different sizes are represented by a coupled set of nonlinear ordinary differential equations. As coagulation proceeds, the number of equations required to describe the size distribution spectrum of an aerosol increases, so that in the extreme one might be required to solve a set of 1000 or more of these equations simultaneously. For example, to determine changes in size distribution of several coagulating monodisperse and polydisperse aerosols, Hidy (1965) solved up to 600 equations simultaneously, and even then there were instances where material was "lost" from the system because it coagulated into sizes larger than the largest size allowed for the particles. As the calculations proceeded, this "loss" limited the accuracy of his numerical solutions.

Use of a Nonlinear Integrodifferential Equation

To circumvent the problem of many equations, it is possible to represent the coagulation equation as a nonlinear integrodifferential equation of the form

$$\frac{\partial (v,t)}{\partial t} = \frac{1}{2}\int_0^v K(\phi, v - \phi) c(\phi, t) c(v - \phi, t)\, d\phi$$

$$- \int_0^\infty K(\phi, v) c(\phi, t) c(v, t)\, d\phi \qquad (18.21)$$

Particle volume is used rather than particle diameter because the volumes are additive whereas the diameters are not. Otherwise, this equation is analogous to the set of ordinary differential equations—in one case a discrete model is used, in the other a continuous model is chosen.

The first term in Eq. 18.21 represents the increase in particles in the volume size range $v + dv$ from the combination of particles of volume ϕ and volume $v - \phi$. The term $c(\phi, t)$ represents the number of particles of volume ϕ at time t; $c(v - \phi, t)$ is the number of particles of volume $v - \phi$ at time t; and the term $K(\phi, v - \phi)$ is the coefficient for coagulation of particles of volume ϕ and volume $v - \phi$.

The second term of this equation represents the loss of particles in the volume size range $v + dv$ resulting from coagulation of particles of volume v and volume ϕ. The term $K(v, \phi)$ is the combined coagulation coefficient for these two particles. Thus this equation gives an expression for the net rate of change of particles whose volumes lie between v and $v + dv$. Solution of the equation with appropriate initial conditions gives the number of particles of volume $v + dv$ at any time t.

To determine the total number of particles per unit volume, it is necessary to integrate over all particle volumes. Symbolically this is

$$C(t) = \int_0^\infty c(v, t)\, dv \tag{18.22}$$

Then the change in the total number of particles with time becomes

$$\frac{dC(t)}{dt} = -\frac{1}{2}\int_0^\infty \int_0^\infty K(v, \phi) c(v, t) c(\phi, t)\, dv\, d\phi \tag{18.23}$$

For the case of constant K and a homogeneous aerosol, Eq. 18.23 reduces to Eq. 18.11.

Terms for Gravity and Deposition Effects

An advantage to using Eq. 18.21 is that other mechanisms which may also remove particles from the coagulating aerosol cloud can be included. For example, removal of particles by gravity settling can be accounted for by adding the term

$$-\vec{v}(v)\, \nabla c(v, t)$$

to the right-hand size of Eq. 18.23; diffusional deposition onto boundary surfaces by the term

$$D(v)\, \nabla^2 c(v, t)$$

$\vec{v}$ is the velocity vector of a particle of volume v and D is its diffusional coefficient (Takahashi and Kasahara, 1968). Thus a generalized equation can be formed which accounts for most, if not all, aerosol behavior with time.

The "Self-Preserving" Size Distribution

It has been suggested by Friedlander (1965), Friedlander and Wang (1966), and Wang and Friedlander (1967) that a coagulating aerosol should, with time, reach the same steady-state size distribution regardless of the aerosol's initial size distribution. This distribution is called the *self-preserving size distribution*. When this steady state is reached, gains by coagulation in the number of particles of a given size are equaled by losses from that size either by coagulation or by sedimentation. For very small particles, sedimentation losses are unimportant; for very large ones, coagulation losses can be neglected. This implies three distinct distribution functions over the entire particle size spectrum. However, with no allowance for a source of particles, these distribution functions are really quasi-steady-state, since eventually the whole system will run out of particles, making a null distribution function.

Hidy (1965) utilized a set of differential equations similar to Eq. 18.19 to test for the existence of self-preservation and found that spectral curves did indeed tend toward an asymptotic value with time. Later Hidy and Brock (1970) showed that the value was itself a function of time and eventually disappears. Hidy's study determined that the quasi-steady-state spectrum would be fully developed in a time given by

$$t = \frac{9\mu}{kTc_0} \tag{18.24}$$

Clark and Whitby (1967) used Friedlander's self-preserving spectrum theory to explain the general shape of the observed size distribution of atmospheric aerosols. Although the formation of the distribution is so slow that there is little likelihood of finding this form in most cases, it is possible that for some global aerosols the quasi self-preserving spectrum of Friedlander is actually developed.

Coagulation of Nonspherical Particles

For nonspherical particles, Muller (1928) postulated that since the diffusion equation applicable to aerosol problems is the same (except for definition of terms) as the general equation for electric fields (Laplace's equation), there should be analogs among the electrostatic terms for various properties of coagulation. For example, the potential should be analogous to particle number concentration, and field strength to particle agglomeration rate. Zebel (1966) pointed out that since field strengths are known to be high at points of sharp curvature, the particle agglomeration rate should also be high at similar places on an irregular particle. The analogy can be carried further to imply that shapes other than spherical tend to enhance coagulation, compared to large spherical particles. However, since particle mobility must also be considered and the decreased mobility of irregularly shaped particles would tend to depress coagulation, Muller's approach should be regarded as only qualitative, giving at best an idea of the effect of particle shape on coagulation rates. It does appear to be true that particle deposition takes place preferentially on sharp protuberances.

External Factors in Coagulation

A number of external physical factors can act to enhance or retard the coagulation of aerosols. These include electrical effects such as the attraction or mutual repulsion of charged particles, polarization effects giving rise to induced forces, sonic agglomeration, gravitational coagulation, and coagulation brought about by turbulence. In the following sections these effects are discussed only briefly since, in general, a detailed treatment is beyond the scope of this text. For more information the reader is referred to the appropriate references.

Electrical effects in coagulation

In coagulation in an electric field, particles diffuse toward one another until they are close enough for electric forces to come into play. Then the motion of the particles assumes an ordered nature.

With electrically charged particles of the same or opposite sign, coagulation may be enhanced or diminished depending on the signs of the two charges and their magnitude. A derivation for the ratio of the coagulation constants for charged and uncharged particles was first given by Fuchs (1934) and is discussed in detail in more recent references by Fuchs (1964) and Zebel (1966). A plot showing the results of Fuchs' derivation is given in Fig. 18.3. When $y < 0$, the forces between the particles are attractive; whereas when $y > 0$, they repel. For a weak bipolar aerosol, the ratio z can be approximated by a straight line through the point $z = 1$ at $y = 0$, indicating that the increase in coagulation brought about by attraction is compensated for by a similar decrease due to repulsion (Zebel, 1966).

On the other hand, with a very strong bipolar aerosol $|y| \gg 1$, the increase in coagulation due to attraction greatly exceeds the decrease due to repulsion, giving a net increase in the coagulation rate.

For a unipolar aerosol it is necessary to consider electrostatic dispersion, i.e., the tendency of charged particles of the same sign to

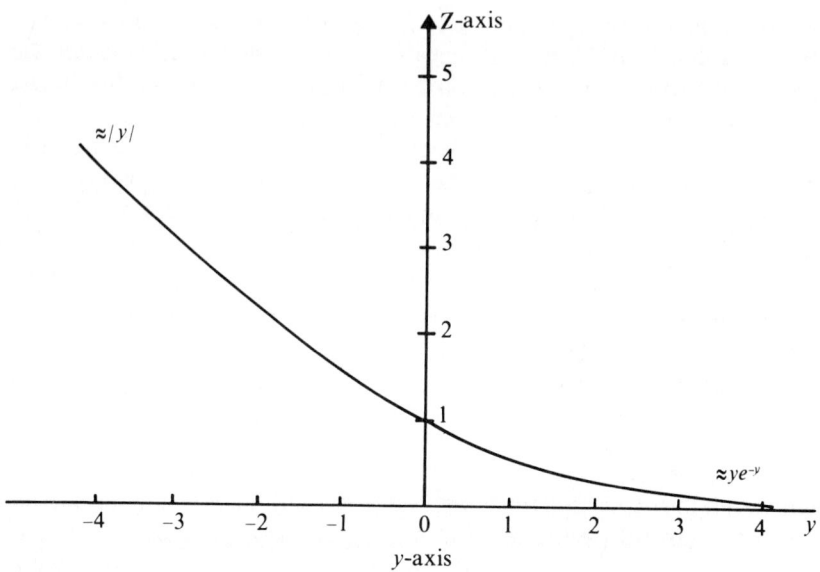

Figure 18.3 Plot of $Z = y/e^y - 1$.

move away from each other. This dispersion tends to reduce the concentration of an aerosol, e.g., by causing particles to deposit on the walls of any containing vessel or nearby surface, and thus can interfere with the measurement of electrical effects in coagulation. Hidy and Brock (1970) have used the Debye-Huckel model to analyze electrostatic effects on coagulation. They showed that when electrostatic dispersion is considered strong, bipolar aerosols will have enhanced coagulation coefficients whereas strongly charged unipolar aerosols will have greatly reduced coagulation coefficients. They caution that these estimates are only approximate since polarization of the electric field can greatly alter the effect of charging in coagulation. Fuchs (1966) pointed out that the coagulation of mists is enhanced only in very strong electric fields (something in excess of 200 V/cm). For solid particles, polarization in an electric field enhances the formation of chainlike aggregates. Particles that are permanent dipoles also form these chainlike structures.

Coagulation in moving atmospheres

Up to this point only brownian motion has been considered. Coagulation can also take place in rapidly moving airstreams, and one might expect that by moving the particles coagulation would be enhanced. This motion could assume two forms. On one hand, there could be an ordered flow of particles in one direction, with the particles moving at different velocities. This could occur, e.g., in a polydisperse aerosol settling by gravity under quiescent conditions. On the other hand, motion could be disordered, as in turbulent mixing.

Coagulation with ordered motion. First consider the case of a large particle moving through a cloud of much smaller particles. What will be the decrease in number concentration of the smaller particles brought about by the larger one? Assuming that the ordered velocity of the smaller particles is negligible compared to the velocity of the larger one, it can be assumed that the large particle removes in unit time all smaller particles contained in a volume given by $(\pi/4)d_L^2 v_L$. The term d_L is the diameter of the larger particle, and v_L is its velocity. However, as discussed in Chap. 5, some of the smaller particles may tend to be displaced from in front of the larger particle by fluid forces set up by the larger particle, so that only a fraction of the particles contained in the volume will actually be collected (the fluid flow pattern is considered to be characterized by the flow around the large particle, taken to be a sphere). A knowledge of this fraction ϵ, known as either the efficiency or capture coefficient, thus permits solution of the problem since the number collected per unit time by a single particle $\mathcal{N}$ becomes

$$\mathcal{N} = \frac{\pi}{4} \epsilon d_L^2 v_L c \qquad (18.25)$$

where c is the number concentration per unit volume of the smaller particles.

Example 18.7 A 50-μm-diameter particle falls through an aerosol of 0.1-μm-diameter spheres containing 10^3 particles per cubic centimeter. If the capture coefficient is 0.01, determine the number of particles collected in unit time.

$$v_g = \tau g = (7.66 \times 10^{-3})(980) = 7.50 \text{ cm/s}$$

$$\mathcal{N} = \frac{\pi}{4}(0.01)(5 \times 10^{-3})^2(7.50)(10^3) = 1.47 \times 10^{-3} \text{ particles/s}$$

Ordered coagulation of particles of approximately the same size. For equal-size particles, prediction of a flow field is extremely difficult because the combined flows of both particles must be considered. These flow fields will vary, as pointed out earlier, as the particles approach each other, in addition to varying with Reynolds number and relative particle size.

An attempt at modeling this situation for viscous flow was made by Hocking (1960), who arrived at the conclusion that collisions between similar-size particles are impossible for particles smaller than 36 μm in diameter. Hocking's computations have been criticized on a number of grounds, and it now appears that a small but finite impaction efficiency does exist. A rough approximation for ϵ in viscous flow has been given by Friedlander (1965) as

$$\epsilon = \left(\frac{d_S}{d_L}\right)^2 - \left(\frac{d_S}{d_L}\right)^4 \tag{18.26}$$

Sonic agglomeration. Particle agglomeration can occur in ordered flow such as that which can be established in a sonic field. Here agglomeration takes place by the different velocities imparted to particles of differing inertia, by aerodynamic attractive forces between the particles, and by radiation pressure which moves the particles toward the vibration antinodes. No complete and adequate theory for acoustic agglomeration exists yet.

Turbulent agglomeration. For turbulent agglomeration two cases should be considered. First, if the inertia of the aerosol particles is approximately the same as that of the medium, the particles will move about with the same velocities as associated air parcels and can be characterized by a turbulence or eddy diffusion coefficient D_T. This coefficient can have a value 10^4 to 10^6 times greater than aerosol diffusion coefficients. Turbulent agglomeration processes can be treated in a manner similar to conventional coagulation except that the larger diffusion coefficients are used.

The second method for aerosol coagulation in turbulent flows arises because of inertial differences between particles of different sizes. The particles accelerate to different velocities by the turbulence depending on their size, and they may then collide with each other. This mechanism is unimportant for a monodisperse aerosol. For a polydisperse aerosol of unspecified size distribution, Levich (1962) has shown that the agglomeration rate is proportional to the basic velocity of the turbulent flow raised to the 9/4 power, indicating that the agglomeration rate increases very rapidly with the turbulent velocity. Since very small particles are rapidly accelerated, this mechanism also decreases in importance as the particle size becomes very small, being most important for particles whose sizes exceed 10^{-5} to 10^{-4} cm in diameter. In all cases brownian diffusion predominates when particles are less than 10^{-6} cm in diameter.

Problems

1 If coagulation is considered to be independent of particle size, how long will it take an aerosol concentration of 5×10^7 particles per cubic centimeter to coagulate to one-half its original value? On average, how many collisions per particle will have taken place?

2 Sulfur dioxide reacts rapidly in the atmosphere to form sulfuric acid droplets with an initial diameter of 0.001 μm. Determine a value for K_0 for coagulation of two of these particles.

3 Compute the value for the coagulation coefficient for 0.07- and 3.0-μm-diameter particles.

4 Particles 1 μm in diameter are coagulating in an atmosphere where the temperature is $-85°C$ and the barometric pressure is 0.5 torr. Compute the value of the coagulation coefficient. Would you expect a greater or lesser rate of coagulation than at ambient conditions?

5 Estimate the time for the concentration of a test aerosol to be reduced by one-half if the aerosol concentration is 50 mg/m^3, its mean particle diameter is 0.1 μm, and $\sigma_g = 2.1$.

Chapter 19

Viable Aerosols

A special class of aerosols are those made up of living biological material (viable aerosols). Pollen spores transported by the wind are in this category, as are the virus-containing droplets of a sneeze or bacteria carried on dust particles. Some of these aerosol materials are benign whereas others can have effects ranging from mildly allergic to deadly as diseases in plants and animals. It is life associated with these aerosols that makes them different from other aerosols. Generally concentrations are so low that these aerosols are important only when they are alive.

An understanding of the behavior of viable aerosols starts with an understanding of the behavior of nonviable aerosols, but there are significant differences which must be taken into account. For example, in sampling, care must be taken to maintain viability throughout the sampling process—this eliminates certain sampling methods from consideration. Otherwise, it is necessary to assume that all material sampled is viable.

Viable aerosol concentrations usually are quite low but can be extremely variable. For example, concentrations on the order of 1000 particles per cubic meter are typical for indoor air (Reponen et al., 1989) or occupational sources (Martinez et al., 1988), but these values can vary by several orders of magnitude. Nevalainen et al. (1992) point out that large fluctuations in viable aerosol concentrations can occur over periods of only a few minutes.

Example 19.1 Typical treated sewage contains 10^7 to 10^9 microorganisms per liter. If sewage containing 10^8 microorganisms per liter is sprayed over land, forming 10-μm-diameter drops, estimate the fraction of droplets that do not contain microorganisms. Hence, estimate the fraction of *sterile* droplets produced.

$$\text{Droplet volume} = \frac{\pi}{6}d^3 = 5.24 \times 10^{-10} \text{ cm}^3$$

$$\text{Droplets in 1 L of liquid} = \frac{1000}{5.24 \times 10^{-10}} = 1.91 \times 10^{12} \text{ drops}$$

$$\text{Drops not containing bacteria} = \frac{1.91 \times 10^{12} - 1 \times 10^8}{1.91 \times 10^{12}} = 99.9948\%$$

Most of the sewage aerosol sprayed is sterile!

Types of Viable Aerosols

Viable aerosols can be viruses, bacteria, or fungi, including both yeasts and molds, alone or attached to inert dust particles, or they can be live pollen spores and fungus spores, slime molds, protozoans, and algae. Because life is an integral part of the definition of this class of aerosols, such "bioaerosols" as plant and animal fragments which are important as aeroallergens are not considered further here, since viability is not essential to the allergic reaction.

Units of Measure

Unlike nonviable aerosols, viable aerosol concentrations cannot simply be reported in terms of particles per unit volume since interest here is in the *number of living particles* per unit volume, and the numbers generally are quite small compared to typical aerosol concentrations.

Viable aerosol samples can be collected on microscope slides and counted directly, with or without staining to reveal or differentiate biological material. Unfortunately this procedure does not show which, if any, of the collected microorganisms are viable. In this case counts are reported as microorganisms or bioaerosol particles.

Collected microorganisms can be transferred to petri dishes for growth and possible isolation of bacteria and fungi on an appropriate nutrient agar or transferred to a cell culture for isolation of viruses (Chatigny et al., 1989). As the bacteria or fungi grow, they form colonies or spots on the agar. These colonies are made up of millions of individual microorganisms. Each colony is considered to arise from one or more individual microorganisms. Hence each aerosol particle which gives rise to a colony on agar is counted as a *colony-forming unit* (CFU), and viable bacteria or fungi are reported as such.

For viruses, host cells are infected with the sample which causes changes in the cells' appearance or bursts the cells and produces plaques (Chatigny et al., 1989). Each plaque is considered to arise from one or more viruses. Hence viruses are reported as *plaque-forming units* (PFUs).

In these procedures, live microorganisms are detected by allowing them to multiply to easily observable numbers. This is both an advantage and a disadvantage to the investigator. Since the microorganisms will grow after collection, very small concentrations of viable aerosol can be detected. At the same time, however, sample results depend on maintaining viability both during and after collection of the sample. If care is not exercised in handling the collected sample, some microorganisms can be killed and the concentration will be measured as less than the true value. Figure 19.1 illustrates the process of viable aerosol sampling.

Example 19.2 An early method of measuring airborne bacteria concentration was to place an agar plate on a flat surface for a specific time, allowing the bacteria to fall on the agar surface. Then if a terminal settling velocity is assumed, the concentration can be determined by dividing the count assumed to have fallen on the agar plate, in microorganisms per square centimeter per second, by the terminal settling velocity.

An agar plate collects 26 viable colonies over a 24-h period. If the plate diameter is 10 cm and 10-μm-diameter particles are assumed, determine the average microorganism concentration over the 24-h period.

Particles collected per square centimeter per second are

$$\frac{26}{(\pi/4)10^2(24 \times 60 \times 60)} = 3.823 \times 10^{-6} \text{ particles/cm}^2$$

Using 0.304 cm/s as the terminal settling velocity for 10-μm drops, we see the estimated concentration is

$$C = \frac{3.823 \times 10^{-6}}{0.304} = 1.26 \times 10^{-5} \text{ CFU/cm}^3$$

$$= 12.6 \text{ CFU/m}^3$$

Factors influencing viable aerosol concentrations

A major difficulty in evaluating the concentration of viable aerosols is that viable materials are extremely sensitive to environmental factors, usually in a negative sense. Thus these aerosols will not exist as

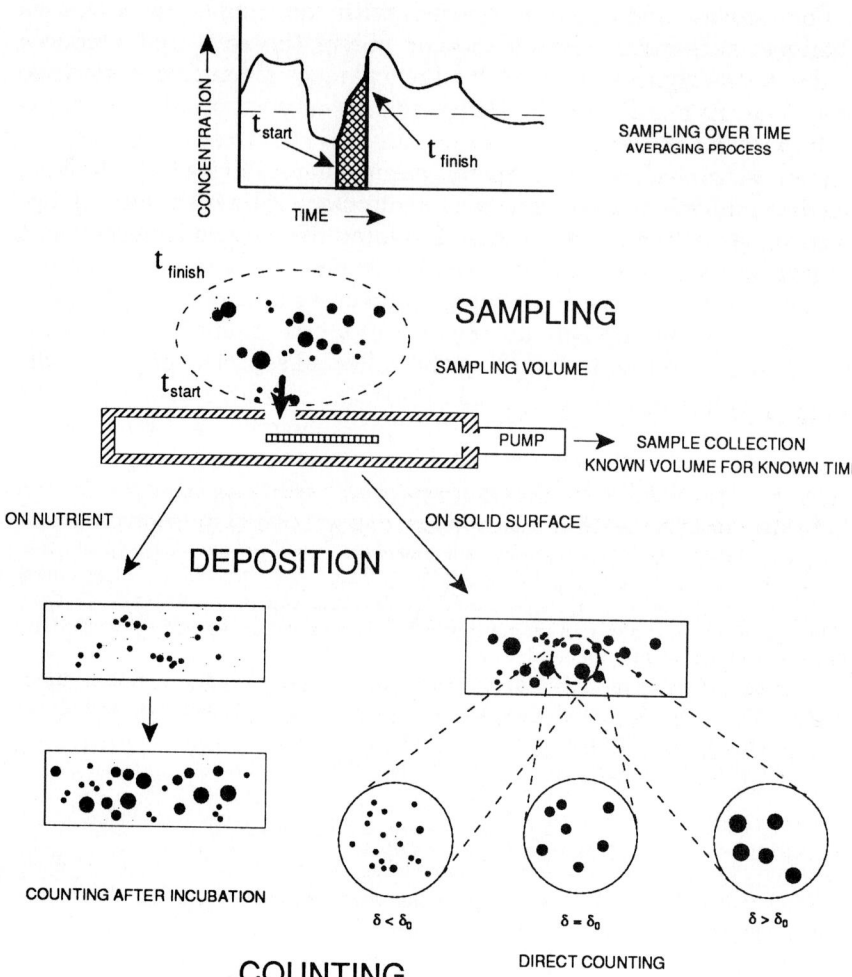

Figure 19.1 The process of bioaerosol sampling. (*After Nevalainen et al., 1992.*)

such for very long, and while they do exist, their concentration will be continuously decreasing.

Details on many of the environmental factors which can cause concentration change are beyond the scope of this book. It is important to realize that such things as species variability, air temperature, humidity, irradiation, or trace materials in the air all can influence microbiological viability, or the effective lifespan of these materials (Reist et al., 1980; Anderson and Cox, 1967).

For some organisms, exposure to oxygen in air represents a toxic gas (Chatigny et al., 1989).

How samples are handled after collection also can influence presumed viable aerosol concentrations. Storage conditions, the culture medium used for sample growth, and mutation or metabolic changes during growth can affect the final viable aerosol concentration (Dimmick and Akers, 1969).

Estimates of Viable Aerosol Concentrations

Despite the difficulties inherent in determining viable aerosol concentrations, it is possible to make a rough concentration estimate by considering loss of viability as a first-order event and assigning a rate constant K_D to it. The value of this rate constant will vary depending on both species and environmental factors, as previously mentioned, but there are some limited data available, e.g., for viruses which show the utility of this approach.

Tyrrell (1967) gives a half-life for "common cold" virus infectivity of about 3 min at 40 to 50 percent relative humidity and 10 min at 90 percent relative humidity. Then Eq. 6.27 can be adapted to account for the decrease in infectivity, to give

$$C = C_0 \exp\left(-K_D t - \frac{v_T t}{H}\right) \tag{19.1}$$

According to Anderson and Cox (1967), this simple first-order kinetic approach may not be followed, and they suggested alternatives such as those given by Monk and Mattuck (1956).

Example 19.3 Figure 19.2 shows the size distribution of a sneeze as given by Lidwell (1967) (10^8 droplets per sneeze). Assuming each of the droplets contains one or more viable microorganisms with a virus survival half-life of 2 min, determine the viable aerosol concentration as a function of time, using Eq. 19.1.

Because v_T is a function of particle size, Eq. 19.1 must be solved for a number of particle diameters. By using 11 size increments, Table 19.1 can be constructed; and a plot of C versus droplet diameter for various times is shown in Fig. 19.3. Then Fig. 19.4 shows the relative decrease in virus concentration as a function of time for the summed data.

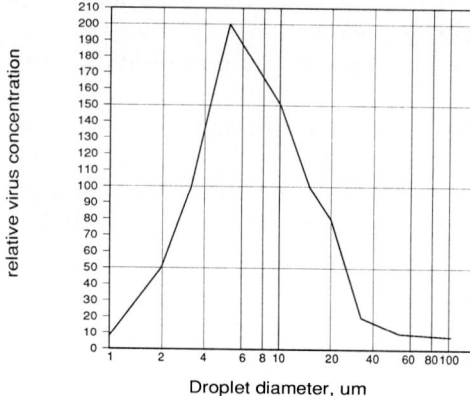

Figure 19.2 Size distribution of a sneeze. (*After Fidwell, 1967.*)

TABLE 19.1 Variation in Viable Aerosol Concentration *C* by Number and Time (Computed from Eq. 19.1)

t	1	2	3	5	10	20	30	50	100	Total
0	8,000	50,000	100,000	200,000	150,000	100,000	80,000	20,000	10,000	726,000
1	5,654	35,273	70,336	139,332	99,914	61,809	44,532	8,255	1,585	466,704
2	3,996	24,884	49,472	97,067	66,552	38,204	24,789	3,407	251	308,621
3	2,824	17,555	34,796	67,623	44,329	23,614	13,799	1,406	40	205,986
4	1,996	12,385	24,474	47,110	29,527	14,595	7,681	580	6	138,356
5	1,411	8,737	17,214	32,820	19,668	9,021	4,276	240	1	93,387
6	997	6,164	12,108	22,864	13,101	5,576	2,380	99	0	63,288
7	705	4,348	8,516	15,929	8,726	3,447	1,325	41	0	43,036
8	498	3,068	5,990	11,097	5,812	2,130	737	17	0	29,349
9	352	2,164	4,213	7,731	3,872	1,317	411	7	0	20,066
10	249	1,527	2,963	5,386	2,579	814	229	3	0	13,748
11	176	1,077	2,084	3,752	1,718	503	127	1	0	9,438
12	124	760	1,466	2,614	1,144	311	71	0	0	6,490
13	88	536	1,031	1,821	762	192	39	0	0	4,470
14	62	378	725	1,269	508	119	22	0	0	3,083
15	44	267	510	884	338	73	12	0	0	2,128
16	31	188	359	616	225	45	7	0	0	1,471
17	22	133	252	429	150	28	4	0	0	1,018
18	15	94	178	299	100	17	2	0	0	705
20	8	47	88	145	44	7	1	0	0	339

Viable Aerosols   325

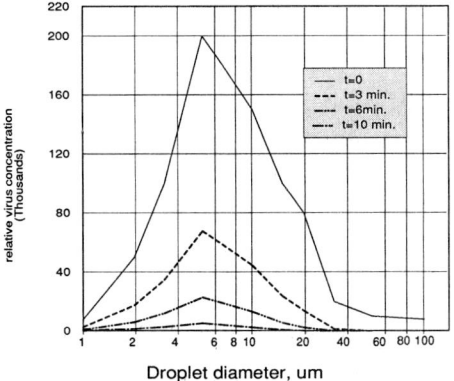

Figure 19.3 Change in viable aerosol concentration as a function of diameter, different times.

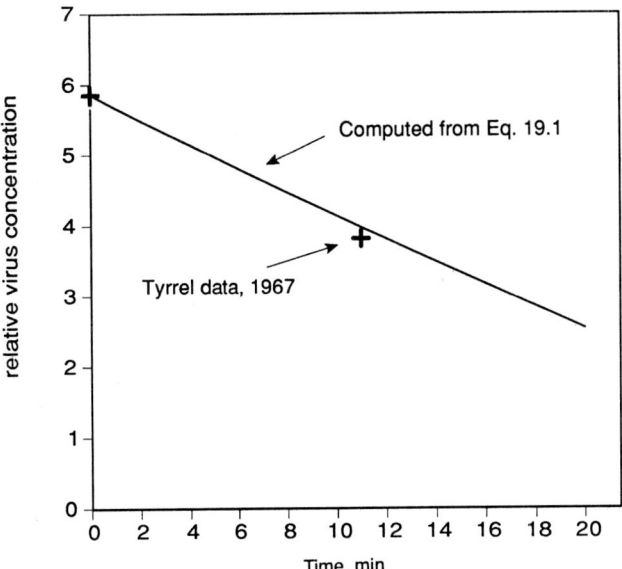

Figure 19.4 Summed virus data, Example 19.3.

Chapter 20

Explosive Aerosols

Any material that will burn in air in solid form may explode if it is dispersed in aerosol form. Explosions of foods, metals, pharmaceuticals, grain products, polymers, and other organic materials all have occurred in the past. Since oxidation is an exothermic reaction, the heat released in burning will rapidly raise the temperature of small particles nearby, and the large surface area presented by these particles encourages more reaction to take place and hence more heat to be produced. A runaway reaction can be the result.

For dust to explode, the following conditions must be met:

1. Dust must be present in suspension in a high enough concentration.
2. There must be sufficient oxygen present to enable combustion to take place.
3. There must be a large enough source of energy to ignite the cloud.

In addition, the severity of the explosion will depend on the nature of the reacting material and the amount and type of confinement of the blast.

In some instances this reaction is brought about on purpose, e.g., rocket fuels which are a mixture of combustible material and oxidant in aerosol form. The close proximity and intimate mixing of large amounts of fuel and oxidant can result in extremely violent explosions.

With settled powders or with very dense dust clouds, there may not be sufficient oxygen for the rapid reaction required for an explosion. Particles must be close enough for the heat generated by one particle to initiate a reaction in its neighbor, but spaced far enough apart that oxygen can easily reach the reacting surface. This implies the existence of both minimum and maximum explosive concentrations. The lower explosive concentration for most dusts ranges from about 20 to 60 g/m³. However, unlike gases, it is very hard to establish a maximum explosive concentration for dusts, but it is somewhere between 2 and 6 kg/m³ (Bartknecht, 1987).

Example 20.1 Given a monodisperse dust cloud made up of 40-μm-diameter spheres with ρ_p = 1.85 g/cm³. If the dust concentration is 30 g/m³, determine the concentration in particles per cubic centimeter.

The volume of one particle is $(\pi/6)(40 \times 10^{-4})^3 = 3.35 \times 10^{-8}$ cm³, and then the mass of 1 particle is

$$3.35 \times 10^{-8} \text{ cm}^3 \times 1.85 \text{ g/cm}^3 = 6.20 \times 10^{-8} \text{ g}$$

With a concentration of 30 g/m³, the number of particles per cubic meter is

$$\frac{30 \text{ g/m}^3}{6.20 \times 10^{-8} \text{ g/particle}} = 4.84 \times 10^8 \text{ particles/m}^3$$

This is 4.84×10^2 particles per cubic centimeter.

Severity of Explosions

Unlike a gas-phase explosion or *detonation* in which the flame speed is close to the speed of sound (340 m/s), the flame speed in a dust explosion is usually relatively slow (about 10 m/s). Here the explosion is described as a *deflagration*. This means that the shock wave in the explosion precedes the flame, resuspending settled dust and enhancing conditions for a second, larger dust explosion. Thus dust explosions tend to propagate as a series of increasingly severe explosions, utilizing resuspended previously settled dust as the explosive material.

Figure 20.1 shows a plot of explosion pressure versus concentration

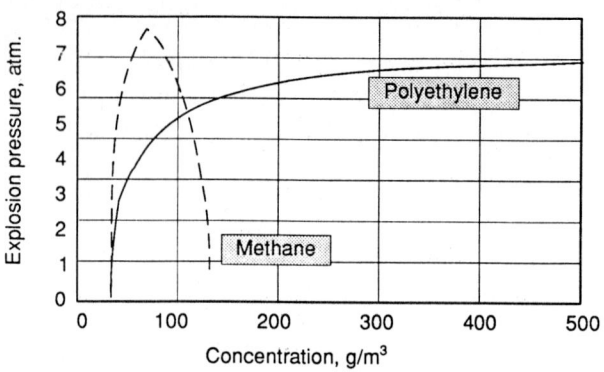

Figure 20.1 Flammability limits for polyethylene dust and methane gas, in air. (*From Hertzberg and Cashdollar, 1987.*)

for polyethylene dust compared to methane, two paraffin hydrocarbons. The maximum explosion pressure developed for the dust is less than that for the methane. Both materials exhibit similar lower explosive limits, consistent with their similar chemical structures. However, whereas the methane shows a clear upper explosive limit, this is not so for the polyethylene dust.

The explosion of a powder or dust can be described by the violence of its fire or explosion (explosion pressure and rate of pressure rise dp/dt) or by the ease with which it ignites. Although it does not necessarily follow that a material which is easy to ignite will burn violently, the two are often related. Since a dust explosion is essentially a rapid oxidation of the dust surface or material emitted from the dust surface, its severity depends on both the reactivity of the material and the surface area available. The severity of the explosion also depends on how heat is transmitted between particles.

Bartknecht (1981) found that the maximum rate of pressure rise $(dp/dt)_{max}$ for dust explosions in spaces ranging from 0.02 to 60 m³ could be expressed as

$$V^{1/3}\left(\frac{dp}{dt}\right)_{max} = K_{St} \quad (20.1)$$

where V is the volume of the space or container in cubic meters and K_{St} a value which is characteristic for a given dust (Cross and Farrer, 1982). It is possible to classify dusts according to their K_{St} value as shown in Table 20.1. For example, Pineau et al. (1976) gave data on the maximum pressure rise for dried milk having d_g = 56.3 μm and σ_g = 2.75, showing a K_{St} value of 63 bar · m/s, placing this particular sample in class St 1.

Types of Explosive Dusts

There are a wide variety of dusts which may explode to give either a detonation or a deflagration. The K_{St} value, if available, can give some indication of the explosiveness of a given material; the higher the K_{St} value, the more explosive the dust. Table 20.2 lists several different materials, showing a range of their K_{St} values.

TABLE 20.1 Dust Classification

Dust explosion class	K_{St}, bar · m/s	Example
St 0	0	Nonexplosive dust
St 1	> 0–200	Grain dust
St 2	200–300	Organic pigments
St 3	> 300	Fine metal dust

SOURCE: Cross and Farrer (1982).

TABLE 20.2 Values of K_{St} for Various Dusts

Dust type	K_{St}, bar · m/s	Reference
Aluminum	2000	Pineau et al. (1976)
Aluminum	16–1900	Cross and Farrer (1982)
Aspirin	190	Eckhoff (1987)
Coal	60–97	Cross and Farrer (1982)
Corn flour	95	Pineau et al. (1976)
Epoxy	200	Pineau et al. (1976)
Flour	80–192	Cross and Farrer (1982)
Phenolic resin	140	Eckhoff (1987)
Polyethylene	138	Eckhoff (1987)
Polyethylene	4–120	Cross and Farrer (1982)
Sugar, granular	110	Geysen et al. (1987)
Sugar, powdered	215	Pineau et al. (1976)
Toner dust	169	Eckhoff (1987)

Note: These values are based on experiments using a 1-m^3 sphere. Previous experimental studies on explosibility were carried out using the Hartmann apparatus which tends to give values for K_{St} lower than the 1-m^3 sphere (see Pineau et al., 1976).

Even the K_{St} values give only a partial picture, since they are dependent on the initial temperature of the material. For example, Wiemann (1987) gives a K_{St} value of 100 bar · m/s for sprayed skim milk with an initial temperature of 50°C and 125 bar · m/s for the same material with an initial temperature of 125°C. For some materials the value of K_{St} decreases with increasing initial temperature. This could be due to removal of some volatiles from the material by the higher temperature prior to the explosion.

If K_{St} values are not known, sometimes the potential explosibility of a dust cloud can still be estimated. Chemical groups such as organic peroxides and some nitro compounds are especially susceptible, as are materials containing significant quantities of oxygen readily available for combustion. With the following basic reaction

$$C_xH_yO_z + \left(x + \frac{y}{4} - \frac{z}{2}\right) O_2 \rightarrow x\ CO_2 + \frac{y}{2}H_2O \quad (20.2)$$

an oxygen balance can be defined as (Cross and Farrer, 1982)

$$\Omega = \frac{-1600(2x + y/2 - z)}{\text{molecular weight}} \quad (20.3)$$

According to Cross and Farrer (1982), if a material has a value of Ω more positive than -200, it should be considered to have the potential to be a highly explosive dust.

Example 20.2 An approximation for the exothermic reaction that takes place during the explosion of polyethylene powder is (Hertzberg and Cashdollar, 1987)

$$8[-CH_2-]_n + 12nO_2 + 45nN_2 \rightarrow 8nCO_2 + 8nH_2O + 45nN_2 + 712 \text{ cal/g}$$

Determine the value of Ω for this reaction.

$$\Omega = \frac{-1600(2x + y/2 - z)}{\text{molecular weight}} = \frac{-1600(24n)}{14n}$$

so $\Omega = -2743$. This is well above the highly explosive limit given by Cross and Farrer, and thus the material should be treated as if there were a potential explosion hazard.

Ignition Sources

As mentioned previously, to initiate a dust explosion, there must be an ignition source as well as a sufficiently high concentration of ignitable dust. Ignition sources can range from obvious ones such as open flames or torches used in welding operations to the careless disposal of burning cigarettes. Sparks from electric discharges represent another ignition source.

Spontaneous combustion can arise from exothermic chemical reactions of some materials, e.g., with water, or through the action of microorganisms. Microorganism growth can heat a medium to about 70°C where, although the heat is sufficient to kill the microorganisms, the increased temperature may be sufficient to rapidly increase the rate of some other chemical reaction which subsequently raises the temperature of the medium to the combustion point (Cross and Farrer, 1982).

Another common source of spontaneous combustion is by oxidation, especially oxidation of finely divided metals or metallic powders or certain fats and oils. Metal powders which have been produced in inert atmospheres are especially susceptible to spontaneous combustion in air because their unoxidized surfaces are highly reactive.

Because of the nature of dust explosions, an ignition source need only cause settled dust to smoulder. Then later resuspended smouldering dust can be the ignition source for an explosion.

For a dust piled on a surface or thick on the walls of a duct, heat is lost by conduction and thus is linear with temperature whereas the reaction rate (and thus heat production) increases exponentially with temperature. This implies the existence of a temperature level, known as the *critical temperature* T_c, above which more heat will accumulate in the reacting material than will be lost from it. Such factors as the size or thickness of the dust deposit, airflow across the dust which may carry away some heat, thermal conductivity and packing of the dust, and temperature distribution within the dust can all influence the ultimate value for T_c.

The effectiveness of any ignition source depends on its size and shape, the amount of energy it delivers to the dust, and its contact time with the dust, as well as with the particular type of dust. For example, Fig. 20.2 shows flammability limits for several dusts as a function of the ignition energy (Hertzberg and Cashdollar, 1987).

Figure 20.3 shows the flammability domains for two different isothermally heated dusts (lycopodium and coal) dispersed in air. As the temperature increases, the minimum dust concentration required for an explosion decreases. Above some temperature the dust becomes *autoignitable*; i.e., it will spontaneously explode if present in high enough concentration. In addition, as the temperature of a dust-air mixture increases toward this autoignitable limit, the likelihood of the mixture's igniting increases. For example, at a cost dust concentration of 400 g/m^3 an increase in initial temperature from 75 to 200°C results in an order-of-magnitude reduction in its spark ignition energy (Hertzberg et al., 1985).

Particle Size

As might be expected, particle size is an important factor in assessing the flammability of dusts. It appears that above some size, dust particles are relatively ineffective as contributors to dust explosions. Figure 20.4 shows this effect for several dusts. In this figure only the mean particle size is given. However, it is expected that particle size distribution would also be a complicating factor in assessing the flammability of dusts.

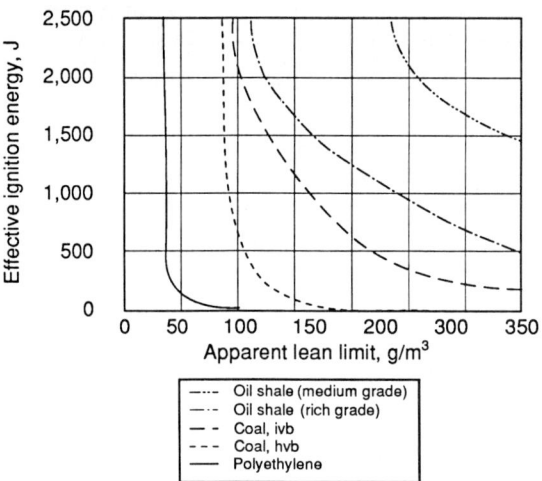

Figure 20.2 Lean flammability limits for five dusts. (*Adapted from Hertzberg and Cashdollar, 1987.*)

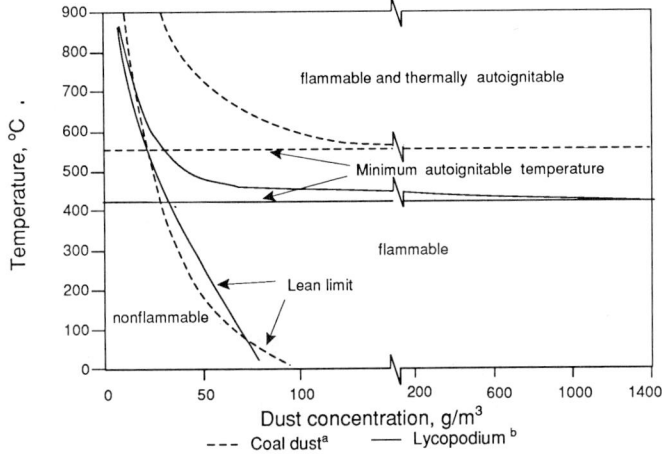

Figure 20.3 Domains of flammability, isothermal heating. [(a) Hertzberg and Cashdollar, 1987. (b) Conti and Hertzberg, 1987.]

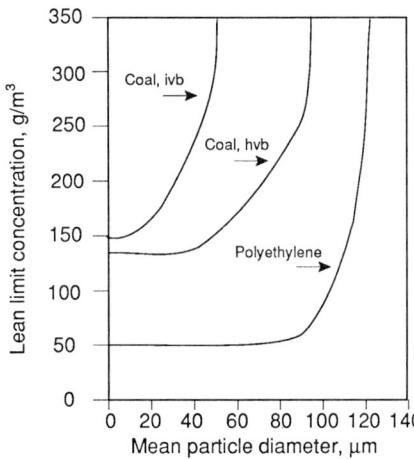

Figure 20.4 Effect of particle diameter on mean flammability limit. (*Hertzberg and Cashdollar, 1987.*)

The data shown in Fig. 20.4 are for tests in which dusts with narrow size distributions were used. Hertzberg and Cashdollar (1987) pointed out that with broad size distributions, strong size dependencies tend to be blurred, but the trends should remain the same. With small mean particle sizes, all flammable material is volatilized from the particles, and the combustion process is "homogeneous." As the particle size increases, only the surface material is volatilized so that more dust is required overall to sustain combustion. Finally, above some upper limit of particle size, the volatilization rate is such that a minimum concentration of combustible volatile material is not achieved.

Example 20.3 If the minimum explosive limit for an organic dust is 50 g/m³, determine the minimum dust surface area for a dust explosion, in square centimeters per cubic centimeters of air for the dried milk described above. Assume $\rho_p = 1.0$.

For the dried milk, $d_g = 56.3$ μm and $\sigma_g = 2.75$. Then from Eq. 2.11

$$d_v = d_g \exp(1.5 \ln^2 \sigma_g) = 56.3 \exp(1.5 \ln^2 2.75) = 261.3 \text{ μm}$$

$$n = \frac{6m}{\pi d^3 \rho} = \frac{6(50 \times 10^{-6}) \text{ g/cm}^3}{(3.14)(261.3 \times 10^{-4})^3 (1.00) \text{ g/cm}^3} = 5.35 \times 10^5 \text{ particles/cm}^3$$

$$d_a = d_g \exp(1.0 \ln^2 \sigma_g) = 56.3 \exp(1.0 \ln^2 2.75) = 156.7 \text{ μm}$$

$$\text{Surface area} = \pi d_a^2 n = (3.14)(156.7 \times 10^{-4})^2 (5.35 \times 10^5) = 4.13 \times 10^{-3} \text{ cm}^2/\text{cm}^3$$

The minimum ignition energy required for a dust explosion is related to particle size. This can be seen in Fig. 20.5a for cornstarch and 20.5b for high-density polyethylene. There appears to be an upper particle size limit of hundreds of micrometers above which there will be no ignition, this limit varying with the type of dust.

Moisture can also act to suppress dust explosions or at least make their occurrence more difficult. Figure 20.6 shows the effect of moisture content on the minimum explosive concentration of grain. Here increasing the moisture content increases the minimum explosive concentration. The effect of moisture on the minimum ignition energy of grain is shown in Fig. 20.7. Increasing the moisture content of the grain increases the ignition energy required in an almost linear fashion. Also the presence of moisture will act to reduce the severity of an explosion if it occurs. This can be seen in Fig. 20.8 where increasing the moisture content of the dust decreases both the maximum and the average rate of pressure rise as well as the overall maximum pressure achieved. Apparently moisture can change the ease with which dust is dispersed as well as altering the oxidation properties of the dust's sur-

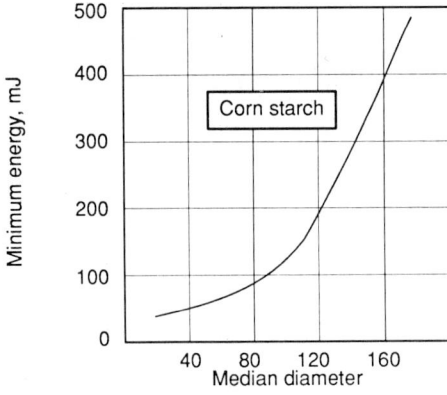

Figure 20.5a Minimum ignition energy of cornstarch as a function of particle diameter. (*Cross and Farrer, 1982.*)

Explosive Aerosols 335

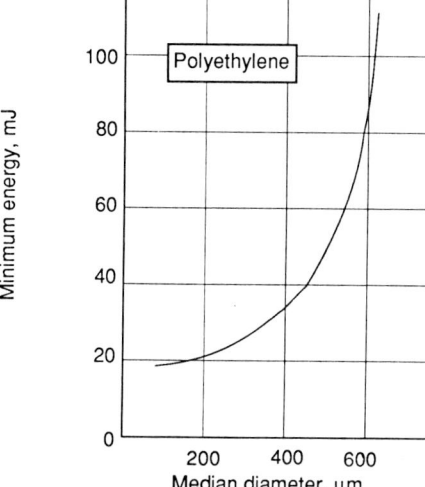

Figure 20.5b Minimum ignition energy of polyethylene as a function of particle diameter. (*Cross and Farrer, 1982.*)

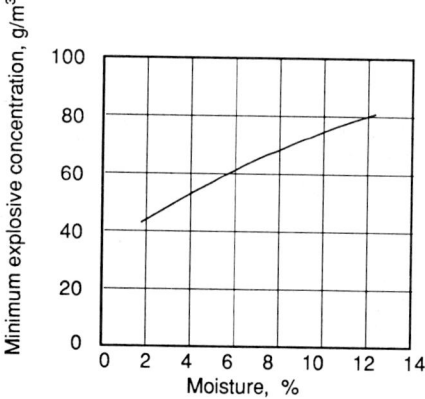

Figure 20.6 Effect of moisture on the minimum explosive concentration of grain. (*Cross and Farrer, 1982.*)

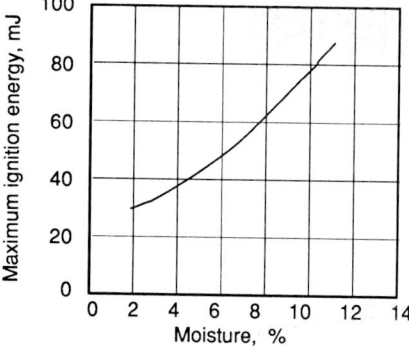

Figure 20.7 Effect of moisture on the minimum ignition energy of grain. (*Cross and Farrer, 1982.*)

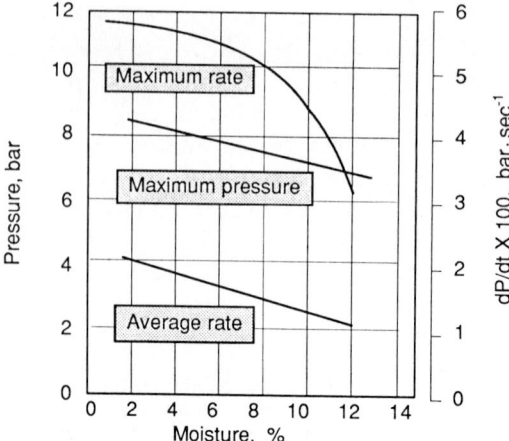

Figure 20.8 Effect of moisture on pressure of an explosion. (*Cross and Farrer, 1982.*)

face. Moisture can act as a heat sink as well as altering the electrical characteristics of some dusts.

Control of Dust Explosions

As mentioned earlier, for dust to explode, at least three conditions must be met:

1. Combustible dust must be present in suspension in a high enough concentration.
2. There must be a large enough source of energy to ignite the cloud.
3. There must be sufficient oxygen present to enable combustion to take place.

Control methods which focus on any of these three conditions will be effective to prevent explosions. For example, probably the most important method for controlling dust explosions is to limit the concentration of dust in the air to below the lower explosive limit (LEL). Besides controlling dust in the air, housekeeping is important as well so that there is no possibility of settled dust's becoming resuspended and producing a dust cloud above the LEL.

Sometimes noncombustible dust can be mixed with combustible dust so that the LEL for the resulting mixture is significantly raised. The noncombustible dust acts to dissipate heat from the burning combustible dust, thereby limiting the spread of the flame front.

If it is not possible to ensure that the LEL for a combustible dust will not be exceeded, then control should focus on potential combustion sources and methods to minimize or eliminate their presence in the dust cloud. Use of nonsparking equipment is one example of combustion source control, as are restrictions on smoking by employees in dusty areas. Because settled dust can smoulder and then, when resuspended, become the source for ignition of a dust cloud, care must be taken to eliminate "hot spots" which could act to start settled dust burning or smouldering. For example, surfaces of light bulbs or other sources of heat need to be protected from settled dust. Electric equipment should be enclosed. Welding or use of open flames must be controlled carefully in dusty areas. Ductwork carrying dust must be well grounded to prevent the buildup of static electricity.

Sometimes it is possible to work with high dust concentrations in spaces where the air is partially deoxygenated or where an inert gas is present in the space. The major problem here comes if the dusty material needs to be transferred from the inert atmosphere to one containing a normal quantity of oxygen.

In addition to reducing the probability of a dust explosion, it is possible to take steps to mitigate the consequences of a dust explosion, should one occur. Use of automatic vents to prevent the buildup of destructive overpressures and constructing areas with blowout partitions are examples of such steps. Also, spaces which may contain these high dust concentrations can be kept fairly open to minimize the development of turbulence in a flame front, should an explosion occur.

Problems

1 Given is a monodisperse dust cloud made up of 40-μm-diameter spheres with ρ_p = 1.85 g/cm^3. If the dust concentration is 30 g/m^3, determine the concentration in particles per cubic centimeter.

2 Using Eq. 20.3, determine the value of Ω for the reaction of an organic aerosol where each molecule has the chemical form $C_{26}H_{36}O_{12}$.

3 If the minimum explosive limit for an organic dust is 30 g/m^3, determine the minimum dust surface area for a dust explosion, in square centimeters per cubic centimeters of air for a powder with d_g = 24 μm and σ_g = 2.1.

4 A graphite mine tunnel has a 3- × 3-m cross-section. How thick must a layer of graphite dust on the floor be to give an explosive concentration of 40 g/m^3 when 15 percent of it is resuspended? Assume the bulk density of the graphite dust layer is 1.4 g/cm^3.

Appendix A

Corrected Sedimentation Velocities

Unit-Density Spheres (NTP)

d, μm	v_g, cm/s	C_c
0.01	7.07×10^{-6}	23.775
0.02	1.45×10^{-5}	12.192
0.04	3.06×10^{-5}	6.422
0.06	4.84×10^{-5}	4.516
0.08	6.80×10^{-5}	3.573
0.10	8.97×10^{-5}	3.015
0.20	2.31×10^{-4}	1.938
0.40	6.88×10^{-4}	1.446
0.60	1.39×10^{-3}	1.294
0.80	2.32×10^{-3}	1.220
1.00	3.50×10^{-3}	1.176
2.00	1.29×10^{-2}	1.088
4.00	4.97×10^{-2}	1.044
6.00	1.10×10^{-1}	1.029
8.00	1.94×10^{-1}	1.022
10.00	3.03×10^{-1}	1.018
20.00	1.20×10^{0}	1.009
40.00	4.78×10^{0}	1.004
60.00	1.07×10^{1}	1.003

Appendix B

Stokes Law

Stokes considered the resistance experienced by a sphere moving uniformly through an incompressible viscous fluid. Viscous flow implies a low Reynolds number Re. He assumed an infinite medium, rigid particle, no slipping at the surface of the particle. Stokes pointed out that his solution was erroneous in the case of a cylinder.

Consider a fluid flowing past a sphere. Take spherical coordinates with the polar axis parallel to u, the direction of motion of the fluid. By symmetry, all quantities of interest will only be functions of r and the polar angle θ.

The resisting force $\vec{F}$ is parallel to the velocity $\vec{u}$. The incremental force acting on a unit of surface area is

$$d\vec{F}_i = p\vec{n}_i - \sigma_{ik}\vec{n}_k$$

where p is the pressure of the fluid, σ_{ik} is the viscosity stress tensor, and $\vec{n}$ is a unit vector. The term σ describes the irreversible "viscous" transfer of momentum in the field. Hence the first term in the above equation is the ordinary pressure of the fluid; the second represents the force of friction, due to viscosity, acting on the surface of the body.

Considering only the components of these forces in the $\vec{u}$ direction and suming over the entire sphere surface give

$$F = \oint (-p \cos\theta + \sigma_{rr} \cos\theta - \sigma_{r\theta} \sin\theta)\, ds$$

where the integration is taken over the whole surface s of the sphere. We must now evaluate p, σ_{rr}, and $\sigma_{r\theta}$. In spherical coordinates the stress tensor is

$$\sigma_{rr} = 2u\frac{\partial v_r}{\partial r}$$

$$\sigma_{re} = \mu\left(\frac{1}{r}\frac{\partial v_r}{\partial \theta} + \frac{\partial v_\theta}{\partial r} - \frac{v_\theta}{r}\right)$$

where v is the velocity of a fluid element ($v = u$ at infinity) and μ is the fluid viscosity. For steady flow of an incompressible fluid, the Navier-Stokes equation is

$$\vec{v} \times \text{grad } \vec{v} = -\frac{1}{\rho}\text{grad } p + \frac{u}{\rho}\nabla\vec{v}$$

the term

$$\vec{v} \times \text{grad } \vec{v} \quad \text{is proportional to} \quad \frac{u^2}{l}$$

and

$$\frac{\mu}{\rho}\nabla\vec{v} \quad \text{is proportional to} \quad \frac{\mu u}{\rho l^2}$$

The ratio of the two is the Reynolds number Re, so if Re is small, we can write

$$\mu \nabla\vec{v} - \text{grad } p = 0$$

which with the equation of continuity

$$\text{div } v = 0$$

completely specifies the motion. Solving these equations gives

$$v_r = (u \cos\theta)\left(1 - \frac{3R}{2r} + \frac{R^3}{2r^3}\right)$$

$$v_\theta = (-u \sin\theta)\left(1 - \frac{3R}{4r} - \frac{R^3}{4r^3}\right)$$

where the surface of the sphere is $r = R$.

The relative pressure is

$$p = -\frac{3u}{2R}u \cos\theta$$

Substituting at $r = R$ gives

$$\sigma_{rr} = 0 \quad \sigma_{r\theta} = -\frac{3\mu}{2R}u \sin\theta$$

thus

$$F = \frac{3\mu u}{2R} \int ds$$

or

$$F = 6\pi R\mu u = 3\pi\mu u d$$

where d is the particle diameter.

For a more complete derivation see Landau and Lifshitz (1969) or Joos (1951).

Appendix C

Error Function

Many integrals of the diffusion equation lead to the error function, which can be written as

$$\mathrm{erf}\, y = \frac{2}{\sqrt{\pi}} \int_0^y \exp(-\xi^2)\, d\xi$$

$$\mathrm{erf}\, \infty = 1$$

$$\mathrm{erf}\, 0 = 0$$

$$\mathrm{erf}(-y) = -\mathrm{erf}\, y$$

The first two derivatives of the error function are

$$\frac{d}{dy}(\mathrm{erf}\, y) = \frac{2}{\sqrt{\pi}} \exp(-y^2)$$

$$\frac{d^2}{dy^2}(\mathrm{erf}\, y) = -\frac{4}{\sqrt{\pi}} y \exp(-y^2)$$

Table C.1 lists $\mathrm{erf}\, y$ for various values of the argument y. In this table y to the nearest tenth is entered in the left-hand column, reading across in hundredths to the argument value of $\mathrm{erf}\, y$. Thus $\mathrm{erf}\, 0.63 = 0.62705$.

TABLE C.1 Error Function erf y

y	0.00	0.01	0.02	0.03	0.04	0.05	0.06	0.07	0.08	0.09
0.0	0.00000	0.01128	0.02256	0.03384	0.04511	0.05637	0.06762	0.07886	0.09008	0.10128
0.1	0.11246	0.12362	0.13476	0.14587	0.15695	0.16800	0.17901	0.18999	0.20094	0.21184
0.2	0.22270	0.23352	0.24430	0.25502	0.26570	0.27633	0.28690	0.29742	0.30788	0.31828
0.3	0.32863	0.33891	0.34913	0.35928	0.36936	0.37938	0.38933	0.39921	0.40901	0.41874
0.4	0.42839	0.43797	0.44747	0.45689	0.46623	0.47548	0.48466	0.49375	0.50275	0.51167
0.5	0.52050	0.52924	0.53790	0.54646	0.55494	0.56332	0.57162	0.57982	0.58792	0.59594
0.6	0.60386	0.61168	0.61941	0.62705	0.63459	0.64203	0.64938	0.65663	0.66378	0.67084
0.7	0.67780	0.68467	0.69143	0.69810	0.70468	0.71116	0.71754	0.72382	0.73001	0.73610
0.8	0.74210	0.74800	0.75381	0.75952	0.76514	0.77067	0.77610	0.78144	0.78669	0.79184
0.9	0.79691	0.80188	0.80677	0.81156	0.81627	0.82089	0.82542	0.82987	0.83423	0.83851
1.0	0.84270	0.84681	0.85084	0.85478	0.85865	0.86244	0.86614	0.86977	0.87333	0.87680
1.1	0.88021	0.88353	0.88679	0.88997	0.89308	0.89612	0.89910	0.90200	0.90484	0.90761
1.2	0.91031	0.91296	0.91553	0.91805	0.92051	0.92290	0.92524	0.92751	0.92973	0.93190
1.3	0.93401	0.93606	0.93807	0.94002	0.94191	0.94376	0.94556	0.94731	0.94902	0.95067
1.4	0.95229	0.95385	0.95538	0.95686	0.95830	0.95970	0.96105	0.96237	0.96365	0.96490
1.5	0.96611	0.96728	0.96841	0.96952	0.97059	0.97162	0.97263	0.97360	0.97455	0.97546
1.6	0.97635	0.97721	0.97804	0.97884	0.97962	0.98038	0.98110	0.98181	0.98249	0.98315
1.7	0.98379	0.98441	0.98500	0.98558	0.98613	0.98667	0.98719	0.98769	0.98817	0.98864
1.8	0.98909	0.98952	0.98994	0.99035	0.99074	0.99111	0.99147	0.99182	0.99216	0.99248
1.9	0.99279	0.99309	0.99338	0.99366	0.99392	0.99418	0.99443	0.99466	0.99489	0.99511
2.0	0.99532	0.99552	0.99572	0.99591	0.99609	0.99626	0.99642	0.99658	0.99673	0.99688
2.1	0.99702	0.99715	0.99728	0.99741	0.99753	0.99764	0.99775	0.99785	0.99795	0.99805
2.2	0.99814	0.99822	0.99831	0.99839	0.99846	0.99854	0.99861	0.99867	0.99874	0.99880
2.3	0.99886	0.99891	0.99897	0.99902	0.99906	0.99911	0.99915	0.99920	0.99924	0.99928
2.4	0.99931	0.99935	0.99938	0.99941	0.99944	0.99947	0.99950	0.99952	0.99955	0.99957
2.5	0.99959	0.99961	0.99963	0.99965	0.99967	0.99969	0.99971	0.99972	0.99974	0.99975
2.6	0.99976	0.99978	0.99979	0.99980	0.99981	0.99982	0.99983	0.99984	0.99985	0.99986
2.7	0.99987	0.99987	0.99988	0.99989	0.99989	0.99990	0.99991	0.99991	0.99992	0.99992
2.8	0.99992	0.99993	0.99993	0.99994	0.99994	0.99994	0.99995	0.99995	0.99995	0.99996
2.9	0.99996	0.99996	0.99996	0.99997	0.99997	0.99997	0.99997	0.99997	0.99997	0.99998

Appendix D

Units, Definitions, and Conversions

Dimensions and Units

ϵ = dielectric constant
m = mass
t = time
l = length
T = temperature

Parameter	Dimensions	cgs Unit
Acceleration	l/t^2	cm/s^2
Angular acceleration	t^{-2}	rad/s
Capacitance	ϵl	statfarad
Charge	$\epsilon^{1/2}m^{1/2}l^{3/2}/t$	esu
Current	$\epsilon^{1/2}m^{1/2}l^{1/2}/t^2$	statamp
Density	m/l^3	g/cm^3
Diffusion coefficient	l^2/t	cm/s^2
Electromotive force	$m^{1/2}l^{1/2}/(\epsilon^{1/2}t)$	statvolt
Energy	ml^2/t^2	erg
Force	ml/t^2	dyn
Heat quantity	ml^2/t^2	cal = 4.185 × 10^7 erg
Kinematic viscosity	l^2/t	St
Mole	m	mol = molecular weight in grams
Momentum	ml/t	g · cm/s
Moment of inertia	ml^2	g · cm^2
Power	ml^2/t^3	W = 10^7 erg/s
Pressure	$m/(lt^2)$	bar = 1 dyn/cm^2
Resistance	$t/(\epsilon l)$	statohm
Surface tension	m/t^2	dyn/cm
Thermal conductivity	$ml/(t^3 T)$	cal/(s · cm^2 · °C)
Velocity	l/t	cm/s
Viscosity	$m/(lt)$	P = dyn · s/cm^2
Work	ml^2/t^2	erg

Conversions

Absolute temperature = 0 kelvins (K) = −273.15 degrees Celsius (°C)
Degrees Fahrenheit (°F) = 32 + 9/5 × °C
1 micrometer (μm) = 10^{-4} centimeter (cm)
1 foot (ft) = 30.5 cm = 12 inches (in)
1 gram (g) = 1/454 pound (lb) = 15.4 grains (gr)
1 cubic meter (m^3) = 1000 liters (L) = 35.3 ft^3
1 m^3/s = 2120 ft^3/min
1 ft^3/min = 28.3 L/min
1 inch water (inH_2O) pressure = 250 pascals (Pa) = 0.036 lb/in^2 absolute
1 atmosphere (atm) pressure = 14.7 psia = 408 inH_2O = 76 cmH_2O
1 atm = 1.0133 × 10^6 dyn/cm^2 = 101,325 N/m^2 = 101,325 Pa
Acceleration of gravity at sea level = 981 cm/s^2
Volume of 1 mol of gas at 0°C is 22.4 L and at 20°C is 24.0 L
When $PV = nRT$ with P = dyn/cm^2, V = cm^3, n = mol, and T = K, R = 8.1432 × 10^7 erg/(K · mol)
When $PV = nRT$ with P = mm mercury Hg, V = cm^3, n = mol, and T = K, R = 62,400 mmHg · cm^3/(K · mol)

Some Properties of Air, Water, and Water Vapor at 1 atm and 20°C

Property	Air	Water	Water Vapor
Density	1.205 × 10^{-3} g/cm^3	1.00 g/cm^3	0.75 × 10^{-3} g/cm^3
Viscosity	1.82 × 10^{-4} P	1.00 × 10^{-2} P	
Molecular weight	28.9 g/mol	18.0 g/mol	
Diffusion coefficient	0.19 cm^2/s		0.24 cm^2/s
Velocity of sound	3.4 × 10^4 cm/s	1.5 × 10^5 cm/s	4.94 × 10^4 cm/s (at 134°C)
Velocity of light	3.00 × 10^{10} cm/s	2.26 × 10^{10} cm/s	

Composition of Dry Air, by Volume

Argon	0.93 percent
Carbon dioxide	0.031 percent
Nitrogen	78.1 percent
Oxygen	20.9 percent
Trace gases	< 0.003 percent

Electrical Units

The force exerted in a vacuum at a distance R on a charge q by a charge Q can be written as

$$F = \frac{1}{4\pi\epsilon_0} \frac{qQ}{R^2} \tag{D.1}$$

Appendix D

Depending on the definition of ϵ_0, there are a number of units which can (and have) been used during the development of electromagnetic theory. There are three specific types of units to consider, the types not necessarily being related.

Rationalization

In Eq. D.1 a factor $1/(4\pi)$ is inserted, somewhat arbitrarily, for ease in solution of many electrical theorems (e.g., Gauss'). Units which are defined as "rationalized" electrical units include this factor of $1/(4\pi)$; "unrationalized" units would omit it.

Physical Units

Using the cgs system with the dyne as the unit of force and erg as the unit of work is impractical for many applications (too large or too small numbers to work with), so the mks system of units is preferred. Then the force is in newtons (1 N = 10^5 dyn), and the unit of work is the joule (1 J = 10^7 erg). A rate of doing work of 1 J/s is 1 W. A watt is also a measure of electrical power, being 1 A flowing with a potential of 1 V.

Electrical Units

Finally, the magnitude of the electrical quantities must be chosen. Often the ampere is taken as the defining unit, and from the definition of the ampere a term $1/(\mu_0 c^2) = \epsilon_0$ arises. In the rationalized mks system $\mu_0 = 4\pi \times 10^{-7}$ N/A^2 and then $\epsilon_0 = 8.8542 \times 10^{12}$.

The basic electrostatic units (esu), however, are the unrationalized cgs units. Hence, to convert from unrationalized cgs units to the rationalized mks system of units, start with the relationship

$$\frac{\text{ESU unit charge [in esu (cgs) units]}}{\text{MKS unit charge (in coulombs)}} = \frac{c}{10}$$

where c is the velocity of light = 2.99793×10^{10} cm/s. Then 4.8×10^{-10} esu = 1.6×10^{-19} C. In Eq. D.1 $\epsilon_0 = 1/(4\pi)$ for the unrationalized cgs system of units, or $\epsilon = 8.8542 \times 10^{-12}$ F/m in the rationalized mks system of units.

To convert to other units, multiply esu or cgs field strength by $\approx 30{,}000$ to get mks field strength; i.e.,

$$F = \frac{F}{q} \qquad E(x) = \frac{F/10^5}{10q/c} \qquad x = \frac{c}{10^6}$$

Or multiply esu (cgs) voltage by 300 to get mks voltage,

$$V = El \quad V(x) = E \times 30{,}000 \times \frac{1}{100} \quad x \approx 300$$

Conversion Factors for Various Electrical Parameters

Electrical unit	Absolute unit	Electrostatic unit
Charge ($\epsilon^{1/2}m^{1/2}l^{3/2}/t$)	1 coulomb (C)	3×10^9 statcoulombs
Current ($\epsilon^{1/2}m^{1/2}l^{3/2}/t^2$)	1 ampere (A)	3×10^9 statamperes
Potential [$m^{1/2}l^{1/2}/(\epsilon^{1/2}t)$]	1 volt (V)	3.34×10^{-3} statvolt
Resistance ([$t/(\epsilon l)$])	1 ohm (Ω)	1.11×10^{-12} statohm

Appendix E

Adiabatic Expansion

The vapor pressure of a substance at temperature T can be estimated from the expression given by Daniels and Alberty (1961) where the factors A and B are constants that depend on the substance under consideration:

$$\log p_\infty (T) = A - \frac{B}{T} \qquad (E.1)$$

Now, letting

p_1 = vapor pressure before expansion

p_2 = partial pressure before expansion

p_3 = partial pressure after expansion

p_4 = vapor pressure after expansion

we see that the saturation ratio S is

$$S = \frac{p_3}{p_4} \qquad (E.2)$$

Furthermore, the temperature of a volume of gas after expansion can be estimated from

$$T_2 = \frac{T_1}{R^{K-1}} \qquad (E.3)$$

where 2 = after, 1 = before, and R is the expansion ratio, defined as

$$R = \text{expansion ratio} = \frac{\text{final volume}}{\text{initial volume}} = \frac{V_2}{V_1} \qquad (E.4)$$

and K is

$$K = \frac{\text{critical pressure specific heat}}{\text{critical volume specific heat}}$$

Since $p_1 V_1 / T_1 = p_2 V_2 / T_2$,

$$p_3 = p_2 \frac{1}{R} \frac{T_2}{T_1} \tag{E.5}$$

If it is assumed that $p_2 = p_1$, that is, saturated conditions exist before expansion, then

$$S = \frac{V_2}{V_1} \exp\left[\frac{B}{T_1}\left(\frac{V_2}{V_1}\right)^{k-1} - 1\right] \tag{E.6}$$

For water, $K = 1.4$ and $B = 5343$, which give the following:

Expansion ratio	Saturation ratio
1.0	1.00
1.1	1.78
1.2	3.08
1.3	5.21
1.4	8.64

Appendix F

Psychrometric Chart

PSYCHROMETRIC CHART
Normal Temperatures
Reproduced Courtesy of Carrier Corporation

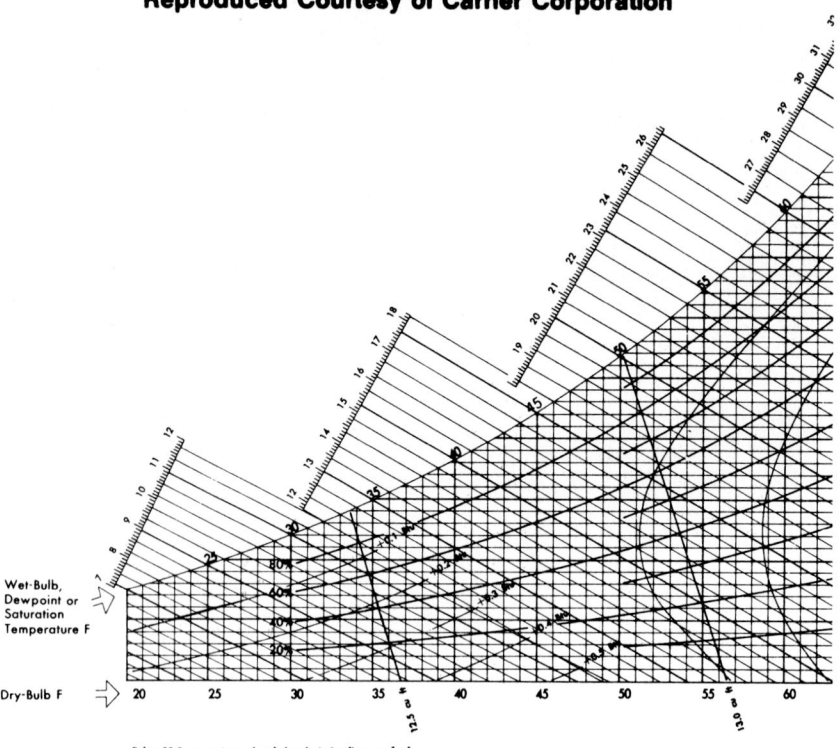

Psychrometric Chart

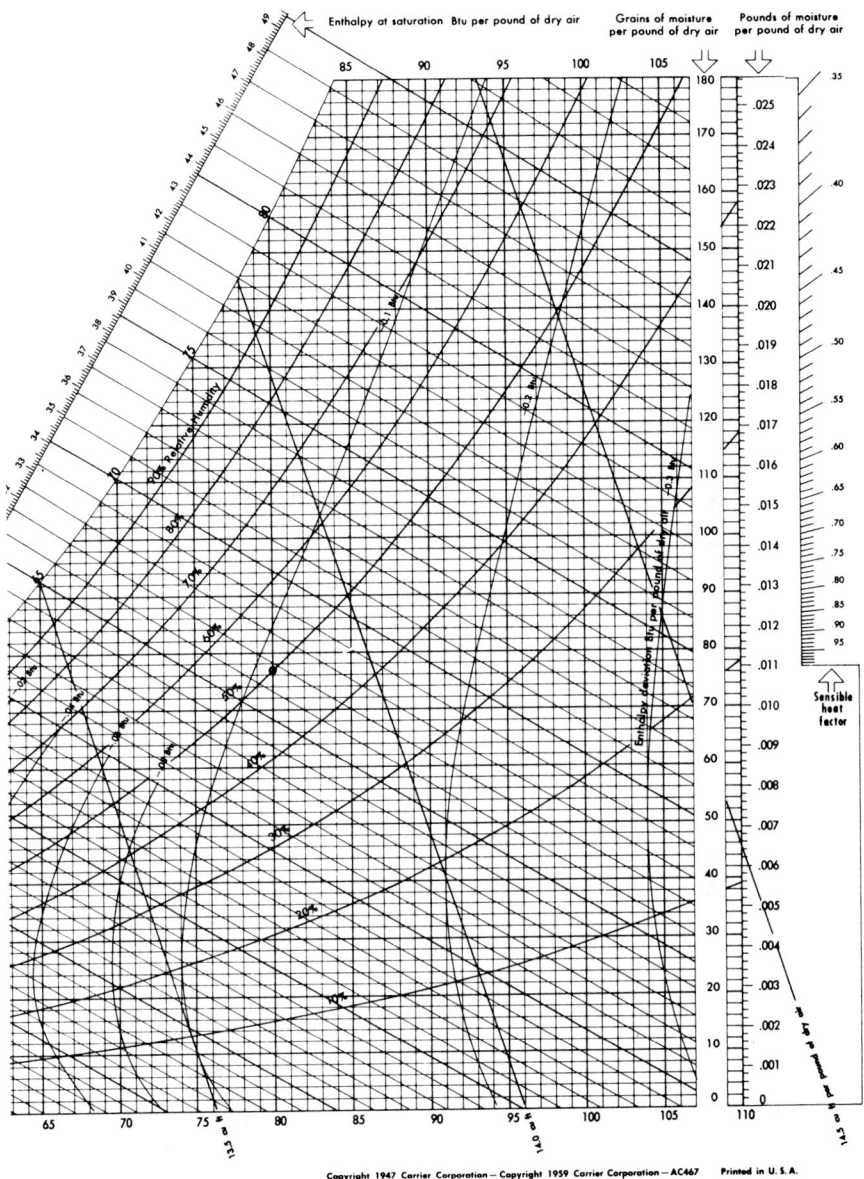

Appendix G

Bessel Functions of Order 1

x	$J_1(x)$	x	$J_1(x)$
0.0	0.00000	2.0	0.57672
0.1	0.04994	2.1	0.56829
0.2	0.09950	2.2	0.55596
0.3	0.14832	2.3	0.53987
0.4	0.19603	2.4	0.52019
0.5	0.24227	2.5	0.49709
0.6	0.28670	2.6	0.47082
0.7	0.32900	2.7	0.44160
0.8	0.36844	2.8	0.40971
0.9	0.40595	2.9	0.37543
1.0	0.44005	3.0	0.33906
1.1	0.47090	3.1	0.30092
1.2	0.49829	3.2	0.26134
1.3	0.52202	3.3	0.22066
1.4	0.54195	3.4	0.17923
1.5	0.55794	3.5	0.13738
1.6	0.56990	3.6	0.09547
1.7	0.57777	3.7	0.05383
1.8	0.58152	3.8	0.01282
1.9	0.58116		

SOURCE: From M. Abramowitz and I. Stegun, *Handbook of Mathematical Functions,* Nat. Bureau of Standards Appl. Math Ser. 55, Washington, 1964, p. 390.

References

Abrahamsen, A. R., in *Industrial Dust Explosions,* ASTM STP 958, K. L. Cashdollar and M. Hertzberg (eds.), Am. Society for Testing Materials, Philadelphia, 1987.
Abrahamson, J., in *Progress in Filtration and Separation 2,* R. J. Wakeman (ed.), Elsevier Sci. Pub. Co., Amsterdam, 1981.
Abramowitz, M., and Stegun, I., *Handbook of Mathematical Functions, AMS 55,* Nat. Bureau of Stds., Washington, 1964.
Adachi, M., Kousaka, Y., and Okuyama, K., *Aerosol Sci. & Tech.,* **7,** 217 (1987).
Aerosol Technology Committee, *Am. Ind. Hyg. Assoc. J.,* **31,** 133 (1970).
Agricola, Georgius, *De Re Metallica,* translated by H. C. Hoover and L. H. Hoover, Dover Publications, New York, 1950.
Ahlberg, M. S., and Hansson, H.-C., *J. Aerosol Sci.,* **14,** 499 (1983).
Aitken, J., *Trans. Royal Soc. Edinb.,* **30,** 337 (1880).
Aitken, J., *Trans. Roy. Soc.,* **32,** 293 (1884a).
Aitken, J., *Nature,* **28,** 139 (1884b).
Aitken, J., *Proc. Royal Soc.,* **A51,** 408 (1892).
Allen, M. D., and Raabe, O. G., *J. Aerosol Sci.,* **13,** 53 (1982).
Allen, M. D., and Raabe, O. G., *Aerosol Sci. & Tech.,* **4,** 269 (1985).
Amelin, A. G., *Theory of Fog Condensation,* Israel Program for Sci. Trans., Jerusalem, 1967.
Bailey, A. G., Balachandran, W., and Williams, T. J., *J. Aerosol Sci.,* **14,** 39 (1983).
Bakanov, S. J., and Roldughin, V. I., *Aerosol Sci. & Tech.,* **7,** 249 (1987).
Banks, D. O., Hall, M. S., and Kurowski, G. J., *J. Aerosol Sci.,* **14,** 87 (1983).
Banks, D. O., and Kurowski, G. J., *J. Aerosol Sci.,* **14,** 463 (1983).
Barrer, R. M., *Diffusion in and through Solids,* Cambridge University Press, New York, 1941.
Barth, W., *Brennstoff-Warme-Kraft.,* **8,** 1 (1956).
Bartknecht, W., *Explosions,* Springer-Verlag, New York, 1981.
Bartknecht, W., in *Industrial Dust Explosions,* ASTM STP 958, K. L. Cashdollar and M. Hertzberg (eds.), American Society for Testing Materials, Philadelphia, 1987.
Bartley, D. L., and Doemeny, L. J., *Am. Ind. Hyg. Assoc. J.,* **47,** 443 (1986).
Battler, J. R., Zhavoronkov, A. A., and Matteson, M. J., *J. Aerosol Sci.,* **13,** 491 (1982).
Berry, J., *J. Opt. Soc. Am.,* **52,** 888 (1962).
Berry, J., *J. Opt. Soc. Am.,* **56,** 460 (1966).
Billings, C. E., et al., *Controlling Airborne Particles,* National Academy of Sciences, Washington, 1980.
Bird, R. B., Stewart, W. E., and Lightfoot, E. N., *Transport Phenomena,* Wiley, New York, 1960.
Bodaszewski, R. W., *Beibl.,* **8,** 488 (1883).
Bodhaine, B. A., and Harris, J. M., *J. Aerosol Sci.,* **13,** 12 (1982).
Boisdron, and Brock, J. R., *Atm. Env.,* **3,** 235 (1969).
Bricard, J., *Geofisica pura e appl.,* **51,** 237 (1962).
Bricard, J., in *Problems of Atmospheric and Space Electricity,* S. C. Coroniti (ed.), Elsevier, Amsterdam, 1965.

Bricard, J., and Pradel, J., in *Aerosol Science*, C. N. Davies (ed.), Academic Press, New York, 1966.
Brock, J. R., *J. Col. Sci.*, **17**, 768 (1962a).
Brock, J. R., *J. Phys. Chem.*, **66**, 1763 (1962b).
Brown, R., *Phil. Mag.*, **4**, 161 (1828).
Brown, R. C., *J. Aerosol Sci.*, **14**, 481 (1983).
Buchholz, T., and Steinwandel, J., *Aerosol Sci. & Tech.*, **3**, 71 (1984).
Burge, H. A., and Feeleyer, J. C., in *Indoor Air Pollution—A Health Perspective*, J. M. Samet and J. D. Baltimore (eds.), Johns Hopkins University Press, 1991, p. 273.
Byers, H. R., *Elements of Cloud Physics*, University of Chicago Press, Chicago, 1965a.
Byers, H. R., *Ind. & Eng. Chem.*, **57**, 11 (1965b).
Cadle, R. D., *Particles in the Atmosphere and Space*, Reinhold Publishing Co., New York, 1966.
Castleman, A. W., Jr., *J. Aerosol Sci.*, **13**, 73 (1982).
Cawood, W., *Trans. Faraday Soc.*, **32**, 1068 (1936).
Chan, T., and Lippmann, M., *Environ. Sci. & Tech.*, **11**, 372 (1977).
Chapman, S., and Cowling, T. G., *The Mathematical Theory of Non-Uniform Gases*, Cambridge University Press, New York, 1961.
Charlson, R. J., Horvath, H., and Pueschel, R. F., *J. Atm. Env.*, **1**, 469 (1967).
Chatigny, M. A., et al., in *Air Sampling Instruments*, ACGIH, Cincinnati, Ohio, 1987, p. 199.
Cheah, K., *J. Aerosol Sci.*, **14**, 47 (1983).
Chen, B. T., and Mercer, T. T., *J. Aerosol Sci.*, **13**, 29 (1982).
Cheng, Y.-S., and Yeh, H.-C., *J. Aerosol Sci.*, **14**, 489 (1983).
Cheng, Y.-S., in *Air Sampling Instruments*, ACGIH, Cincinnati, Ohio, 1987, p. 405.
Cheng, Y.-S., et al., *Aerosol Sci. & Tech.*, **8**, 199 (1988).
Chomiak, J., and Gupta, A. K., *J. Aerosol Sci.*, **20**, 1 (1989).
Clark, W. E., and Whitby, K. T., *J. Atm. Sci.*, **26**, 603 (1967).
Clarke, A. D., et al., *Atm. Env.*, **21**, 1455 (1987).
Colbeck, I., Hardman, E. J., and Harrison, R. M., *J. Aerosol Sci.*, **20**, 765 (1989).
Colbeck, I., et al., *Aerosol Sci. & Tech.*, **7**, 841 (1987).
Coombs, M. A., and Cuddihy, R. G., *J. Aerosol Sci.*, **14**, 75 (1983).
Connor, W. D., and Hodkinson, J. R., *Observations on the Optical Properties and Visual Effects of Smoke Plumes*, PHS Pub. 949-AP-30, Washington, 1967.
Conti, R. S., and Hertzberg, M., in *Industrial Dust Explosions*, ASTM STP 958, K. L. Cashdollar and M. Hertzberg (eds.), American Society for Testing Materials, Philadelphia, 1987.
Cooper, D. G., and Reist, P. C., *J. Col. Int. Sci.*, **45**, 17 (1973).
Cooper, D. W., *J. Aerosol Sci.*, **13**, 11 (1982).
Corr, D., et al., *J. Aerosol Sci.*, **14**, 70 (1983).
Coulier, M., *J. Pharm. Chim.*, **22**, 165 (1875).
Crawford, M., *Air Pollution Control Theory*, McGraw-Hill, New York, 1976.
Cross, J., and Farrer, D., *Dust Explosions*, Plenum Press, New York, 1982.
Cunningham, E., *Proc. Roy. Soc. A*, **83**, 357 (1910).
Daniels, F., and Alberty, R., *Physical Chemistry*, 2d ed., Wiley, New York, 1961.
Davies, C. N., *Ann. Occ. Hyg.*, **8**, 239 (1965).
Davies, C. N., in *Aerosol Science*, C. N. Davies (ed.), Academic Press, New York, 1966, p. 393.
Davies, C. N., in *Fundamentals of Aerosol Science*, D. T. Shaw (ed.), Wiley, New York, 1978.
Davies, C. N., and Subard, M., *J. Aerosol Sci.*, **13**, 59 (1982).
Davies, C. N., Jayasekera, P. N., and Cheung. K. K., *J. Aerosol Sci.*, **14**, 163 (1983).
Davison, S. W., and Gentry, J. W., *Aerosol Sci. & Tech.*, **4**, 157 (1985).
Deirmendjian, D., *Electromagnetic Scattering on Spherical Polydispersions*, American Elsevier, New York, 1969.
Denman, H. H., Heller, W., and Pangonis, W. J., *Angular Scattering Functions for Spheres*, Wayne State University Press, Detroit, 1966.
Dennis, R., *Handbook on Aerosols*, TID-26608, NTIS, Springfield, Va., 1976.
Derjaguin, B. V., and Bakanov, S. P., *Kolloidn. Zh.*, **21**, 377 (1959).

Derjaguin, B. V., and Yalamov, Yu., *J. Col. Sci.*, **20**, 555 (1965).
Derjaguin, B. V., Storozhilova, A. I., and Rabinovich, Y. I., *J. Col. Int. Sci.*, **21**, 35 (1966).
Deutsch, W., *Ann. d. Phys.*, **68**, 335 (1922).
Dietz, P. W., *Am. Inst. Chem. Eng. J.*, **27**, 288 (1981).
Drinker, P. D., and Hatch, T., *Industrial Dusts*, 2d ed., McGraw-Hill, New York, 1954.
DuBard, J. L., McDonald, J. R., and Sparks, L. E., *J. Aerosol Sci.*, **14**, 5 (1983).
Eckhoff, R. K., in *Industrial Dust Explosions*, ASTM STP 958, K. L. Cashdollar and M. Hertzberg (eds.), American Society for Testing Materials, Philadelphia, 1987.
Einbinder, H., *J. Chem. Phys.*, **26**, 948 (1957).
Einstein, A., *Z. Phys.*, **27**, 1 (1924).
Einstein, A., *Investigations on the Theory of Brownian Movement*, Dover, New York, 1956.
Emi, H., Kanaoka, C., and Kuwabara, Y., *J. Aerosol Sci.*, **13**, 40 (1982).
Epstein, P., *Z. Phys.*, **54**, 537 (1929).
Exner, *Sitzungsber Akad. Wiss. Wien*, **56**, Part II, 116 (1867).
Fjeld, R. A., Gauntt, R. O., and McFarland, A. R., *J. Aerosol Sci.*, **14**, 541 (1983).
Flagan, R. C., *Aerosol Sci. & Tech.*, **7**, 783 (1987).
Flanagan, V. P. V., and O'Connor, T. C., *Geofisica pura e applicata*, **50**, 148 (1961).
Fleagle, R. G., and Businger, J. A., *An Introduction to Atmospheric Physics*, Academic Press, New York, 1963.
Fletcher, F. W., *J. Chem. Phys.*, **31**, 1136 (1958a).
Fletcher, F. W., *J. Chem. Phys.*, **29**, 572 (1958b).
Foitzik, I., Hebermehl, G., and Spankuch, D., *Optik*, **23**, 274 (1965/1966).
Fox, D. L., Kuhlman, M. R., and Reist, P. C., *Colloid and Interface Science*, vol. 2, M. Kerker (ed.), Academic Press, New York, 1976.
Friedlander, S. K., in *Aerosols, Physical Chemistry and Applications*, C. Spurny (ed.), Gordon and Breach, New York, 1965.
Friedlander, S. K., and Wang, C. S., *J. Col. Int. Sci.*, **22**, 126 (1966).
Friedlander, S. K., *Smoke, Dust and Haze*, Wiley, New York, 1977, p. 43.
Fuchs, N. A., *Z. Phys.*, **89**, 736 (1934).
Fuchs, N. A., *Izv. Akad. Nauk. SSSR, Ser. Geogr. Geogfiz.*, **11**, 341 (1957).
Fuchs, N. A., *Evaporation and Droplet Growth in Gaseous Media*, Pergamon Press, London, 1959.
Fuchs, N. A., *Geofisica pura e appl.*, **56**, 185 (1963).
Fuchs, N. A., *The Mechanics of Aerosols*, Pergamon Press, New York, 1964.
Fuchs, N. A., in *Fundamentals of Aerosol Science*, David T. Shaw (ed.), Wiley, New York, 1978.
Fuchs, N. A., *J. Aerosol Sci.*, **13**, 327 (1982).
Fuchs, N. A., *J. Aerosol Sci.*, **13**, 32 (1982).
Fuzzi, S., Orsi, G., and Mariotti, M., *J. Aerosol Sci.*, **14**, 135 (1983).
Gäggeler, H. W., et al., *J. Aerosol Sci.*, **20**, 557 (1989).
Geysen, et al., in *Industrial Dust Explosions*, ASTM STP 958, K. L. Cashdollar and M. Hertzberg (eds.), American Society for Testing Materials, Philadelphia, 1987.
Gibbs, W. E., *Clouds and Smoke*, P. Blakiston Son & Co., Philadelphia, 1924, p. 63.
Giese, R., De Bary, E., Bullrich, K., and Vinnemann, C. V., *Tabellen der Streufunktionen* $i_1(\theta)$, $i_2(\theta)$, Akademie-Verlag, Berlin, 1962.
Gillespie, T., *J. Col. Sci.*, **18**, 562 (1963).
Goetz, A., and Pueschel, R., *Atmos. Env.*, **1**, 287 (1967).
Goren, S. L., *J. Col. Int. Sci.*, **61**, 77 (1977).
Gormley, P. G., and Kennedy, M., *Proc. Royal Irish Acad.*, **52A**, 163 (1949).
Green, H. L., and Lane W. R., *Particulate Clouds: Dusts, Smokes and Mists*, 2d ed., D. Van Nostrand, New York, 1964.
Gucker, F. T., and O'Konski, C. T., *J. Col. Sci.*, **4**, 541 (1949).
Hangal, S., and Willeke, K., *Atm. Env.*, **24A**, 2379 (1990a).
Hangal, S., and Willeke, K., *Env. Sci. & Tech.*, **24**, 688 (1990b).
Happel, J., and Brenner, H., *Low Reynolds Number Hydrodynamics*, Prentice-Hall, Englewood Cliffs, N.J., 1965.
Hautojarvi, A., Hautala, M., and Raunemaa, T., *J. Aerosol Sci.*, **14**, 507 (1983).
Henry, J. F., Peters, L. K., and Kermode, R. I., *J. Aerosol Sci.* **13**, 34 (1982).

Herdan, G., *Small Particle Statistics*, 2d ed., Academic Press, New York, 1960.
Hering, S. V., and Marple, V. A., in *Cascade Impactor, Sampling and Data Analysis*, American Industrial Hygiene Association, Akron, Ohio, 1986, p. 103.
Hertzberg, M., and Cashdollar, K. L., in *Industrial Dust Explosions*, ASTM STP 958, K. L. Cashdollar and M. Hertzberg (eds.), American Society for Testing Materials, Philadelphia, 1987.
Hidy, G. M., *J. Col. Sci.*, **20**, 123 (1965).
Hidy, G. M., and Brock, J. R., *The Dynamics of Aerocolloidal Systems*, Pergamon Press, Oxford, 1970.
Hinds, W., *Aerosol Technology*, Wiley, New York, 1982.
Hochrainer, D., and Zebel, G., *J. Aerosol Sci.*, **12**, 49 (1981).
Hocking, L. M., in *Aerodynamic Capture of Particles*, E. G. Richardson (ed.), Pergamon Press, Oxford, 1960.
Hodkinson, J. R., in *Aerosol Science*, C. N. Davies (ed.), Academic Press, New York, 1966, p. 287.
Hofschreuder, P., Vrins, E., and van Boxel, J., *J. Aerosol Sci.*, **14**, 65 (1983).
Hogan, A., et al., *J. Aerosol Sci.*, **16**, 391 (1985).
Holland, A. C., and Draper, J. S., *Applied Optics*, **6**, 511 (1967).
Hoppel, W. A., and Frick, G. M., *Aerosol Sci. & Tech.*, **5**, 1 (1986).
Horvath, H., *Aerosol Sci. & Tech.*, **7**, 837 (1987).
Hussin, A., et al., *J. Aerosol Sci.*, **14**, 671 (1983).
Iozia, D. L. and Leith, D., *Filtration and Separation*, **26**, 272 (1989).
Ivchenko, I. N., and Yalamov, Y., *Russian J. Phys. Chem.*, **45**, 317 (1971).
Jacobsen, S., and Brock, J. R., *J. Col. Int. Sci.*, **20**, 544 (1965).
Janka, K., Kivosto, T., and Makynen, J., *J. Aerosol Sci.*, **13**, 45 (1982).
Jantunen, M. J., and Reist, P. C., *Aerosol Science*, **14**, 127 (1983).
Jennings, S. G., *Applied Optics*, **25**, 2499 (1986).
Jennings, S. G., *J. Aerosol Sci.*, **19**, 159 (1988).
John, W., *Annals of the American Conference of Government Industrial Hygiene*, **11**, 75 (1986).
Johnson, D. L., and Martonen, T. B., *Am. Ind. Hyg. Assoc. J.*, **50**, 408 (1989).
Johnson, I., and LaMer, V. K., *J. Am. Chem. Soc.*, **69**, 1184 (1947).
Johnston, A. M., *J. Aerosol Sci.*, **14**, 643 (1983).
Johnston, A. M., Vincent, J. H., and Jones, A. D., *Aerosol Sci. & Tech.*, **6**, 115 (1987).
Joos, G., *Theoretical Physics*, 2d ed., Blackie and Son, Glasgow, 1951.
Jost, W., *Diffusion in Solids, Liquids and Gases*, Academic Press, New York, 1952.
Junge, C., *Tellus*, **5**, 1 (1953).
Junge, C., *Air Chemistry and Radioactivity*, Academic Press, New York, 1963.
Keefe, D., Nolan, P. J., and Rich, T. A., *Proc. Royal Irish Acad.*, **60A**, 27 (1959).
Keng, E. Y. H., and Orr, C., *J. Col. Int. Sci.*, **22**, 107 (1966).
Kerker, M., *Electromagnetic Scattering*, Pergamon Press, New York, 1963.
Kerker, M., *The Scattering of Light and Other Electromagnetic Radiation*, Academic Press, New York, 1969.
Kim, C. S., Trujillo, D., and McDonald, R., *J. Aerosol Sci.*, **14**, 633 (1983).
Kitani, S., *J. Col. Sci.*, **15**, 287 (1960).
Koch, W., *J. Aerosol Sci.*, **13**, 415 (1982).
Koschmieder, H., *Beitr. Phys. freien Atm.*, **12**, 33 (1924a).
Koschmieder, H., *Beitr. Phys. freien Atm.*, **12**, 171 (1924b).
Kuhlman, M. R., "A Model of Sulfuric Acid Aerosol Processes," Ph.D. diss., University of North Carolina, Chapel Hill, 1982.
Landau, L. D., and Lifshitz, E. M., *Fluid Mechanics*, Pergamon Press, New York, 1959.
Lanza, A. J., in *Silicosis and Asbestosis*, Oxford University Press, New York, 1938.
Lawton, J., and Weinberg, F. J., *Electrical Aspects of Combustion*, Clarendon Press, Oxford, 1969.
Leigh, D., and Licht, W., *A.I.Ch.E. Symp. Series*, **126**, 196 (1972).
Lenard, P., *Ann. der Phys.*, **47**, 413 (1915).
Leonard, G. L., Mitchner, M., and Self, S. A., *J. Aerosol Sci.*, **13**, 27 (1982).
Leong, K. H., *J. Aerosol Sci.*, **13**, 603 (1982).
Leong, K. H., et al., *J. Aerosol Sci.*, **14**, 23 (1983a).
Leong, K. H., Hopke, K., and Stukel, J. J., *J. Aerosol Sci.*, **14**, 611 (1983b).

Levich, V. G., *Physicochemical Hydrodynamics,* Prentice-Hall, Englewood Cliffs, N.J., 1962.
Levin, L. M., *Physics of Coarse Aerosols,* Trans. 815, Department of Commerce TT 64 19550, June 1963.
Licht, W., *Air Pollution Control Engineering,* Marcel Dekker, New York, 1980.
Lippman, M., *Am. Ind. Hyg. Assoc. J.,* **31,** 138 (1970).
Lippman, Morton, in *Air Sampling Instruments for Evaluation of Atmospheric Contaminants,* 7th ed., ACGIH, Cincinnati, Ohio, 1989, p. 163.
Liu, B. Y. H., and Pui, D. Y. H., *J. Col. Int. Sci.,* **58,** 142 (1977).
Liu, B. Y. H., Whitby, K. T., and Yu, H. S., *J. Col. Int. Sci.,* **23,** 367 (1967a).
Liu, B. Y. H., Whitby, K. T., and Yu, H. S., *J. Appl. Phys.,* **38,** 1592 (1967b).
Liu, C. S., and Gentry, J. W., *J. Aerosol Sci.,* **13,** 12 (1982).
Lodge, O. J., *Nature,* **28,** 297 (1883).
Lodge, O. J., and Clark, J. W., *Phil. Mag.,* **17,** 214 (1884).
Lopez, A., et al., *J. Aerosol Sci.,* **14,** 99 (1983).
Lowan, A. N., *Tables of Scattering Functions for Spherical Particles,* NBS Appl. Math Ser. 4, Washington, 1948.
Ludlum, F. H., *Clouds and Storms,* Pennsylvania State University Press, University Park, 1980.
Makynen, J., et al., *J. Aerosol Sci.,* **13,** 52 (1982).
Maron, S. H., and Elder, M. E., *J. Col. Sci.,* **18,** 199 (1963).
Marple, V. A., "A Fundamental Study of Inertial Impactors," Ph.D. diss., University of Minnesota, Part. Tech. Lab., Pub. No. 144, 1970.
Marple, V. A., and Willeke, K., in *Aerosol Measurement,* D. A. Lundgren, et al. (eds.), 1979, p. 90.
Marple, V. A., and Rubow, K. L., *Cascade Impactor, Sampling and Data Analysis,* American Industrial Hygiene Association, Akron, Ohio, 1986, p. 79.
Martonen, T. B., *Am. Ind. Hyg. Assoc. J.,* **43,** 154 (1982).
Martonen, T. B., *J. Aerosol Sci.,* **14,** 11 (1983).
Martonen, T., *ES&T,* xx, xxx (1989).
Mason, B. J., *The Physics of Clouds,* 2d ed., Clarendon Press, Oxford, 1971.
Mason, E., and Chapman, S. J., *J. Chem. Phys.,* **36,** 627 (1962).
Maxwell, J. C., *Royal Soc. London, Phil. Trans.,* 170 1, 231 (1880).
Maxwell, J. C., *Scientific Papers,* **2,** 639 (1890).
May, K. R., *J. Aerosol Sci.,* **13,** 37 (1982).
Mayya, Y. S., Dua, S. K., and Kotrappa, P., *J. Aerosol Sci.,* **13,** 33 (1982).
McDaniel, E. W., *Collision Theory in Ionized Gases,* Wiley, New York, 1964.
Mercer, T. T., *Aerosol Technology and Hazard Evaluation,* Academic Press, New York, 1973.
Middleton, W. E. K., *Vision through the Atmosphere,* University of Toronto Press, Toronto, 1963.
Mie, G., *Ann. Physik.,* **25,** 377 (1908).
Miller, F. C., and Loeb, L. B., *J. Appl. Phys.,* **22,** 494 (1951a).
Miller, F. C., and Loeb, L. B., *J. Appl. Phys.,* **22,** 614 (1951b).
Miller, F. C., and Loeb, L. B., *J. Appl. Phys.,* **22,** 740 (1951c).
Miller, J. G., and Heinemann, H., *Science,* **107,** 144 (1948).
Miller, S. W., and Bodhaine, B. A., *J. Aerosol Sci.,* **13,** 48 (1982).
Millikan, R. A., *Science,* **32,** 436 (1910).
Millikan, R. A., *Phys. Rev.,* 2d series, **22,** 1 (1923).
Miyake, S., *Aero Res. Inst., Tokyo,* **10,** 85 (1935).
Mokler, B., personal communication, 1969.
Monchick, L., and Blackmore, R., *Aerosol Sci. & Tech.,* **5,** 27 (1986).
Moss, O., and Kenoyer, J. L., *Cascade Impactor, Sampling and Data Analysis,* American Industrial Hygiene Association, Akron, Ohio, 1986, p. 23.
Mozurkewich, M., *Aerosol Sci. & Tech.,* **5,** 223 (1986).
Muller, H., *Kolliodchem. Beih.,* **27,** 223 (1928).
Napper, D. H., and Ottewill, R. H., *Trans. Faraday Soc.,* **60,** 1466 (1964).
Natanson, G. L., *Soviet Phys. Tech. Phys.,* **5,** 538 (1960).
National Materials Advisory Board, *Prevention of Grain Elevator and Mill Explosions,*

NMAB 367-2, Commission on Engineering and Technical Systems, NAS, National Academy Press, 1982.
Nevalainen, A., et al., *Atm. Env.*, **26A,** 000 (1992).
Newman, S., *J. Aerosol Sci.*, **14,** 69 (1983).
Niven, W. D. (ed.), *The Scientific Papers of James Clerk Maxwell*, vol. 2, Dover Publications Inc., New York, 1965, p. 703.
Nolan, P. J., and Doherty, D. J., *Proc. Royal Irish Acad.*, **53A,** 163 (1950).
Nolan, P. F., and Jennings, S. G., *J. Atm. & Oceanic Tech.*, **4,** 391 (1987).
O'Connor, T. C., and Sharkey, W. P., *Proc. Royal Irish Acad.*, **61A,** 15 (1960).
Ohta, S., *J. Aeosol Sci.*, **13,** 13 (1982).
Orr, C., Hurd, F. K., and Corbett, W. J., *J. Col. Sci.*, **13,** 472 (1958a).
Orr, C., et al., *J. Met.*, **15,** 240 (1958b).
Orr, C., and Keng, E. *J. Atmos. Sci.*, **21,** 475 (1963).
Orr, C., *Particulate Technology*, Macmillan, New York, 1966.
Orsi, G., and Fuzzi, S., *J. Aerosol Sci.*, **14,** 581 (1983).
Pauthenier, M. M., and Moreau-Hanot, M., *J. Phys. et Rad.*, Ser. 7, **3,** 590 (1932).
Pendorf, R., *J. Opt. Soc. Am.*, **52,** 797 (1962).
Perry, R. H., and Chilton, C. H., *Chemical Engineers' Handbook*, 5th ed., McGraw-Hill, New York, 1973.
Peterson, T. W., Stratmann, F., and Fissan, H., *J. Aerosol Sci.*, **20,** 683 (1989).
Pineau, et al., *Cahiers de Notes Documentaires*, Institute Nationale de Recherché et de Securitié, Paris, 1976, p. 33.
Pinnick, R. G., and Chỳlek, P., *J. Geophys., Res.*, **85,** 4059 (1980).
Pinnick, R. G., et al., *J. Geophys. Res.*, **88,** 6787 (1983).
Porstendorfer, J., et al., *J. Aerosol Sci.*, **15,** 47 (1984).
Prandtl, L., and Tietjens, O. G., *Fundamentals of Hydro- and Aeromechanics*, Dover Publications, New York, 1957.
Prodi, F., and Tomasi, C., *J. Aerosol Sci.*, **14,** 517 (1983).
Pruppacher, H. R., and Klett, J. D., *Microphysics of Clouds and Precipitation*, D. Reidel Pub. Co., Boston, 1978.
Quindos, L. S., Soto, J., and Villar, E., *J. Aerosol Sci.*, **14,** 495 (1983).
Raabe, O. G., *J. Aerosol Sci.*, **2,** 289 (1971).
Rader, D. J., and Marple, V. A., *Aerosol Sci. & Tech.*, **4,** 141 (1985).
Rader, D. J., McMurry, P. H., and Smith, S., *Aerosol Sci. & Tech.*, **6,** 247 (1987).
Raes, F., Kodas, T. T., and Friedlander, S. K., *Aerosol Sci. & Tech.*, **12,** 856 (1990).
Ranz, W. E., and Wong, J. B., *Ind. Eng. Chem.*, **44,** 1371 (1952).
Ranz, W. E., *Aerosol Sci. & Tech.*, **4,** 417 (1985).
Rao, A. K., and Whitby, K. T., *J. Aerosol Sci.*, **9,** 77 (1978).
Rayleigh, Lord, *Proc. Roy. Soc.*, **34,** 414 (1882).
Rayleigh, L., *Nature*, **28,** 139 (1884).
Rayleigh, Lord, *Phil. Mag.*, **47,** 375 (1899).
Robinson, R. A., and Stokes, R. H., *Electrolyte Solutions*, Butterworths, London, 1959.
Rodes, C., et al., *Aerosol Sci. & Tech.*, **13,** 220 (1990).
Roedel, W., *J. Aerosol Sci.*, **13,** 597 (1982).
Rohmann, H., *Zeit. fur Phys.*, **17,** 253 (1923).
Rose, H. E., and Wood, A. J., *An Introduction to Electrostatic Precipitation in Theory and Practice*, Constable and Company, London, 1966.
Rosenblatt, P., and LaMer, V. K., *Phys. Rev.*, **70,** 385 (1946).
Rosinski, J., and Lecinski, A., *J. Aerosol Sci.*, **14,** 49 (1983).
Sartor, J. D., and Atkinson, W. R., *Science*, **157,** 1267 (1967).
Schadt, C. F., and Cadle, R. D., *J. Col. Sci.*, **12,** 356 (1957).
Schadt, C. F., and Cadle, R. D., *J. Phys. Chem.*, **65,** 1694 (1961).
Scheibel, H. G., and Porstendorfer, J., *J. Aerosol Sci.*, **14,** 113 (1983).
Schlichting, H., *Boundary Layer Theory*, 6th ed., McGraw-Hill, New York, 1968.
Schmitt, J. L., Kassner, J. L., Jr., and Podzimek, J., *J. Aerosol Sci.*, **13,** 37 (1982).
Schmitt, K. H., *Z. Naturf.*, **14a,** 589 (1959).
Schneider, T., and Holst, E., *J. Aerosol Sci.*, **14,** 139 (1983).
Seaver, A. E., *Aerosol Sci. & Tech.*, **3,** 177 (1984).
Seddig, R., *Phys. Zeit.*, **9,** 465 (1908).

Sedunov, Y. S., *Physics of Drop Formation in the Atmosphere*, Halsted Press, Wiley, New York, 1974.
Sehmel, G. A., *Aerosol Deposition from Turbulent Airstreams in Vertical Conduits*, BNWL-578, Richland, Wash., 1968.
Seinfeld, J. H., and Warren, D. R., *Aerosol Sci. & Tech.*, **4**, 31 (1985).
Shaw, G. E., *J. Aerosol Sci.*, **14**, 475 (1983).
Sinclair, D., *Handbook on Aerosols*, H. F. Johnstone (ed.), Government Printing Office, Washington, 1950.
Sinclair, D., *A.I.H.A.J.*, **33**, 729 (1972).
Sinclair, D., *Aerosol Sci. & Tech.*, **3**, 125 (1984).
Sliney, D., and Wolbarsht, M., *Safety with Lasers and Other Optical Sources*, Plenum Press, New York, 1980.
Smolik, J., and Vitovec, J., *J. Aerosol Sci.*, **13**, 587 (1982).
Smoluchowski, M., *Bull. Acad. Sci., Cracow*, **1a**, 28 (1911).
Smoluchowski, M., *Zeit. Phys. Chemie*, **92**, 129 (1918a).
Smoluchowski, M., *Zeit. Phys. Chemie*, **92**, 167 (1918b).
Spurny, K. R., Gentry, J. W., and Stober, W., in *Fundamentals of Aerosol Science*, D. T. Shaw (ed.), Wiley, New York, 1978.
Stauffer, D., *J. Aerosol Sci.*, **7**, 319 (1976).
Stauffer, D., *J. Aerosol Sci.*, **14**, 173 (1983).
Stöber, W., and Flachsbart, H., *Env. Sci. & Tech.*, **3**, 1280 (1969).
Stöber, W., in *Assessment of Airborne Particles*, T. T. Mercer, T. E. Morrow, and W. Stöber (eds.), Charles Thomas, Philadelphia, 1972.
Stratmann, F., et al., *Aerosol Sci. & Tech.*, **9**, 115 (1988).
Sutton, O. G., *Mathematics in Action*, 2d ed., G. Bell and Sons, London, 1957.
Sutugin, A. G., *Russian Chem. Rev.*, **38**, 79 (1969).
Svedberg, Z. f. *Electroch.*, **12**, 853 (1909).
Takahashi, K., and Kasahara, A., *Atm. Env.*, **2**, 441 (1968).
Talbot, L., Cheng, R. K., Schefer, R. W., and Willis, D. R., *J. Fluid Mech.*, **101**, 737 (1980).
Tikhomirov, M., Tunitskii, N., and Petryanov, I., *Dokl. Akad. nauk SSSR*, **94**, 865 (1942).
Tillery, M. I., in *Aerosol Measurement*, Dale Lundgren et al. (eds.), University of Florida Press, Gainesville, 1979, p. xx.
Tompson, R. V., and Loyalka, S. K., *J. Aerosol Sci.*, **17**, 723 (1986).
Tong, N. T., and Bird, G. A., *J. Col. Int. Sci.*, **35**, 403 (1971).
Tricker, R. A., *Introduction to Meteorological Optics*, American Elsevier, New York, 1970.
Tsang, T. H., and Korgaonkar, N., *Aerosol Sci. & Tech.*, **7**, 317 (1987).
Tulis, J. J., *Introduction to Aerosols, Aerobiology and Infectious Disease*, 1989.
Turner, J. R., and Hering, S. V., *J. Aerosol Sci.*, **18**, 215 (1987).
Tyndall, J., *Proc. Roy. Inst.*, **6**, 3 (1870).
U.S. Environmental Protection Agency, *Ambient Air Quality Standards for Particulate Matter*, 40 CFR Parts 51, 52, 53, and 58, Federal Register, 52(126), 1987, pp. 24634–24750.
Van de Hulst, H. C., *Light Scattering by Small Particles*, Wiley, New York, 1957.
Van der Meulen, A., *J. APCA*, **35**, 383 (1986).
Wait, G. R., *Carnegie Inst. Wash. News Serv. Bull.*, **3**, 1 (1934).
Waldmann, L., *Z. Naturf.*, **14a**, 589 (1959).
Walstra, P., *Brit. J. Appl. Phys.*, **15**, 1545 (1964).
Wang, C. S., and Friedlander, S. K., *J. Col. Int. Sci.*, **24**, 170 (1967).
Wang, H. C., et al., *J. Aerosol Sci.*, **14**, 703 (1983).
Watson, H. H., *Trans. Faraday Soc.*, **32**, 1073 (1936).
Weast, R. C. (ed.), *Handbook of Chemistry and Physics*, 54th ed., CRC Press, Cleveland, Ohio, 1973.
Wen, H. Y., Reischl, G. P., and Kasper, G., *J. Aerosol Sci*, **15**, 89 (1984).
Went, F. W., *Proc. Natl. Acad. Sci.*, **46**, 212 (1960).
Wheeldon, J. M., and Burnard, G. K., *Filtr. and Separat.*, **24**, 178 (1987).
Whitby, K. T., *Rev. Sci. Inst.*, **32**, 1351 (1961).

Whitby, K. T., and Peterson, C. M., *Ind. & Eng. Chem. Fundamentals*, **4**, 66 (1965).
Whitby, K. T., and Liu, B. Y. H., in *Aerosol Science*, C. N. Davies (ed.), Academic Press, New York, 1966.
Whitby, K. T., and Liu, B. Y. H., *J. Col. Int. Sci.*, **25**, 537 (1967).
Whitby, K. T., and Willeke, K., in *Aerosol Measurement*, D. A. Lundgren, et al. (eds.), University of Florida Press, Gainesville, 1979.
White, H. J., *Industrial Electrostatic Precipitation*, Addison-Wesley, Reading, Mass., 1963.
White, H. J., *J. APCA*, **27**, 206 (1977).
Whytlaw-Gray, R., and Patterson, H. S., *Smoke*, E. Arnold, London, 1932.
Wiedensohler, A., et al., *J. Aerosol Sci.*, **17**, 413 (1986).
Wiedensohler, A., *J. Aerosol Sci.*, **19**, 387 (1988).
Wiemann, in *Industrial Dust Explosions*, ASTM STP 958, K. L. Cashdollar and M. Hertzberg (eds.), American Society for Testing Materials, Philadelphia, 1987.
Wildi, J., and Thomann, H., *J. Aerosol Sci.*, **14**, 615 (1983).
Willeke, K., and Brockman, J. E., *Atm. Env.*, **11**, 995 (1977).
Wilson, C. T. R., *Phil. Trans. Royal Soc.*, **A189**, 265 (1897).
Yaglou, C. P., and Benamin, L. C., *Heating, Piping and Air Conditioning*, January 1934, p. 25.
Yalamov, Yu, I., Vasiljeva, L. Yu, and Schukin, E. R., *J. Col. Int. Sci.*, **62**, 503 (1977).
Zak, E. G., *Zh. Goefiz.*, **6**, 452 (1936).
Zebel, G., in *Aerosol Science*, Academic Press, New York, 1966.
Zimmerman, N. J., Reist, P. C., and Turner, A. G., *Appl. and Env. Microbiology*, **53**, 99 (1987).

Index

Abrahamson, J., 120
Absorbance, 265
Absorption, 263, 266–267
Acceleration, 75–76, 82–84
   forces for, 91
   radial, 91–92, 115
   time-dependent cases of, 82
Acoustic agglomeration, 316
Adiabatic expansion, 351–352
Adsorption, 241, 247
Aerodynamic boundary layers, 155–159
Aerodynamic diameters, 6, 71–72, 106–107
Aerodynamics, 31
Aerosol owl, 295
Aerosol spectrometers, 117
Agglomerates, 8, 10–11
   (*See also* Coagulation)
Air:
   coagulation constant for, 305
   composition of, 348
   currents in, 91
   density of, 47
   diffusion in, 150–152
   in ducts, 48
   ion concentrations in, 200–201
   mean free path in, 39–40
   mobility of ions in, 193
   pollution of, 225
   resisting force of, 59–60, 66–67, 75
   scattering efficiency factor for, 271
   spinning of, 91
   temperatures of, 225
   terminal velocities in, 66
   and thermophoretic velocity, 175
   viscosity of, 41, 46–47
Air pressure and dust-free spaces, 165
Airstreams, 109, 150–152, 314–315
Aitken, J., 165, 225, 293
Aitken nuclei, 238
Albedo, 265
Alberty, R., 32, 41, 351
Algae, 320

Allen, M. D., 61
Alveolar region, 7, 125
Amelin, A. G., 226–227, 233
American Conference of Governmental Industrial Hygienists (ACGIH), 126–128
Angstrom units, 1
Angular intensity functions, 283–285
Angular scattering, 281–282
   applications of, 294–299
   Mie theory, 283–288
   of polydisperse particles, 289–294
Anisokinetic sampling, 121–124
Anterior unciliated nares, 125
Apparent mean free path, 142
Arithmetic mean, 26
Asbestos fibers, 4
   aerodynamic diameter of, 71–72
   settling velocity of, 81
Atkinson, W. R., 199
Atomic Energy Commission (AEC), 126–127
Attenuation coefficient, 267–368
Attrition, 4, 13
Autoignitable dust, 332
Automatic vents, 337
Average charges per particle, 206
Average diameters, 13, 25
Average displacement, 136, 140–141
Average kinetic energy, 38
Average surface, 26
Average velocity, 38, 116, 139
Average volume, 26
Avogadro's number, 32, 134, 229

Backscatter, 291–292
Bacteria, 319–320
Bakanov, S. P., 168
Bar charts, 15–16
Barometric distribution, 139–140
Barrer, R. M., 146
Bartknecht, W., 328–329

Baseballs, 31–32
Basketballs, 1
Beer, bubbles in, 67
Beer's law, 267
Bernoulli, J., 33
Berry, J., 264
Bessel functions, 288, 355
Billings, C. E., 241
Bioaerosols, 319–325
Blue moons, 293
Blue sky, 290
Bodaszewski, R. W., 131
Boilers, thermal deposition in, 165
Boisdron, Y., 189
Boltzmann approach, 204–205
Boltzmann equation, 139
Boltzmann velocity, 141
Boltzmann's constant, 36, 135, 229
Boltzmann's law, 201
Bouguer's law, 266–270
Boundary layers, 155–159
Brenner, H., 65
Bricard, J., 189
British Medical Research Council (BMRC), 125, 127
Brock, J. R., 169, 171–175, 189, 312, 314
Brock's equation, 171–172
Brown, Robert, 131
Brownian motion, 131–132
  and apparent mean free path, 142
  barometric distribution by, 139–140
  and diffusion coefficient, 140–141
  displacement by, 136–138
  Einstein's theory of, 133–136
  and Fick's laws, 132–133
  of rotation, 138–139
Bubbles in beer, 67
Buffer zones, 157
Burnard, G. K., 120
Businger, J. A., 201
Byers, H. R., 244

Cadle, R. D., 171, 240, 293
Calibration of impactors, 107–108
Cannonballs, flight of, 49–50
Capture of particles, 75
Carbon particles, 75, 266
Cascade impactor arrangement, 105
Cashdollar, K. L., 330, 332–333
Cawood, W., 167
CCN (cloud condensation nuclei), 239
Centimeters, 1
Centrifugation, 115–117

Chains, Stokes' law with, 68–70
Chan, T., 120
Chapman, S. J., 168–169
Characteristic charge, 187
Characteristic length and diffusion, 155
Charging mechanisms, 179
  collisions with ions, 185
  contact electrification, 182–183
  corona discharge, 195–198
  diffusion, 185–189, 195
  electric, 179–183
  equilibrium with, 200–201
    steady-state theory of, 201–207
    transient approach to, 207–208
  field charging, 185, 189–195
  flame ionization, 184–185
  and force, 179–180
  frictional, 184
  ionization, 181–182, 184–189, 195–198, 213–214
  maximum charge with, 198–200
  and mobility, 180–181
  speed of, 187, 194
  spray electrification, 183
  static electrification, 182
Charlson, R. J., 278
Charts, histograms, 15–19
Chatigny, M. A., 320–321, 323
Chemical groups, explosions with, 330
Cheng, Y. -S., 81, 153
Chilton, C. H., 41, 176
Cigarette smoke particles, 136, 150
Ciliated nasal passages, 125
Circular jet impactors, 96
Circular paths, 91–92, 115
Circulation of particles, 66–67, 86
Clark, W. E., 165, 312
Clausius-Clapeyron equation, 227
Cloud chambers, 226, 236
Cloud condensation nuclei (CCN), 239
Clouds and cloud droplets, 225
  formation of, 236
  turbidity coefficient for, 269
CN (condensation nuclei), 226, 238–241
Coagulation, 301
  differential equation form for, 309–310
  electrical effects in, 313–314
  external factors in, 313–316
  monodisperse spherical, 301–306
  in moving atmospheres, 314–315
  nonlinear integrodifferential equations for, 310–311
  nonspherical, 312–313

Coagulation (*Cont.*):
  with ordered motion, 315
  and particle size, 306–309, 315–316
  rate of, 308
  sonic, 316
  turbulent, 316
Coagulation constant, 304–307
Coastlines, length of, 9–10
Collection efficiency, 94, 96–100, 221–222
Collisions, 32, 38
  elastic, 32, 49
  with ions, 185
  and thermal diffusion, 302
Colloidal solutions, 301
Colony-forming units (CFUs), 320
Color of sky, 290, 293–295
Combustion sources, 337
  (*See also* Explosions and explosive aerosols)
Concentration boundary layers, 157–159
Concentration change, 149
Concentration gradient, 145
Condensation and evaporation, 3, 13, 225
  concentrations of, 64
  hysteresis in, 246–248
  and isokinetic sampling, 124
  in Langmuir's equation, 257–259
  of liquid drops, 225
  Maxwell's equation for, 251–256
  of moving droplets, 260–262
  and nucleation, 226
    formation rate of, 232–233
    heterogeneous, 238–248
    homogeneous, 228–232
    ions for, 233–238
  observations of, 225–226
  and saturation ratios, 227–231
  surface for, 226
  time of, in saturated media, 259–260
Condensation nuclei (CN), 226, 238–241
Conductivity, thermal, 40–42, 168, 171
Confined sedimentation, 65
Constant field strength, 213
Constant velocity, 75, 81, 92
Contact electrification, 182–183
Continuous curves, 17
Continuous medium, 31, 60–63
Continuum, 45
Contrast, 264, 276–279
Convection currents, 165, 171
Convective diffusion, 150–152
Conversion factors, 348, 350

Cooper, D. G., 185, 208
Coordinate systems for motion, 76, 79–80
Corona discharge, 195–198, 214–217
Cosmic radiation, 201
Coulier, M., 225
Coulomb effect, 188
Coulomb's law, 211
Counting rate in particle counters, 298
Crawford, M., 52
Creep flow, 170
Critical drop size, 230
Critical nuclei, 232–233
Critical Stokes number, 113
Critical temperature, 331
Cross, J., 329–331
Cross-sectional diameter of fibers, 71
Crystallization, 248
Cumulative distribution percentages, 106
Cumulative lognormal distributions, 24
Cunningham, E., 60
Cunningham correction factor, 60–64, 69
  and coagulation, 307
  for settling velocity, 81–82
  and thermophoretic velocity, 169
Currents:
  air, 91
  convection, 165, 171
  corona, 215
  density of, 193
  diffusion, 135, 145, 303
  ion, 190, 192–193
Curves, 17, 19
Curvilinear motion, 75, 91–92
Cyclone particle collectors, 117–120
Cylinders:
  diffusion in, 133–136
  settling in, 86–87
  thermophoresis in, 167

Daniels, F., 32, 41, 351
Davies, C. N., 61, 122, 155, 158, 255, 257
Debye-Huckel model, 314
Deceleration of particles, 75, 83–84
Definitions, 2–3, 347–350
  of average diameters, 26
  of force, 179–180
  of means, 25–29
Deflagration, 328–329
Deformation drag, 50–51
Deirmendjian, D., 266, 272
Denman, H. H., 283, 288
Dennis, R., 296

Index

Density:
  of air, 47
  of currents, 190, 192–193
  and toxicity, 125
Deposition, 121
  in boilers, 165
  in coagulation, 311–312
  by diffusion, 149, 152–160
  by inertia, 121–122
  in respiratory system, 125
  velocity of, 160
  on walls, 102, 146–148
Derjaguin, B. V., 168, 172–175
Detonation, 328–329
Deutsch, W., 222
Diameters, 4–7
  aerodynamic, 6, 71–72, 106–107
  for air, 175
  of average mass, 25
  of cyclones, 119
  equilibrium, 234
  of gases, 39
  geometric, 23
  mean and median, 13–15, 23–29, 106–107
  of respirable dust, 126
  and saturation ratios, 230
  for settling particles, 55
  of small particles, 103
Dielectric constants of liquids, 182
Diffraction, 294–296
Diffusion, 131
  and brownian displacement, 136–138
  charging by, 185–189, 195
  and collisions, 302
  convective, 150–152
  current in, 135, 145, 303
  in cylinders, 133–136
  deposition by, 149
  Fick's laws of, 132–133, 135, 137, 145–147, 151
  force of, 134, 159
  of gases, 40–42, 132–133, 253
  non-steady-state, 146–150
  steady-state, 145–146
  and tube deposition, 152–160
  velocity of, 158–160
Diffusion coefficients, 132, 155, 304
  effects of mass on, 140–141
  gas, 40–41
  for water vapor, 252–253
Diffusion constants, 136

Dimensional analysis, 46
Direction of motion, 75
Disintegration of droplets, 198–200
Dissipation by wind, 68
Distance:
  in cyclones, 118
  between molecules, 33
  stop, 83, 95, 142
Distributions:
  barometric, 139–140
  of charges, 201–207
  with coagulation, 312
  cumulative, 24, 106
  histograms for, 15–19
  log probability paper for, 24–25
  lognormal, 20, 22–24, 27, 106–107
  mathematical representation of, 19–24
  Maxwell-Boltzmann, 35–37, 139
  of mean and median diameters, 13–15, 25–29
  normal, 20–22
  in respirable sampling, 125
Doherty, D. J., 207
Drag, 49
  deformation, 50–51
  form and friction, 71
  and Reynolds number, 50–56
  surface, 64
Drag coefficient, 50–56
Drift, 193, 219–222
Drinker, P. D., 27
Droplets:
  disintegration of, 198–200
  evaporation of, 251, 260–262
  growth of, 251, 260–262
  lifetime of, 256–260
  moving, 260–262
  (*See also* Water and water droplets)
Dry air:
  composition of, 348
  mean free path in, 40
  viscosity in, 41
Ducts:
  air flow in, 48
  and dust explosions, 337
Dust explosions, 179, 184, 327
  control of, 336–337
  ignition sources for, 331–335
  and particle size, 332–336
  severity of, 328–329
  types of particles in, 329–331
Dust-free space, 163–165, 175–177

Dust particles:
  autoignitable, 332
  and condensation nuclei, 238
  in cyclones, 117
  definition of, 2
  explosions of (*see* Dust explosions)
  flammability of, 332–336
  respirable, 125–128
Dynamic shape factor, 68–71

Eddy diffusion coefficient, 158
Effective migration velocity, 222
Efficiency:
  of boilers and heat exchangers, 165
  of cyclones, 119–120
  of electrostatic precipitators, 221–222
  of impactors, 94, 96–100, 111
Einbinder, H., 185
Einstein, Albert, 133–136, 167
Elastic collisions, 32, 49
Elder, M. E., 296
Electric charges (*see* Charging mechanisms)
Electric discharges, sparks from, 331
Electric fields, 179–180, 190–191
  drift in, 219–221
  for geometrics, 213–218
  gravity in, 76–77
  with particles present, 216–218
  perturbations in, 218–219
  strength of, 180–181, 209–210, 212–218
Electric flux, 191, 193, 210
Electric forces, 179, 181, 348–349
Electrical effects in coagulation, 313–314
Electrical units, 211–212, 348–350
Electrolytes, 182, 243
Electronegative gases, 198
Electrostatic controlled kinetics:
  charge saturation in, 195
  and Coulomb's law, 211
  drift in, 219–221
  efficiency of, 221–222
  electric fields for, 209–210, 213–218
  electrical units for, 211–212
  field perturbations in, 218–219
  field strength for, 210, 212, 214–218
  potential for, 212–213
Electrostatic units (esu), 349–350
Embryos, 226, 228, 232
Energy, transfer of, 40
Epstein, P., 173
Epstein's equation, 170–171

Equation of motion, 76–78, 220
Equilibrium charge distribution, 201–207
Equilibrium diameters, 234
Equilibrium times, 82, 207
Equilibrium vapor pressure, 243
Error function, 147, 345–346
Errors:
  from anisokinetic sampling, 122
  with impactor data, 107–108
Evaporation (*see* Condensation and evaporation)
Expansion, adiabatic, 351–352
Expansion ratios, 228, 351–352
Explosions and explosive aerosols, 179, 184, 327
  control of, 336–337
  ignition sources for, 331–332
  and particle size, 332–336
  severity of, 328–329
  types of, 329–331
Extinction, 263
  and Bouger's law, 266–270
  coefficient of, 267, 270–276
  and contrast, 276–279
  efficiency factor for, 268, 273–275
  particle size in, 294–295
  terms for, 264–266
Extinction paradox, 271–272

Factor of proportionality, 210
Falling particles, 81
Farrer, D., 329–331
Feret's diameter, 4–6
Fibers, 4
  aerodynamic diameter of, 71–72
  persistence of, 73
  settling velocity of, 81
  Stokes' law with, 68–71
Fick's laws of diffusion, 132–133, 135, 137, 145–147, 151
Field charging, 185, 189–195
Field of force, 180
Field strength, 180–181, 209–210, 212–218
50 percent Stokes number, 103
Filter fan model, 153–155
Final velocity, 53–55, 66, 75, 81–82
Flachsbart, H., 69, 117
Flame ionization, 184–185
Flammability of dusts, 332–336
Flanagan, V. P. V., 207
Fleagle, R. G., 201
Fletcher, F. W., 242
Flocculated particles, 68

Flocs, 8
Fluid dynamics, 31
Fluid mechanics, 45
Fluid properties, 31
  drag, 49–56
  gas behavior, 33–40
  kinetic theory for, 32–33
  macroscopic, 45–56
  Reynolds number for, 45–49
  viscosity, conductivity, and diffusion, 40–42
Flux, 191, 193, 210
Fly ash, 8
  and drift velocity, 221
  and field strength, 217–218
Fog, 3
Foitzik, I., 289
Force, 75, 179–180
Force balance equations, 76
Force vector, 180
Form drag, 71
Forward scattering, 281–282, 291–293, 296–297
Fox, D. L., 233
Fractal geometry, 8–11
Fraunhofer diffraction, 296
Free energy and ions, 233–237
Free molecule region, thermophoresis in, 166–169
Friction drag, 71
Frictional electrification, 183–184
Frictional resistance, 50
Friedlander, S. K., 312, 316
Fuchs, N. A., 61
  and brownian rotation, 138
  and coagulation, 313–314
  and diffusion charging, 188
  and diffusion and collisions, 304
  and equilibrium charge distributions, 204–205
  and free mean path, 142
  and impaction, 93
  and lifetime of drops, 258–259, 261
  and mean square displacement, 141
  and thermophoresis, 165, 173
Fumes, 2, 8
  removal of, 217
  sampling, 124
Fungi, 320

Gas-phase reactions, 226
Gases, 31
  adsorption of, 241

Gases (*Cont.*):
  behavior of, 33–40
  in cyclones, 117
  diameters of, 39
  diffusion in, 40–42, 132–133, 253
  electronegative, 198
  inertia of, 60
  molecules present in, 32–33
  in pipes, 47
  transfer of momentum, energy and mass within, 40
  two-dimensional trajectories of, 140–141
  velocity of, 35, 37–38
  viscosity and conductivity of, 40–42
  volume of, 32
Geometric mean, 26
Geometric mean diameter, 23
Geometric median, 24
Geometric standard deviation, 23–24
Geometrics, electric fields for, 213–218
Geometry, fractal, 8–11
Giant nuclei, 238
Gibbs, W. E., 165
Giese, R., 287–288
Gillespie, T., 306
Glycerol particles, 171
Goetz, A., 240
Goren, S. L., 176
Graph paper, 24
Graphs, histograms, 15–19
Gravity:
  in coagulation, 311–312
  deposition from, 121
  in electric fields, 76–77
  and particle motion, 77–81
Green, H. L., 95, 296
Grinding wheels, particles thrown from, 86
Grounding of ductwork, 337
Growth and evaporation, 251–262
Gucker, F. T., 297

Happel, J., 65
Hatch, T., 27
Hatch-Choate equation, 27–29
Haze, definition of, 3
Heat (*see* Temperature)
Heat exchangers, deposition in, 165
Heineman, H., 183
Helium gas, viscosity of, 42
Herdan, G., 24
Hering, S. V., 97, 103–104

Hering design, 97
Hering impactors, 104
Hertzberg, M., 330, 332–333
Heterogeneous nucleation, 226, 238–248
Hidy, G. M., 310, 312, 314
Higher-order Tyndall spectra, 295–296
Histograms, 15–19
Hochrainer, D., 219
Hocking, L. M., 315–316
Hodkinson, J. R., 270, 296
Homogeneous aerosols, 3
Homogeneous nucleation, 226, 228–232
Hoppel, W. A., 188, 204–205
Horizontal range of particles, 83
Hot bodies, 163–165
Hot spots, 337
Humidity:
  and condensation, 246–248
  and dust explosions, 184, 334–336
  and extinction measurements, 279
Hydrated protons, speed of, 189
Hydrogen:
  and corona, 198
  speed of, 35
Hyperbolic streams, 108
Hysteresis, 246–248

Ideal stirred settling, 86–88
Ignition sources, 331–332
Image effect, 188
Image force, 218–219
Impaction, 92–96
  and bounce, 101–102
  curvilinear motion, 91–92
  data analysis with, 106–113
  from inertia, 121–122
  operation of, 96–101
  phase trajectories with, 108–113
  and pressure drop, 104–105
  and small size, 102–104
Impaction parameter, 95
Impactors:
  analysis of, 105–107
  configuration of, 95
  cut points for, 91, 96–97, 100, 103
  efficiency of, 94, 96–100, 111
  errors with, 107–108
  operation of, 96–101
  particle bounce in, 101–102
  phase trajectories for analyzing, 108–113
  pressure drop in, 104–105
  for small particle sizes, 102–104

Impingers, 91
Induction, 210, 218
Inelastic impacts, 49
Inertia, 45, 47, 91
  deposition by, 121–122
  in diffusion, 141
  of gas molecules, 60
Infinite medium, 64–65
Inorganic composition of condensation
  nuclei, 241
Insoluble nuclei, 242
Intensity of light, 282–283, 285, 290–291
Intermediate flow, 47–48
Internal circulation, 66–67
Interval size, 15
Ions and ionization, 181–182
  and charge equilibrium, 200–201
  collisions with, 185
  concentrations of, 186, 200–201, 203
  from corona discharge, 195–198
  current density of, 190, 192–193
  diffusion charging with, 186–189
  in flames, 184–185
  mobility of, 193, 215
  negative, 183
  as nuclei, 233–238
  space charges from, 213–216
Isokinetic sampling, 120–124
Isometric particles, 3
  brownian rotation of, 139
  persistence of, 73
  shape of, 68
Ivchenko, I. N., 172

Jacobsen, S., 169
Jennings, S. G., 39–41, 61
Jets, 96, 101, 103
Johnson, I., 71–72, 200, 295
Joos, G., 343
Jost, W., 132, 146
Joules, 349
Junge, C., 241

Kasahara, A., 311
Keefe, D., 201, 205
Kelvin's equation, 228–232
Keng, E. Y. H., 172
Kenoyer, J. L., 101
Kerker, M., 263, 283, 288, 290, 294
Kinematic viscosity, 45–47, 169
Kinetic energy, average value of, 38
Kinetics, 75
  acceleration and deceleration, 83–84

Kinetics (*Cont.*):
  of centrifugation, 115–117
  of cyclones, 117–120
  electrostatic controlled (*see* Electrostatic controlled kinetics)
  equation of motion, 76–77
    without external forces, 77–80
  ideal stirred settling, 86–88
  impaction (*see* Impaction)
  isokinetic sampling, 120–124
  limitations on, 84
  one-dimensional motion, 84–86
  respirable sampling, 124–128
  stop distance, 83
  terminal settling velocities, 81–82
  theory of, 31–33, 42
Kirsch, A., 153
Kitani, S., 296
Klett, J. D., 213, 232, 253, 261
Knudsen number, 61
  and evaporation, 254
  and thermophoresis, 166, 168, 170, 175
Köhler curves, 244
Koschmieder, H., 277
Krakatoa, eruption of, 240

LaMer, V. K., 171–172, 295
Laminar boundary layers, 155–157
Laminar flow, 47–48, 119
Landau, L. D., 155, 343
Lane, W. R., 95, 296
Langevin ions, 201
Langmuir's equation, 256–259
Laplace's equation, 212–214
Laplacian operator, 133
Large ions, 201
Large nuclei, 238
Larynx, 125
Lawton, J., 184–185
Leith, D., 71–72, 120
Lenard, P., 183
Length ratio, 72
Levich, V. G., 151–152, 157, 316
Licht, W., 120
Lifetime of droplets, 256–260
Lifshitz, E. M., 155, 343
Light (*see* Angular scattering; Extinction)
Line charts, 15
Lines of force, 209
Lippman, M., 120, 125, 127
Liquid aerosols:
  charge limitations on, 198

Liquid aerosols (*Cont.*):
  condensation of, 225
  dielectric constants of, 182
  shape of, 68
  supercooling of, 226
Liu, B. H. Y., 188–189, 192, 199, 298
Living biological material, 319–325
Lodge, O. J., 165
Loeb, L. B., 195
Log probability paper, 24–25, 106–107
Logarithms, 17, 22
Lognormal distributions, 20, 22–24, 27, 106–107
Los Angeles smog, 241
Low-pressure impactors, 103
Lowan, A. N., 283
Lower explosive limit (LEL), 336
Ludlum, F. H., 255
Lumergs, 265
Lumination, 264
Luminators, 264
Lungs:
  removal of particles by, 93
  toxic doses to, 125

McDaniel, E. W., 193
Macroscopic fluid properties:
  drag, 49–56
  Reynolds number for, 45–49
Mandelbrot, Benoit, 8–11
Maron, S. H., 296
Marple, V. A., 96–100, 103–104
Martin's diameter, 4–6
Mason, B. J., 168–169, 213, 232
Mass:
  average, diameter of, 25
  concentrations of, and toxic doses, 125
  effect of, on diffusion coefficients, 140–141
  of large particles, 19
  median diameters of, 25, 106–107
  transfer of, 40
  (*See also* Volume)
Mass respirable sampling, 125
Mathematical representation of distributions, 19–24
Maximum explosive concentrations, 328
Maxwell, James, 170, 251
Maxwell-Boltzmann distributions, 35–37, 139
Maxwell's equation, 251–256
Mean diameter, 13–15, 23

Mean free path, 7, 38–40
  apparent, 142
  in Stokes equation, 61
Mean square displacement:
  and brownian motion, 137
  and diffusion, 141
Mean values, 20–21, 25–29
Median diameters, 13–15, 24, 26, 28, 106–107
Medians, 21, 24–25
Metal fumes, 2, 8
Metal powders, 331
Meteorological range, 278
Meters, 1
Micrometers, 1
Microorganism growth as ignition source, 331
Midpoints, 14–15, 21
Mie, G., 263
Mie region, 293
Mie theory, 263, 283–293, 295
Miller, F. C., 183, 195
Millikan, R. A., 60
Millikan oil drop experiment, 59
Millikan resistance factor, 60–64
Minimum explosive concentrations, 328
Mists:
  coagulation of, 314
  definition of, 3
Mixed nuclei, 241
Mixing:
  in boundary layers, 157
  and settling, 86–88
Miyake, S., 165
Mks system of units, 349
MMAD (mass median aerodynamic diameter), 106–107
Mobility, 135
  and charging mechanisms, 180–181
  of ions, 193, 215
Mode, 26
Moist air:
  and dust explosions, 334
  mean free path in, 40
  viscosity in, 41
  (See also Humidity)
Molds, 320
Molecules, 32
  average distance between, 33
  average energy of, 38
  behavior of, 32
  collision of, 38
  diameters of, 39

Molecules (Cont.):
  in small volumes of gas, 32
  velocities of, 33–38
Momentum, 34, 40, 49, 76, 166, 168, 341
Momentum accommodation coefficient, 172
Monodisperse particles, 3, 13
  barometric distribution of, 139
  and charging, 185
  coagulation of, 301–306
  size of, 4, 296
Moon, 293–294
Moreau-Hanot, M., 190
Morphological properties, 3–11
Moss, O., 101
Most probable velocity of gases, 37
Motion, 135
  and coagulation, 313
  curvilinear, 91–92
  in electric fields, 180–181
  equation of, 76–77, 220
  without external forces, 77–80
  one-dimensional, 84–86
  rectilinear, 75
  (See also Brownian motion; Thermophoresis)
Motor oil, viscosity of, 42
Mount St. Helens, eruption of, 240
Moving atmospheres, coagulation in, 314–315
Moving droplets, growth and evaporation of, 260–262
Muller, H., 312–313

Napper, D. H., 264
Natanson, G. L., 189
Navier-Stokes equation, 46, 63, 342
Negative corona, 197
Negative ions:
  mobility of, 193
  saturation ratios with, 237
Net displacement in brownian motion, 136
Nevalainen, A., 319
Newton, I., 49
Newtons, 349
Newton's second law of motion, 34
Nitrogen, ions with, 197–198
Niven, W. D., 170
Nolan, P. J., 207
Nonrigid particles and Stokes' equation, 65–68
Nonsparking equipment, 337

Nonspherical particles, coagulation of, 312–313
Non-steady-state diffusion, 146–150
Normal curve, 20
Normal distributions, 20–22
Nozzles, 96
Nucleation, 226
　heterogeneous, 238–248
　homogeneous, 228–232
Nuclei:
　insoluble, 242
　ions as, 233–238
　rate of formation of, 232–233
　soluble, 242–246
　utilization of, 241–242
Number of living particles per unit volume, 320

Oceans and ion concentrations, 201
O'Connor, T. C., 207
Oil, viscosity of, 42
O'Konski, C. T., 297
Optical properties (see Angular scattering; Extinction)
Oral cavity, 125
Ordinate scale, 17
Ordinate value, 17
Organic composition of condensation nuclei, 240–241
Orr, C., 52, 172, 247
Osmotic coefficients, 243
Osmotic pressure, 134
Ottewill, R. H., 264
Oxidation process, 2, 331
Oxygen:
　and explosions, 327, 336
　ions with, 197–198
Ozone, 197–198

Parabolic flow profile, 156–157
Particles:
　acceleration and deceleration of, 83–84
　centrifugation of, 115–117
　charge on, 180–181, 198
　coagulation of (see Coagulation)
　diameters of (see Diameters)
　diffusion of (see Diffusion)
　distribution of (see Distributions)
　fractal dimensions of, 9
　inertia of, 91, 122
　interactions induced by, 64–65
　kinetics of (see Electrostatic controlled kinetics; Impaction; Kinetics)

Particles (*Cont.*):
　mass of (see Mass; Volume)
　motion of (see Brownian motion; Motion; Thermophoresis)
　refractive indices of, 266
　shape of, 3–4, 6
　size of (see Size of particles)
　in Stokes' law, 68–73
　velocity of, 83
Particles displaced per unit area, 137
Paths, 7, 38–40, 61, 111, 142
Patterson, H. S., 302
Pauthenier, M. M., 190
Peclet numbers, 151–152, 155
Penndorf, R., 284
Perfect gases, velocity of, 35
Perry, R. H., 41, 176
Perturbations in electric fields, 218–219
Peterson, C. M., 172, 208
Pharynx, 125
Phase diagrams, 111–112
Phase trajectory analysis, 108–113
Photochemical production, 240
Photometric units, 265
Photooxidation, 240
Physical units, 349
Pipes, turbulence in, 47
Planar streams, 108
Plane polarized light, 282
Plane vertical wall, diffusion on, 146–148
Plaque-forming units (PFUs), 321
Platelets, 3
Point charge, field strength of, 210
Poiseuille flow, 156–157
Poisson's equation, 212, 215–216
Polarization force, 218
Polarization of light, 290
　incident, 282
　scattered, 284
Pollen spores, 319–320
Pollution, air, 225
Polonium-210 particles, 140
Polydisperse aerosols, 3, 13
　and coagulation, 308
　concentration of, 88
　and scattering, 289–294
　size of, 4
Polyethylene dust, 328–329
Position, equation for, 79–80
Positive corona, 197–198
Positive ions:
　mobility of, 193
　saturation ratios with, 237

Powder, explosions of, 329
Prandtl, L., 45
Prandtl number, 176
Pressure, 33
    and corona currents, 215
    and diffusion, 41
    drag from, 50–51
    on droplets, 229
    and dust-free spaces, 165
    and mean free path, 40
    osmotic, 134
    on sphere surfaces, 71
    and temperature, 351
    and thermophoretic velocity, 175
    on vapors, 243, 252–253, 351
Pressure drop in impactors, 104–105
Probability function, 147
Problem of radiative transfer, 293–294
Problems, visualization of, 31
Projected area diameters, 5–6
Protozoans, 320
Pruppacher, H. R., 213, 232, 253, 261
Psychometric chart, 353
Pueschel, R., 240
Pui, D. Y. H., 188–189

Quartz particles, 183–184

Raabe, O., 26, 61, 97, 99
Radial acceleration, 91–92, 115
Radial velocity, 92–93, 116, 118
Radiation scattering, 283–293
Radiative transfer, 293–294
Radioactive sources, 208
Radioactivity and ion production, 201
Raindrops, 3, 13, 225
    ion concentrations in, 201
    terminal velocities of, 66
Ranz, W. E., 95, 120
Rao, A. K., 96
Rationalized electrical units, 349
Rayleigh, Lord, 163, 290
Rayleigh limit, 198–199
Rayleigh region, 290
Rayleigh scattering, 271, 282–283, 289–291
Rebound, 102, 170
Receptors, 264, 276
Rectangular jets, 96, 101
Rectangular slot impactors, 95
Rectilinear motion, 75
Red light, 293, 295
Reflectance, 265

Reflection coefficient, 172
Refractive indices, 266–267, 296
Regional deposition, 125
Reist, P. C., 8, 185, 208, 323
Relative humidity (*see* Humidity)
Relaxation times for spheres, 82
Resistance, 49–51, 59–64, 66–67, 75–76
Respirable dust, 125–128
Respirable fraction, 127–128
Respirable sampling, 124–128
Respiratory system:
    deposition in, 125
    toxic doses to, 124–125
    upper, 7
Retention time for gases within cyclones, 117–118
Reynolds number, 45–49, 135
    and drag, 50–56
    and impactor efficiency, 97–99
    for jets and nozzles, 96
    one-dimensional motion in, 84–86
    and Peclet number, 152
    and Stokes' equation, 63
    for viscous flow, 341–342
Rigid particles and Stokes' equation, 65–68
Robinson, R. A., 243
Rods, aerodynamic diameter of, 71
Rohmann, H., 190
Root-mean-square displacement, 136–137, 141
Root-mean-square velocity, 37–38
Rose, H. E., 222
Rosenblatt, P., 171–172
Rotation:
    brownian, 138
    of coordinate systems, 79
Round jet impactors, 96, 99
Rubow, K. L., 97

Sampling:
    devices for, 91
    isokinetic, 120–124
    respirable, 124–128
    thermophoresis for, 166
Sartor, J. D., 199
Saturation charges, 190–192, 194–195
Saturation ratios, 227–238, 352
Sauter diameter, 26, 275
Sawdust, 13
Scattering, 7, 263
    applications of, 294–299
    coefficient of, 267

Scattering (*Cont.*):
  diagrams of, 293
  efficiency factor for, 268
  forward angle of, 281–282, 291–293, 296–297
  intensity coefficient of, 282
  Mie, 283–293
  for particle sizing, 297
  patterns in, 291–293
  of polydisperse particles, 289–294
Schadt, C. F., 171
Schlicting, H., 157
Schmitt, K. H., 168–169, 172
Schmitt number, 261
Seddig, R., 133
Sedimentation, 65
Sedimentation velocities, 7, 339
Sehmel, G. A., 158
Selective collection, 121
Self-preserving size distribution, 312
Settling, 75
  stirred, 86–88
  velocities in, 53–55, 64, 66, 81–82
Sewage, 319–320
Shapes of particles, 3–4, 6, 68
Sharkey, W. P., 207
Silica:
  Cunningham correction factor for, 63
  settling velocity of, 81–82
Simple coagulation theory, 308
Simple diffusion, 131–143
Single-particle scattering measurements, 297–299
Single scattering in Bouger's law, 270
Singular-node phase diagram type, 112
Size of particles, 1, 4–7, 11, 297
  and acceleration, 82
  and centrifuges, 115
  and coagulation, 306–309, 315–316
  and Cunningham correction factor, 62
  in cyclones, 119
  distributions of (*see* Distributions)
  and eddy diffusion coefficient, 158
  in extinction and scattering of light, 294–295
  and extinction efficiency factors, 271
  and flammability of dusts, 332–336
  with impactors, 96–97
  intervals for, 19
  monodisperse, 4, 296
  scattering technique for, 297
  and toxic doses, 125

Size of particles (*Cont.*):
  and turbidity coefficients, 275
  and viscosity, 60
Skewed curves, 22
Skim milk, 330
Skin friction, 50
Sky, color of, 290, 293–295
Slime molds, 320
Slip, 60–64, 169–170, 172
Small ions, 201
Small particles:
  anisokinetic sampling with, 123–124
  in cyclones, 119
  diffusion for, 131
  impactors for, 102–104
  measuring, 103
  number of, 19
Small volumes, 33
Smog, 3, 241
Smoke particles, 3
  brownian motion of, 131
  concentration of, 87
  definition of, 2
  and diffusion, 136, 150
Smoking and dust explosions, 337
Smoluchowski, M., 302
Smooth curves, 19
Sneezes, 319, 323–324
Snow, 225
Sodium chloride:
  condensation on, 242–248
  diameter of, 7
  solutions of, 248
Soluble nuclei, 242–246
Sonic agglomeration, 316
Source of radiant energy, 264
Space charge effects, 217
Sparks from electric discharges, 331
Spherical particles, 3, 139
  brownian rotation of, 138
  charging, 185
  deposition velocity of, 160
  diffusion coefficients for, 136
  electric charge on, 192
  electric force on, 180
  extinction efficiency factor for, 272
  flux through, 210
  ions striking, 186, 188
  maximum charge on, 198
  mobility of, 181
  one-dimensional motion of, 85
  Peclet number for, 152
  pressure on, 71

Spherical particles (*Cont.*):
  relaxation times for, 82
  repelling forces between, 211
  resisting force on, 60
  sedimentation velocities of, 339
  in Stokes' law, 68–73
  stop distance of, 83
  turbidity coefficient for, 268–269
Spinning air, 91
Spontaneous combustion, 331
Spontaneous condensation, 232
Spontaneous nucleation, 226
Spray electrification, 183
Spurny, K. R., 72
Standard deviation, 20–21, 23–24
Static electricity, 182, 337
Statistical assumptions, 33
Statistical mechanics, 31
Steady-state conditions, 83
Steady-state diffusion, 145–146
Steady-state size distribution, 312
Steady-state theory of charge equilibrium, 201–207
Stechkina, I., 153
Stirred settling, ideal, 86–88
Stöber, W., 69–70, 117
Stöber spectrometer design, 117
Stokes, G. G., 59
Stokes' diameter, 6
Stokes' flow, 47
Stokes' law, 59, 135, 341–343
  continuous medium assumption in, 60–63
  incompressible medium assumption in, 63
  rigid particle assumption in, 65–68
  spherical particle assumption in, 68–73
  and thermophoretic velocity, 169
Stokes number, 95
Stokes region, evaporation in, 261
Stokes resistance, 76
  for tangential velocities, 92
  and thermal force, 173
Stop distance, 83, 95, 142
Stratmann, F., 166, 176
Stream velocity, 83
Streamline flow, 47
Stress tensor, 341–342
Structures, aerosol, 8
Sun, color of, 293
Supercooling of liquids, 226
Supersaturations, 228, 232–233
Surface area, 11–12, 18–19
Surface-area fractal dimension, 10

Surface-area median diameter, 26, 28
Surfaces:
  for condensation, 226
  mean diameter of, 26, 275
  properties of, 11–12
Sutton, O. G., 50
Sutugin, A. G., 232
Svedberg, Z., 133
S/W ratios for impactors, 97, 100

Takahashi, K., 311
Talbot, L., 168–169, 175
Tangential velocity, 91, 93, 116
Temperature:
  and conductivity, 40–42, 168, 171
  and corona currents, 215
  critical, 331
  and diffusion, 41, 253
  and dust-free spaces, 165, 175–177
  jump in, 173
  and lifetime of droplets, 258
  in Maxwell's equation, 254–256
  and mean free path, 40
  and pressure, 351
  and viscosity, 42
  (*See also* Thermophoresis)
Terminal velocities, 53–55, 66, 75, 81–82
Thermal coagulation, 301
Thermal conductivity, 40–42, 168, 171
Thermal creep, 170
Thermal deposition, 165
Thermal diffusion and collisions, 302
Thermal forces, 163, 168–173
Thermal gradients, 163
Thermal slip flow, 169–170, 172
Thermophoresis, 163–164
  Brock's equation for, 171–172
  Derjaguin and Talamov's equation for, 172–173
  dust free space from, 175–177
  Epstein's equation for, 170–171
  in free molecule regions, 166–169
  observations of, 165–166
  in slip-flow regions, 169–170
  theory of, 166
  velocity in, 173–176
Threshold of brightness contrast, 278
Thunderstorms, 201
Tietjens, O. G., 45
Tikhomirov, M., 306
Time factors:
  in acceleration, 82
  in charging, 194

Time factors (*Cont.*):
  in coagulation, 305
Total mass, 25
Total scattering efficiency, 282
Toxic doses, estimation of, 124–125
Tracheobronchial tree, 125
Trajectories:
  equation for, 80
  phase, 108–113
  two-dimensional, 140–141
Transfer of energy, 40
Transfer of mass, 40
Transfer of momentum, 34, 40, 49, 166, 168, 341
Transient approach to charge equilibrium, 207–208
Transmittance, 265
Tribo electrification, 183–184
Tube deposition, 152–160
Turbidity coefficient, 267–269, 275
Turbulence:
  eddy diffusion coefficients for, 158
  in impactors, 104
Turbulent agglomeration, 316
Turbulent boundary layer, 155, 157
Turbulent flow, 47–49, 119
Twinkling effect, 139
Two-dimensional trajectories, 140–141
Tyndall, J., 163
Tyndall spectra, higher-order, 295–296

Unconfined sedimentation, 65
Unipolar ions, 186–189
Units, 1, 347–350
  electrical, 211–212
  photometric, 265
  for viable aerosols, 320–323
Unpolarized incident light, 282
Upper respiratory system, 7

Van de Hulst, H. C., 270, 288, 290
Vapor:
  condensation of, 3
  pressure of, 243, 252–253, 351
  saturation ratio for, 227
  and temperature, 351
Vector notation, 76–79
Velocities, 76, 79
  average, 38, 116, 139
  Boltzmann, 141
  and brownian motion, 133
  diffusion, 158–160
  drift, 220–222

Velocities (*Cont.*):
  jump in, 170
  Maxwell-Boltzmann distribution of, 36–37, 139
  and mobility, 135
  molecular, 33–38
  radial, 92–93, 116, 118
  sedimentation, 7, 339
  settling, 53–55, 64, 66, 81–82
  tangential, 91, 93, 116
  thermal, 189
  thermophoretic, 169, 173–176
  vectors for, 79
Ventilation systems, 75
Vertical walls, diffusion on, 146–150
Viable aerosols, 319
  estimates of, 323–325
  factors affecting, 321–323
  types of, 320
  units of measure for, 320–321
Viruses, 319–321, 323, 325
Viscosity:
  of air, 41, 46–47
  of gases, 40–42
  and internal circulation, 66–67
  and particle size, 60
  and temperature, 42
Viscosity drag, 50
Viscous flow, 47, 59, 341–342
Viscous forces, 45, 47
Viscous motion and Stokes' law, 59
  continuous medium assumption in, 60–63
  incompressible medium assumption in, 63
  rigid particle assumption in, 65–68
  spherical particle assumption in, 68–73
  viscous medium assumption in, 63
Viscous sublayers, 157
Viscous transfer of momentum, 341
Visibility and contrast, 276
Visual range, 278–279
Visualization of problems, 31
Volatization rate, 333
Volume, 25
  diameter of, 25–26, 28
  distributions of, 18–19
  of gases, 32–33
  (*See also* Mass)

Wait, G. R., 201
Waldmann, L., 168
Waldmann's equation, 175

Walls:
  deposition on, 102, 146–148
  diffusion on, 146–150
Wang, C. S., 312
Water and water droplets, 228
  diffusion coefficients for, 252–253
  equilibrium diameter of, 234
  evaporation rates for, 253, 255–256
  extinction efficiency factor for, 271
  free energy of, 229, 232
  growth and evaporation of, 235
  properties of, 348
  resistance force of air on, 66–67
  saturation ratios for, 228, 231, 234–238
  settling velocity in, 55
  supercooling of, 226
  surface area of, 11–12
  terminal velocities of, 66
  (*See also* Droplets)
Water embryos, 232
Watson, H. H., 165
Wavelength in Bouger's law, 270

Weinberg, F. J., 184–185
Well-designed impactors, 97
Went, F. W., 240
Wet-bulb temperature, 255
Wetted perimeter concept, 96
Wheeldon, J. M., 120
Whitby, K. T., 96, 192, 199, 208, 297–298, 312
White, H. J., 186–189, 195, 215–216
Whytlaw-Gray, R., 302
Wiedensohler, A., 205–206
Willeke, K., 98, 297
Wilson, C. T. R., 226, 233
Wind, dissipation by, 68
Wire-in-cylinder arrangement, 197
Wong, J. B., 95
Wood, A. J., 222

Yalamov, Yu, 172–175
Yeasts, 320

Zebel, G., 312–313

ABOUT THE AUTHOR

Parker C. Reist is Professor of Air and Industrial Hygiene Engineering in the Department of Environmental Sciences and Engineering at the University of North Carolina at Chapel Hill. An engineering graduate of Pennsylvania State University, he holds S.M. degrees from the Massachusetts Institute of Technology and the Harvard School of Public Health, and a doctor of science degree from the Harvard School of Public Health. His current research interests include nucleated condensation processes, aerosol formation mechanisms, testing the efficiency of pleated media filters, rainborne pollution, the aerodynamic behavior of platelet aerosols, and the application of fractal geometry to the prediction of aerosol dynamics.